VOLUME NINETY

ADVANCES IN BOTANICAL RESEARCH

Metabolism, Structure and Function of Plant Tetrapyrroles: Introduction, Microbial and Eukaryotic Chlorophyll Synthesis and Catabolism

ADVANCES IN BOTANICAL RESEARCH

Series Editor

Jean-Pierre Jacquot
Membre de l'Institut Universitaire de France, Unité Mixte de Recherche INRA, UHP 1136 "Interaction Arbres Microorganismes", Université de Lorraine, Faculté des Sciences, Vandoeuvre, France

VOLUME NINETY

ADVANCES IN BOTANICAL RESEARCH

Metabolism, Structure and Function of Plant Tetrapyrroles: Introduction, Microbial and Eukaryotic Chlorophyll Synthesis and Catabolism

Volume Editor

BERNHARD GRIMM

Institute of Biology/Plant Physiology,
Faculty of Life Science,
Humboldt-Universität zu Berlin,
Berlin, Germany

ELSEVIER

ACADEMIC PRESS

An imprint of Elsevier

Academic Press is an imprint of Elsevier
125 London Wall, London, EC2Y 5AS, United Kingdom
The Boulevard, Langford Lane, Kidlington, Oxford OX5 1GB, United Kingdom
50 Hampshire Street, 5th Floor, Cambridge, MA 02139, United States
525 B Street, Suite 1650, San Diego, CA 92101, United States

First edition 2019

ISBN: 978-0-08-102752-3
ISSN: 0065-2296

For information on all Academic Press publications
visit our website at https://www.elsevier.com/books-and-journals

Publisher: Zoe Kruze
Acquisition Editor: Sam Mahfoudh
Editorial Project Manager: Peter Llewellyn
Production Project Manager: James Selvam
Cover Designer: Christian Bilbow

Typeset by SPi Global, India

Contents

Contributors

Donald A. Bryant
Department of Biochemistry and Molecular Biology, The Pennsylvania State University, University Park, PA; Department of Chemistry and Biochemistry, Montana State University, Bozeman, MT, United States

Daniel P. Canniffe
Institute of Integrative Biology, University of Liverpool, Liverpool, United Kingdom

Min Chen
School of Life and Environmental Sciences, University of Sydney, Sydney, NSW, Australia

Jaromír Cihlář
Faculty of Science, University of South Bohemia; Institute of Parasitology, Biology Centre CAS, České Budějovice, Czech Republic

Leszek Fiedor
Faculty of Biochemistry, Biophysics and Biotechnology, Jagiellonian University, Kraków, Poland

Zoltán Füssy
Faculty of Science, University of South Bohemia; Institute of Parasitology, Biology Centre CAS, České Budějovice, Czech Republic

Mareike Hauenstein
Institute of Plant and Microbial Biology, University of Zurich, Zurich, Switzerland

Stefan Hörtensteiner
Institute of Plant and Microbial Biology, University of Zurich, Zurich, Switzerland

Bernhard Kräutler
Institute of Organic Chemistry & Center for Molecular Biosciences (CMBI), University of Innsbruck, Innsbruck, Austria

Dariusz Latowski
Department of Plant Physiology and Biochemistry, Faculty of Biochemistry, Biophysics and Biotechnology, Jagiellonian University in Krakow, Krakow, Poland

Beata Myśliwa-Kurdziel
Department of Plant Physiology and Biochemistry, Faculty of Biochemistry, Biophysics and Biotechnology, Jagiellonian University in Krakow, Krakow, Poland

Miroslav Oborník
Faculty of Science, University of South Bohemia; Institute of Parasitology, Biology Centre CAS, České Budějovice, Czech Republic

Mariusz Pilch
Faculty of Biochemistry, Biophysics and Biotechnology; Faculty of Chemistry, Jagiellonian University, Kraków, Poland

Kazimierz Strzałka
Department of Plant Physiology and Biochemistry, Faculty of Biochemistry, Biophysics and Biotechnology; Malopolska Centre of Biotechnology, Jagiellonian University in Krakow, Krakow, Poland

Ayumi Tanaka
Institute of Low Temperature Science, Hokkaido University, Sapporo, Japan

Ryouichi Tanaka
Institute of Low Temperature Science, Hokkaido University, Sapporo, Japan

Jennifer L. Thweatt
Department of Biochemistry and Molecular Biology, The Pennsylvania State University, University Park, PA, United States

Robert D. Willows
Department of Molecular Sciences, Macquarie University, North Ryde, NSW, Australia

Mateusz Zbyradowski
Faculty of Biochemistry, Biophysics and Biotechnology, Jagiellonian University, Kraków, Poland

Preface

It is not long time ago that the green pigment chlorophyll was considered to be an insignificant organic molecule. Photosynthesis was self-evident an important process, but attracted only spectroscopists or "traditionally classic" plant physiologists, when they studied total plants and algae. The metabolic pathway of tetrapyrroles held interest for plant biochemists only, who reported their results—with due modesty—in not particularly spectacular journals. The pathway and the mode of action of the enzymes involved in tetrapyrrole biosynthesis were considered to be too complex. It caused already headaches to pronounce enzymes and intermediates correctly.

It is already 32 years ago that I was introduced in 1987 as a postdoctoral fellow to 5-aminolevulinic acid and chlorophyll biosynthesis by Gamini Kannangara and Simon Gough (for example: Kannangara, C. G., Gough, S. P., Bruyant, P., Hoober, J. K., Kahn, A., & von Wettstein, D. (1988). tRNA(Glu) as a cofactor in delta-aminolevulinate biosynthesis: steps that regulate chlorophyll synthesis. *Trends in Biochemical Sciences, 13*, 139–143) in the Department of Diter von Wettstein at the Carlsberg Laboratory, Copenhagen. Due to apparently inexhaustible resources of plant material, manpower, and time, we harvested every week 10 kg of 5-day-old greening barley seedlings for the partial purification of protein fractions of crude plastids for enzyme assays of the initial enzymes of 5-aminolevulinic acid. We spent generally more than 15 h of the first day of our preparation with harvest of the leaves, chloroplast extraction, column chromatography, and protein concentration and desalting. Then on day 2 we quickly assayed the few milliliters of purified protein solution, before the enzyme activity quite rapidly degraded. It remained for us to enjoy a final good Carlsberg beer and to soak finally the new seeds for germination.

Since that time, a burst of new exciting findings related to chlorophyll were released, and chlorophyll metabolism and its structure and function have attracted far more scientists than in the 1990s. Chlorophyll became not only interesting for organic chemists and spectroscopists but also for molecular and cell biologists, micro biologists, geneticists, and evolutionary biologists.

Therefore, it seemed to be the appropriate time to give a resume on several aspects of chlorophyll. Fortunately, may researchers were willing to contribute to the lively discussion on function, structure and metabolism of the green pigment. Numerous manuscripts were collected that we can offer two volumes on this subject matter. Chlorophyll has been considered not only in the context of its main function in photosynthesis (more details in the introductory Chapter 1, Vol. 90), but almost each chlorophyll molecule in single proteins of the photosynthetic complexes, including the huge light-harvesting antenna (Chapter 9, Vol. 91: Evolution and function of light harvesting antenna in oxygenic photosynthesis), was separately investigated in their contribution to the transfer of absorbed excitation energy and to the assembly to photosynthetic complexes. While initially only the proteins of photosynthetic complexes were separately characterized, their assembly and mutual functional impact with the pigments and cofactors were thoroughly investigated (Chapter 7, Vol. 91: Chlorophyll-binding subunits of photosystem I and II: biosynthesis, chlorophyll incorporation and assembly). The first chlorophyll catabolites were identified and reported and more catabolic intermediates and the involved enzymes of chlorophyll degradation were identified and explored (Chapter 7, Vol. 90). Not only the multiple functions of chlorophyll in energy transfer and charge separation to promote the photosynthetic electron transfer have been specified, but also the impact of tetrapyrrole metabolites were explored up to the point that also intermediates and tetrapyrrole end products were proposed to act as emitters of plastid-localized retrograde signaling (Chapter 8, Vol. 91: Retrograde signaling: The discovery of gun mutants and the consequence Current state of retrograde signaling derived by tetrapyrrole biosynthesis). More and more variants of chlorophyll and bacteriochlorophyll were detected and their specific functions and metabolism described (Chapters 2–4, Vol. 90).

The predominant number of genes encoding proteins of chlorophyll synthesis has been initially identified in photosynthesis-related operons of bacterial genomes before the homologous plant genes were identified and by means of the physiological and genetic analysis of mutants of single open reading frames within these operons the function of each gene product in chlorophyll biosynthesis was unraveled. At a higher level of complexity, researcher specifically define nuclear transcription factors which control genes encoding enzymes and factors of tetrapyrrole metabolism (Chapter 5, Vol. 91: Transcriptional control for the chlorophyll metabolism). I personally pursued the enormous progress on the characterization of tetrapyrrole biosynthesis. I was fascinating to follow how much care photosynthetic organisms evolved to control the complex metabolic pathway and to ensure with multifaceted regulatory mechanism a fine-tuned formation of tetrapyrroles for multiple light-dependent processes (photomorphogenesis, light harvesting, and photosynthesis) but also light-independent reactions (redox reaction, detoxification, electron transport, sulfur, and nitrogen assimilation) (see further details in the Chapters 5, 6, Vol. 90; Chapter 3, Vol. 91: Posttranslational control of 5-aminolevulinic acid synthesis; Chapter 4, Vol. 91: Porphyrin and heme synthesis; Chapter 6, Vol. 91: Posttranslational control of tetrapyrrole biosynthesis: Interacting proteins, chaperones, auxiliary factors). Ultimately, the intriguing question about the origin of the two photosystems of oxygenic photosynthesis has stimulated research in evolutionary biology to unravel the phylogeny of the tetrapyrrole biosynthetic pathway and to decipher the mosaic composition of genes involved in this metabolism of organisms of derived from multiple endosymbiotic processes (see for details in Chapter 8, Vol. 90).

Since the first cyanobacterial high resolution X-ray crystallographic structures of photosystem I and II were published, enormous progress in the 3-D structure elucidation of light-sensitive pigment containing photosynthetic complexes has been reported when femto second crystallography studies were published (Chapter 2, Vol. 91: Time-resolved femto second crystallography studies of photosynthetic chlorophyll-binding protein complexes). Single molecule spectroscopy enabled specific examination of a spectroscopic view on the chlorophyll-binding proteins and opened another dimension of elucidation on the individual properties of pigment-proteins (Chapter 1, Vol. 91: Studying conformational changes of proteins via single molecule spectroscopy: Cryogenic temperatures versus room temperature).

This development in the studies of function of chlorophyll and the control of chlorophyll metabolism was possible due to interdisciplinary activity of biologists, biochemists, and physicists. This enormous prog ress in pigment synthesis deserves a compilation of manuscripts which survey all achievements on structure, function, and mechanisms of tetrapyrroles.

Many of us are still aware of the former monographic compilations on chlorophyll, such as Scheer, H. (Ed.). (1991). *Chlorophylls*. CRC Press: Boca Raton; and Grimm, B., Porra R. J., Rüdiger W., & Scheer H. (Eds.). (2006). *Chlorophylls and bacteriochlorophylls, biochemistry, biophysics, functions and applications*. Springer Nature Switzerland AG. The aim of the two volumes 90 and 91 are to provide at the right time an up to date overview on chlorophyll metabolism, function, and structure. It was intended to summarize the progress on multiple forms of chlorophyll metabolism in unicellular and multicellular photoautotrophs, to highlight new methods for the elucidation of structure and function of chlorophyll-binding proteins and to emphasize how complex the posttranslational control of synthesis, catabolism, and assembly of chlorophyll is for a fine-tuned supply of pigments in photosynthesis.

I am very thankful for each manuscript and I wish to acknowledge all contributors for their valuable reviews, which made this two volumes to an enormous compendium for many aspects of chlorophyll. We exceeded with our manuscripts the size of a single volume of the Advances in Botanical Research and fill two volumes with an exciting survey on multiple aspects of chlorophyll. I am quite confident that the two volumes will reach a broad visibility among plant science and will be another essential reference for students and scientists, who wish to receive a pronounced and detailed monographic overview.

Bernhard Grimm
May 1, 2019

Acknowledgments

I wish to acknowledge my scientific coworkers, technical assistants, the gardener, and the students, who all contributed to the exciting results of our research. I also acknowledge the funding by the Deutsche Forschungsgemeinschaft, which currently enables my research on tetrapyrrole biosynthesis: Gr936 18-2 (FOR2092), TRR175/1, TPC04, and Gr936 20-.

CHAPTER ONE

Tetrapyrrole pigments of photosynthetic antennae and reaction centers of higher plants: Structures, biophysics, functions, biochemistry, mechanisms of regulation, applications

Leszek Fiedor[a,*], Mateusz Zbyradowski[a], Mariusz Pilch[a,b]
[a]Faculty of Biochemistry, Biophysics and Biotechnology, Jagiellonian University, Kraków, Poland
[b]Faculty of Chemistry, Jagiellonian University, Kraków, Poland
*Corresponding author: e-mail address: leszek.fiedor@uj.edu.pl

Contents

Abstract

The chapter gives an overview about chlorophylls involved in oxygenic photosynthesis and summarizes the most recent findings about these key photosynthetic pigments. First, a general description of structural, stereochemical and photophysical features of chlorophylls is given, as this is relevant to their functioning. Next, the details of the electronic structure and the mechanisms contributing to fine tuning of chlorophylls

Advances in Botanical Research, Volume 90
ISSN 0065-2296
https://doi.org/10.1016/bs.abr.2019.04.001

in photosynthetic apparatus are discussed, including the roles of conservative structural elements of these molecules. Briefly, also the use and results of quantum chemical approaches are described. Further, the biosynthesis and biodegradation of chlorophylls is shortly reviewed, followed by a survey of the photosensitizing and photoprotective functions of chlorophylls, the commercial and therapeutic applications of the pigments and their impact on animal organisms.

Abbreviations

ABCG2	binding cassette super-family G member 2
ALA	δ-aminolevulinic acid
ATP	adenosine triphosphate
BChl	bacteriochlorophyll
BCRP	breast cancer resistance protein
CAO	chlorophyll *a* oxygenase
CASPT2	complete active space perturbation theory
CBR	chlorophyll *b* reductase
CC	coupled-cluster theory
CCG	chlorophyll catabolic gene
Chl	chlorophyll
Chlase	chlorophyllase
Chlide	chlorophyllide
chlorin e_6	*N*-aspartyl-functionalized chlorophyllin
CMI	central metal ion
Crt	carotenoid
DFT	density functional theory
DMAPP	dimethylallyl-pyrophosphate
DNCC	dioxobilin-type non-fluorescent chlorophyll catabolite
DPOR	light-independent protochlorophyllide oxidoreductase
geranyl-PP	geranyl-pyrophosphate
GGPP	geranylgeranyl pyrophosphate
HCAR	hydroxymethyl chlorophyll *a* reductase
HOMO	highest occupied molecular orbital
hydroxyl-pFCC	C33-hydroxylated primary fluorescent chlorophyll catabolite
IPP	isopentenyl-pyrophosphate
IR	infrared
ISC	intersystem crossing
LHC	light-harvesting complex
LPOR	light-dependent protochlorophyllide oxidoreductase
LUMO	lowest unoccupied molecular orbital
MgDVP	divinyl protochlorophyllide
MgPMe	Mg-protoporphyrin IX monomethyl ester
MgPPIX	Mg-protoporphyrin IX
MM	molecular mechanics
MO	molecular orbital
MP2	Møller–Plesset perturbation theory order 2
NADPH	nicotinamide adenine dinucleotide phosphate
NCC	non-fluorescent chlorophyll catabolite

NYC	non-yellow coloring
NYE	non-yellowing
P680	photosystem II primary donor
P700	photosystem I primary donor
PAO	pheophorbide *a* oxygenase
PBG	porphobilinogen
PChlide	protochlorophyllide
PDT	photodynamic therapy
pFCC	primary fluorescent chlorophyll catabolite
Pheide	pheophorbide
Pheo	pheophytin
phytyl-PP	phytyl pyrophosphate
POR	protochlorophyllide oxidoreductase
PPH	pheophytinase
PPIX	protoporphyrin IX
Protogen	protoporphyrinogen
PS	photosystem
QM	quantum mechanics
RC	reaction center
RCC	red chlorophyll catabolite
RCCR	red chlorophyll catabolite reductase
ROS	reactive oxygen species
SAC-CI	symmetry adapted cluster/configuration interaction
SAM	*S*-adenosyl-L-methionine
TIC55	translocon at the inner chloroplast envelope 55

1. Introduction

Chlorophylls (Chls) are a class of green or blue-green tetrapyrrolic pigments, which can be referred to as the colors of life on our planet. They are widely distributed in the biosphere and the seasonal changes in their content within vegetation are the only biological process visible from outer space. The unique system of π-electrons delocalized over their macrocyclic skeleton and the variety of side groups located around it enable them to function as the principal photoactive and structural components of the photosynthetic apparatus of all photosynthetic organisms. Since they are designed by nature to interact with light, they are responsible for capturing solar photons and converting their energy into the energy of chemical bonds in a form accessible to virtually all life forms. The functional versatility of Chls in the photosynthetic apparatus is intriguing; they function as antenna pigments, active in the absorption of photons, the short-term storage and transfer of excitation energy, and in the photosynthetic reaction centers

(RCs) as both the electron donors and acceptors. In addition, in virtually all photosynthetic Chl-binding proteins they play a key structural role. The photosynthetic activity of Chls, which has already been going on for billions of years, has resulted in the release of gigantic amounts of molecular oxygen into the atmosphere, which has greatly contributed to the geology as well as the evolution of the entire biosphere. These features render Chls indispensible for photosynthesis, and these fascinating pigments have been the subject of intensive investigations for over a century.

2. Structures, occurrence and functions of chlorophylls

2.1 Tetrapyrrolic macrocycle

The structural formulae of Chl *a*, Chl *b*, Chls c_1, c_2 and c_3 (Chls *c*), Chl *d* and Chl *f*, i.e., the Chls associated with oxygenic photosynthesis, are shown in Fig. 1. Chl *a* is the most important and universally abundant photosynthetic pigment on Earth, functioning both as an energy absorber in the light-harvesting complexes (LHCs) and an electron transfer cofactor in RCs. It occurs in all plants, cyanobacteria and algae. Chl *b*, a product of Chl *a*

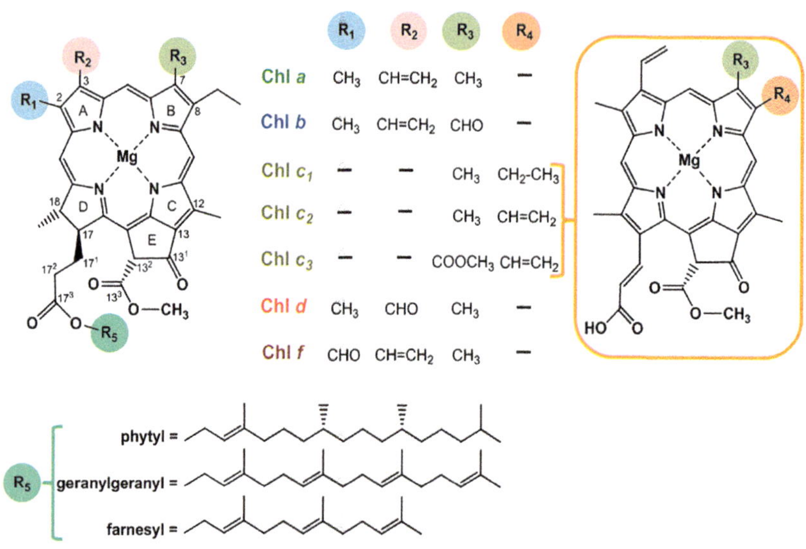

	R1	R2	R3	R4
Chl *a*	CH_3	$CH{=}CH_2$	CH_3	–
Chl *b*	CH_3	$CH{=}CH_2$	CHO	–
Chl c_1	–	–	CH_3	CH_2-CH_3
Chl c_2	–	–	CH_3	$CH{=}CH_2$
Chl c_3	–	–	$COOCH_3$	$CH{=}CH_2$
Chl *d*	CH_3	CHO	CH_3	–
Chl *f*	CHO	$CH{=}CH_2$	CH_3	–

Fig. 1 The chemical formulae of chlorophylls identified in organisms performing oxygenic photosynthesis, with the atom numbering according to the IUPAC system. Formally, chlorophylls *c1—c3*, due to the presence of free carboxylic group in the 17^3 position and the double C—C bond in the 17^1 position should be classified as modified chlorophyllides. The structures of the side chains esterifying the C-17^3 propionic residue, which may be found in chlorophylls are also shown.

oxygenation with molecular oxygen, which is mainly found in higher plants and algae, functions as an accessory antenna pigment in the LHCs associated with photosystem I and II (PSI, PSII). Both Chl *a* and Chl *b* are essential structural cofactors required in the formation and folding of Chl-protein complexes. Chls *c* are light-harvesting pigments, which rather augment the functioning of Chl *a*. Chl *d* was first found in red algae as a minor pigment or isolation artifact. Recently, traces of Chl *c* have been found in an extremophilic cyanobacterium *Chroococcidiopsis thermalis*, grown under far-red light. Later it was identified as the major Chl in *Acaryochloris marina*, a widespread cyanobacterium, where it acts not only as an antenna pigment but also replaces Chl *a* in the PSI reaction center (Hu et al., 1998; Larkum & Kühl, 2005; Nürnberg et al., 2018). Chl *f*, the most recently identified Chl, was discovered in methanolic extracts of samples containing a cyanobacterium, at first uncharacterized, which was obtained from stromatolites (Chen et al., 2010). Afterward it was also found as a pigment that accompanies Chl *a* and Chl *d* in more wide-spread cyanobacterial species, grown either under far-red or natural light, such as the KC1 strain from Lake Biwa in Japan, *Chlorogloeopsis fritschii* and *Chroococcidiopsis thermalis*. In the latter organism, Chl *f* has been shown to participate in light-harvesting as well as in the primary photochemistry of both photosystems, thus demonstrating that the spectral limits of oxygenic photosynthesis extend up to ~740 nm (Airs et al., 2014; Allakhverdiev et al., 2016; Chen, Li, Birch, & Willows, 2012; Miyashita et al., 2014; Nürnberg et al., 2018) (More information found on Chls *c* and *d* can be found in the chapters "Chlorophylls *c*—Occurrence, synthesis, properties, photosynthetic and evolutionary significance" and "Chlorophylls *d* and *f*: Synthesis, occurrence, light-harvesting, and pigment organization in chlorophyll-binding protein complexes" of this volume by Myśliwa-Kurdziel et al. and Chen, respectively).

The molecular framework of Chls consists of a five-member phorbin macrocycle, i.e., an aromatic tetrapyrrolic ring derived from a porphyrin, fused with an isocyclic ring. With the addition of a carbonyl functionality in the 13^1 position it makes up a molecular skeleton common to practically all Chls. In naturally occurring pigments, the central cavity formed by the core four nitrogen atoms hosts either the Mg^{2+} ion (Chls) or two protons (pheophytins, Pheos). Due to a combination of the various degrees of macrocycle saturation and the diversity of the peripheral side groups that can be found around the macrocycle, the number of naturally occurring Chls and bacteriochlorophylls (BChls) scores 100 variants (Scheer, 2006) including newly discovered species, such as Chl *f*. The molecules of naturally occurring Chls, which possess at least several stereochemical centers (e.g., three in the Chl *a* macrocycle), are chiral and stereochemically occur always in pure forms (Fig. 1). Chls, either in their free state in organic solvents or protein-bound, permanently carry at least one axial ligand (Scheer, 2006) and in the diastereotopic monoligated state (Balaban, Fromme, Holzwarth, Krauss, & Prokhorenko, 2002), due to the out-of-plane shift of the central Mg, they are asymmetric. The stereochemistry of Chls is an important biochemical and structural aspect. The high resolution crystal structures of the various Chl-binding proteins available show that the strict stereoselectivity is also retained during their formation, and in general, there is a high degree of stereoselectivity at all levels of Chl-protein interactions. Intriguingly, the stereochemistry of Chls that forms the special pair P700 of the PSI reaction center is of particular importance for its functioning, the highly conserved heterodimer of Chl *a* and Chl *a*′, its 13^2-*R* diastereoisomer (Jordan et al., 2001; Watanabe et al., 1985).

2.2 The aliphatic side chain

With the exception of Chls *c*, the C-17^3 propionic acid side chain of Chls is usually esterified by the phytyl moiety, a C_{20} (aliphatic) isoprenoid residue which makes up as much as one-third of the molecule in terms of the molecular mass. In rare cases, other isoprenoid alcohols, such as geranylgeraniol or farnesol, are found in this position (Fig. 1). The phytyl moiety endows Chls with a considerable hydrophobicity enhancing their lipophilic character and rendering them practically water-insoluble. Due to this side moiety, Chls can be classified as lipids. This hydrophobic residue also determines the aggregation and other physicochemical properties of Chls and their ability to interact with a hydrophobic environment (Agostiano, Catucci, Colafemmina, & Scheer, 2002; Fiedor, Stąsiek, Myśliwa-Kurdziel, & Strzałka, 2003), often providing an anchor that properly positions the pigments within the hydrophobic domains of photosynthetic ensembles (Jordan et al., 2001; Liu et al., 2004). Regarding the influence of the phytyl moiety on the π-electron system of Chls, it is commonly accepted that phytyl is neutral and spectroscopically silent. However, a detailed spectroscopic study on model Chls has revealed small but significant differences in the spectral characteristics of the phytylated and non-phytylated pigments (Fiedor, Kania, Myśliwa-Kurdziel, & Stochel, 2008; Fiedor et al., 2003). The phytyl moiety, in spite of its size and hydrophobicity, does not directly influence the solvatochromic and coordinative properties of Chls, which permits a good coupling between the chromophore and its first solvation shell and evens out the probabilities of ligand binding to both faces of the macrocycle. The phytyl chain slightly affects other properties of the pigment, such as interactions of the H-bond donors with core hydrogens in the free bases, which is consistent with the occurrence of steric interactions between this bulky residue and the periphery of the macrocycle near ring V. Such a steric strain, in turn, contributes to the stabilization of Chl conformation. For this reason, the presence of the phytyl chain seems to be required to fine-tune the pigment in the apoproteins. This implies that truncated chlorophylls, often used for simplicity in theoretical studies, are not adequate as models of native chromophores, especially when their specific properties are modeled. On the other hand, the esterification of the C-17^3 propionate seems necessary because the carboxyl group may act as a strong chelator of the central Mg and hence in Chls *c* this side chain undergoes desaturation to prevent such intramolecular chelation.

2.3 Central Mg^{2+} ion

The central cavity in Chls is able to accommodate various divalent metal ions, some of which are able to spontaneously and irreversibly replace the native Mg^{2+} (Küpper, Setlik, & Spiller, 2002; Schunck & Marchlewski, 1894). In spite of this, for a variety of reasons, Mg^{2+} was selected by nature as the central metal ion in Chls (Fiedor et al., 2008; Kotkowiak, Dudkowiak, & Fiedor, 2017; Scheer, 1991, 2006). The central Mg^{2+} has no effect on the solvatochromism of the Chl π-electron system and is light enough not to cause heavy atom effects in the molecule (Clarke & Frank, 1977; Drzewiecka-Matuszek, Skalna, Karocki, Stochel, & Fiedor, 2005; Kotkowiak et al., 2017). As a closed shell element (frontier $3s^2$ configuration), Mg^{2+} does not electronically perturb the pigment π-electron system but its binding in the core contributes to the conformational stabilization of the macrocycle (Fiedor et al., 2008). But most importantly, a coordinatively unsaturated Mg^{2+} acts as the coordination center for axial ligand binding, which constitutes the strongest bond between Chls and polypeptides (Balaban et al., 2002; Kania & Fiedor, 2006). This versatile design permits Chls to retain both high photochemical and redox activity in photosynthetic complexes. And thus, a single type of pigment is serving in energy absorption and excitation transfer (light-harvesting) as well as in electron transfer (charge separation), apart from being a key structural cofactor for the Chl-binding proteins.

Despite the high charge density on the Mg^{2+} ion, it is only weakly bound in the central cavity of Chls and can be readily removed in the presence of weak acids, at relatively high pH values, yielding the corresponding Pheos or pheophorbides (Pheides) from chlorophyllides (Chlides). The Mg-Pheo complexes are much more labile than complexes with other divalent metal ions, such as Zn^{2+}, Cu^{2+} or Ni^{2+}. The results of recent investigations on bacteriochlorophyll *a* (BChl *a*) and a series of its metallosubstituted analogs, with the use of the time-resolved optoacoustic technique and various electronic spectroscopies combined with the high pressure technique, challenge the classical picture of Chls as simple coordination complexes of Mg^{2+}. These studies show that the type of bonding between the central metal ion (CMI) and the nitrogen atoms that form the central cavity in Chls is influenced by several factors, such as electron-electron repulsion in the central cavity, the occupancy, and the geometric, symmetry and energetic properties of particular orbitals involved in bond formation. In addition, the character of Mg^{2+}-Pheo bonding largely determines the photophysical features of Chls, crucial for photoprotection (Kania et al., 2014; Kotkowiak et al., 2017; Pilch et al., 2013). According to this refined model, the central Mg^{2+} ion, due to its very stable electronic neon configuration and high energy of the 3s and 3p orbitals, can only form weak bonds with the core nitrogens of a mostly ionic (electrostatic) character. In contrast, the Zn^{2+} ion (closed d shell) binds via stronger, classical coordinative bonds, while the divalent transition metal ions (unfilled d shell) form mixed covalently-coordinative bonds of exceptional strength (Fig. 2A). This gradual change in the bonding character and strength, from weak electrostatic to strong coordination and even stronger mixed coordination-covalent bonds, agrees very well with the chemical stability of Chls and their metalloderivatives, and explains the varying influence of the CMI on Chl photophysics (Drzewiecka-Matuszek et al., 2005; Fiedor et al., 2000; Kotkowiak et al., 2017; Musewald et al., 1998; Pilch et al., 2013). Thus, Mg^{2+} complexes

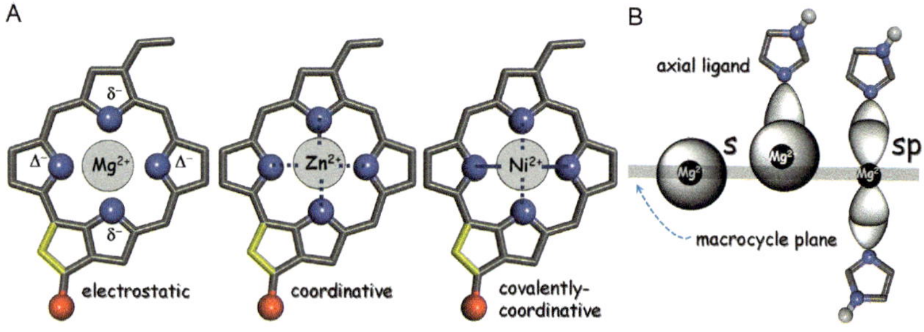

Fig. 2 (A) The character of chemical bonding between the central metal ion and the nitrogens in the binding pocket of chlorophylls varies from weak electrostatic (Mg) to purely coordinative (Zn) and mixed covalently-coordinative (Ni), depending on the metal ion type. (B) Schematic presentation of the acceptor orbitals of the central Mg^{2+} ion and the donor ligand orbital involved in coordination of a single or two axial ligands (see the text for details).

are labile and readily undergo pheophytinization, at a pH as low as 6 (Mazaki, Watanabe, Takahashi, Struck, & Scheer, 1992), and are prone to spontaneous metal ion exchange (Küpper et al., 2002; Schunck & Marchlewski, 1894). This model also rationalizes the interactions of Chls with axial ligands (Fig. 2B). Since the 3s and 3p orbitals of Mg^{2+} are not involved in bonding to the core nitrogens, the bond to a single axial ligand can be formed via electron pair donation to the acceptor 3s orbital. The binding of two axial ligands requires the hybridization of the 3s and $3p_z$ orbitals yielding two sp hybrids, each able to accommodate a pair of electrons. In either case, the classical coordination bonds are formed, and formally the coordination number of the central Mg^{2+} is 1 and 2, respectively. Due to the electron repulsion in the molecule center, when a single ligand is coordinated, the Mg^{2+} ion is pushed out the molecular plane, and two axial ligands are bound via weaker bonds than the single ligand. The symmetric biligated state, in which the two ligands force Mg^{2+} back into the plane, is thermodynamically less favored (Callahan & Cotton, 1987; Kania & Fiedor, 2006). A strong single coordination bond to the spherical s orbital may be formed with equal effectivity at different directions, whereas the two sp hybrids have considerable directional properties and form effective bonds only along the normal to the macrocycle plane. As a consequence, the Chl molecules are bound to their apoproteins predominately via single axial coordination of nucleophilic amino acid residues (most frequently a histidine) while biligation is seen very rarely (Ben-Fredj, Ben Lakhdar, & Ruiz-Lopez, 2009; Fiedor, 2006; Jordan et al., 2001; Liu et al., 2004; Oba & Tamiaki, 2002; Umena, Kawakami, Shen, & Kamiya, 2011). Importantly, the water molecules are not preferred as a second axial ligand to Chls, mainly due to entropic constrains. Therefore, water molecules do not saturate the axial ligand binding sites, which facilitates a controlled and strong coordination of amino acid residues.

The functioning of Chl clusters in LHCs seems to be somehow related to the asymmetric monoligation of Chl molecules (Balaban, 2005), but the s and sp. orbitals involved in axial ligand binding do not discriminate between the two diastereotopic faces of the Chl macrocycle. Thus, any asymmetric or preferential axial ligation must result from steric interactions between voluminous polypeptidic ligands and the peripheral side groups of Chl (Fiedor et al., 2008; Kania & Fiedor, 2006; Oba & Tamiaki, 2002, 2014). The coordination of the histidine side residue, or less frequently other amino acid residues, to the central Mg^{2+} is not only the strongest Chl-protein interaction (Fiedor et al., 2008; Kania & Fiedor, 2006; Oba & Tamiaki, 2014), it is also the major driving force for the folding of the photosynthetic protein complexes (Davis et al., 1996; Hoober, Eggink, & Chen, 2007; Horn & Paulsen, 2002; Paulsen, Finkenzeller, & Kühlein, 1993). The Mg^{2+}-Pheo bonding and ligation of axial ligands are crucial in the stability and functioning of all photosynthetic pigment-proteins and also in their biosynthesis and biodegradation. In particular, the balance between stability and lability of the Mg complexes is of global importance to the biosphere because it facilitates the circulation of organic matter and elements (Christ & Hörtensteiner, 2014).

3. Photophysical features of chlorophylls

3.1 Electronic structure of chlorophylls

The unique photophysical features which underlie the diversity of the functions of Chl and the efficiency of Chls in photosynthesis stem from the properties of their electronic structure. The electronic absorption spectra of Chl *a*, Chl *b*, Chl *d*, and Chl *f* and two of their precursors, protoporphyrin IX (PPIX, free base) and protochlorophyllide (PChlide) *a*, recorded in organic solvents, are shown in Fig. 3. The absorption spectra of Chls, with three well separated major electronic transitions, differ considerably from that of a porphyrin, whereas the absorption spectrum of PChlide still resembles it. Ordered on the energy scale, the Q_Y, Q_X, and Soret bands correspond to electronic levels S_1, S_2 and S_3 of the chromophore. The spectra of Chls have a shape typical of the chlorin spectrum, with two intense absorption bands on the blue (the Soret) and red (Q_Y) edges of the visible region of the spectrum, and several bands of low intensity located in between (the Q_X region).

With respect to the spectrum of Chl *a*, the Q_Y transition in Chl *d* and Chl *f* is red-shifted by at least 30 and 40 nm, respectively, and in Chl *b*, on the contrary, it is blue-shifted by about 12 nm. The Soret maxima show the opposite trend. The relative positions of the S_1 level energies correspond well to the redox potentials of monomeric Chls in organic solvents (Chl *b* + 1.96 V, Chl *a* + 1.93 V, Chl *d* + 1.79 V, Chl *f* + 1.67 V, all in acetonitrile (Miyashita et al., 2014)). Another effect seen on the spectra of the series is a gradual strong increase in the oscillator strength of the Q_Y transition, apparently due to the intensity flow from the Soret transitions. Thanks to this symmetry-related effect (see above) Chls show extremely high molar absorptivity in the red region, from 6.37×10^4 in Chl *d* to 7.00×10^4 and 7.11×10^4 [$L mol^{-1} cm^{-1}$] in Chl *a* and Chl *f*, respectively, as determined in methanol (Li, Scales, Blankenship, Willows, & Chen, 2012; Porra, Thompson, & Kreidmann, 1989). Intriguingly, the relative increase in the Q_Y transition intensity does not necessarily translate into an increase in the cross section for light absorption.

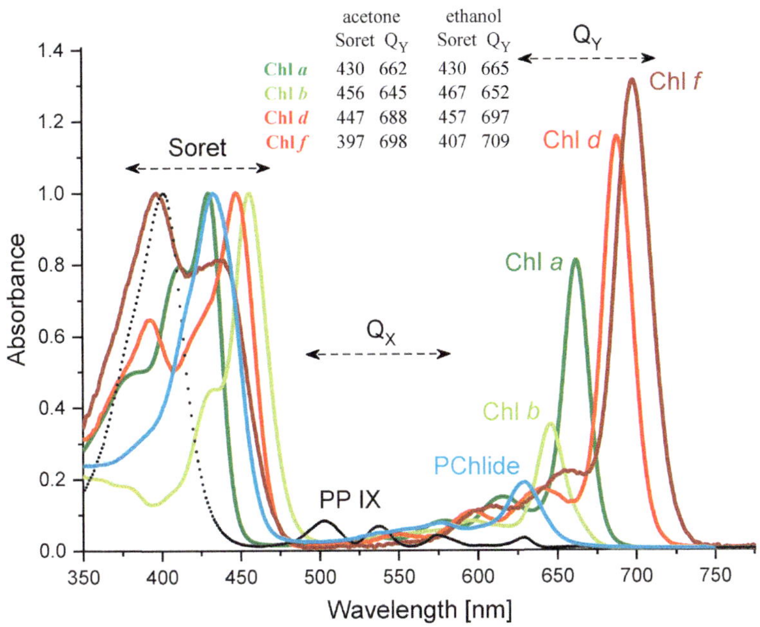

Fig. 3 Electronic absorption spectra of protoporphyrin IX (PPIX), protochlorophyllide *a* (PChlide) and various chlorophylls in acetone, normalized to their maxima in the Soret region. In the inset, the positions of the major absorption maxima (in nm) of Chls in acetone and ethanol. Apparently, the energies of the electronic transitions in Chls are to a large extent determined by substituents capable of coupling to the central π-electron system (compare the structures in Fig. 1) and only slightly are affected by the interactions with solvents. The spectrum of PChlide: the courtesy of Dr. B. Myśliwa-Kurdziel (Myśliwa-Kurdziel, Kruk, & Strzałka, 2004); spectra of Chl *d* and Chl *f*: the courtesy of Prof. M. Chen (Chen & Blankenship, 2011).

The macrocyclic skeleton of Chls is host to a large aromatic system of delocalized π-electrons, whose size and shape (symmetry) may vary depending on the coupling to peripheral substituents that possess π-electrons, such as vinyl or carbonyl residues (see Fig. 1). The basic properties of this π-electron system can be explained well within the so called four-orbital model developed by Gouterman, Wagniere, and Snyder (1972), which considers a porphyrin whose valence molecular orbitals (MOs) are perturbed due to changes in molecular symmetry. In an unsubstituted porphyrin (D_{4h} symmetry), the valence molecular orbitals, i.e., the highest occupied MO (HOMO) and HOMO − 1, correspond to the a_{1u} (π) and a_{2u} (π), and lowest unoccupied MO (LUMO) and LUMO + 1 to e_{gx} (π*) and e_{gy} (π*). The energies of the two π orbitals just happen to be close to each other, and the HOMO can be either one, whereas the two π* orbitals constitute a

Glossary of basic quantum chemistry terms

Orbital symmetry. The group theory is often used in quantum chemistry to classify the wavefunctions (and molecules as well) according to their symmetry properties. The **a** and **e** symbols and the *g* (gerade) and *u* (ungerade) indices are used for orbitals to denote how they behave upon rotation and with respect to an inversion center, respectively. In addition, the a and e symbols carry information about orbital degeneracy (**a**—non degenerated, **e**—doubly degenerated).

Computational methods. The wavefunction of a molecule is complicated and contains more information than is usually needed and could possibly be directly interpreted in physical terms. Therefore, it can be replaced by a simpler function that involves fewer variables and from which it is easier (and faster) to calculate the energy and other properties of the molecule. In the most commonly used approach, called *density-functional theory* (**DFT**), the total energy and other ground state properties of molecules are calculated from their ground state electron densities, described by respective functionals. The *coupled-cluster* (**CC**) *method* is a more advanced approach, in which an exact form of the ground state molecular electronic wavefunction is applied, as obtained by using the cluster operator acting on the ground state Hartee-Fock wavefunction. The *configuration interaction* (**CI**) approach takes into account the electron correlation effects by including linear combinations of the excited and ground state wavefunctions in the description of the total wavefunction.

In order to implement the Hartee-Fock or DFT methods in the numerical calculations, complex mathematical expressions (the *potentials*), which describe the physical forces and interactions between the components of atoms (electrons and nuclei), are used to calculate the energy of the system (an atom, a molecule) on the basis of functions which determine electron density distribution. These *basis functions* (basis set) are composed of functions which describe electrons on the individual atoms and represent the overall electronic wavefunction of a molecule.

Modern ab initio computational methods enable, via the so called population analysis (introduced by Mulliken, developed further by Loewdin), to estimate the chemical reactivity of atoms or molecules by calculating the electron density distribution on the atoms (atomic charges). This helps to bring the outcomes of the calculations closer to intuitively understood chemical properties, such *hardness* and *softness* of atoms and molecules, taken from the theory of soft and hard acids and bases. In this respect, often *Fukui functions* are used, which describe the electron density in frontier orbitals (**HOMO/LUMO**).

degenerative pair. In symmetric porphyrins, the transitions between the HOMO and LUMO are either additive or cancel each other out, as reflected in their electronic absorption profiles (Fig. 3) which comprise a very intense high-energy band (the Soret) and a series of weak bands situated at lower energies (the Q bands). Consecutive saturations of the C_β-C_β bonds in the pyrrole rings in Chls (and bacteriochlorophylls) remove this degeneracy and lead to discrimination between the in-plane *X*-*Y* molecular axes. This shift in orbital energies affects the electronic transitions that result from interactions between one-electron promotions from the a_{1u} (π) and a_{2u} (π) orbitals to the e_{gx} (π^*) and e_{gy} (π^*) orbitals of differing *X*-*Y* polarizations. The pairwise combinations of these promotions give rise to four transitions of varying oscillator strengths, namely the intense B_X, B_Y (the Soret bands) and Q_Y bands, and a weak Q_X band.

The properties of Chls as the key photoactive cofactors of photosynthesis have for a long time been challenged by quantum-chemistry computational approaches, contributing to a better understanding of the mechanisms underlying various aspects of the functioning of these unique pigments (Scherer & Fischer, 2009). The calculations intended at first to explain the electronic structure and spectral properties of Chls on a semi-quantitative level, fully confirmed the validity of the Gouterman four-orbital model and were at the same time very conclusive and helpful in the interpretation of the optical spectra (Gudowska-Nowak, Newton, & Fajer, 1990; Hanson, 1991; Linnanto & Korppi-Tommola, 2000, 2004, 2006; Parusel & Grimme, 2000). Over the last 20 years, due to the development of density functional theory (DFT) and coupled-cluster theory (CC), as well as the increasing computational power, the computations of the photophysical and physicochemical features of Chls and BChls have become much more quantitative (Linnanto & Korppi-Tommola, 2000, 2004; Sundholm, 1999). However, no universal computational method exists and the applicability, precision and fidelity of the calculations depend on how a particular combination of the potential and basis set is able to reproduce the particular features of the system. In DFT-based methods, the Becke three-parameter Lee-Yang-Parr (B3LYP), CAM-B3LYP, and Minnesota Functionals (M06, M06L, M06-2X, M06-HF) potentials are used most frequently. In more advanced approaches, in order to take the electron correlation effects into account, Møller–Plesset perturbation theory order 2 (MP2), complete active space perturbation theory (CASPT2), and symmetry adapted cluster/configuration interaction (SAC-CI) are used. The selection of a proper function basis is another important element of the calculations and the 6-31G*, cc-pVTZ and cc-pV5Z, DGDZVP bases are among those most often used. For the analysis of Chl-environment interactions with explicit solvents and in larger Chl-protein assemblies the quantum mechanics and molecular mechanics methods (QM/MM) are more appropriate (Linnanto & Korppi-Tommola, 2006; Saito, Suzuki, & Ishikita, 2018).

In general, ab initio DFT-based methods predict quite well the spectroscopic features of Chls, but even in more accurate computations, the energies of the Q_Y, O_X, B_X, and B_Y states in Chls are considerably overestimated and only their relative positions are fairly well reproduced (Cai, Crossley, Reimers, Kobayashi, & Amos, 2006; Dahlbom & Reimers, 2005). Nevertheless, recent measurements of Chl absorption spectrum in vacuo seem to somewhat validate these computational results, showing that the impact of the environment on the energy levels in Chls is greater than previously assumed (Milne, Toker, Rubio, & Nielsen, 2015).

Nowadays, ab initio computations are used, almost on a routine basis, to tackle more detailed questions concerning excited state properties, charge transfer states, the spectral and redox tuning of Chls, solvation effects, and the impact of various external factors on electronic transitions, such as interactions with axial ligands, amino acid residues and metal ions (Bechaieb, Fredj, Akacha, & Gerard, 2016; Cai et al., 2006; Heimdal, Jensen, Devarajan, & Ryde, 2007; Rutkowska-Zbik, Witko, & Fiedor, 2013). Ab initio computations employed to examine the chemical reactivity of the pigments and interactions of Chls and Pheos with metal ions, provide reactivity descriptors such as hardness, softness, electrophilicity, atomic Loewdin charges, and Fukui functions, useful in explaining the rates of metalation reactions or in identifying sites susceptible to electrophilic or nucleophilic attacks (Kania et al., 2014; Orzeł et al., 2010). Other studies, on metallosubstituted Chls, have showed large differences in the macrocycle-CMI bonding energies between the Mg complexes and complexes with transition metals ions (Pd, Pt, Cr, Ni, etc.), corroborating experimental results (Kotkowiak et al., 2017). The computational approach was also used in the vibrational analysis of Chl molecules, undertaken in order to simulate their IR and high-resolution fluorescence and phosphorescence spectra (Etinski, Petkovic, & Ristic, 2016), and to accurately predict changes in normal mode frequency upon the one electron oxidation or reduction of Chls (O'Malley, 2000).

The changes in the spectral properties, seen along the sequence of the tetrapyrrolic pigments PPIX-PChlide-Chl *a*-Chl *b*-Chl *d*-Chl *f*, reflects the major evolutionary strategy exploited by photosynthetic organisms in order to increase their photosynthetic performance and to maintain the synergic cooperativity of two photosystems under light of different spectral quality (Schliep, Cavigliasso, Quinnell, Stranger, & Larkum, 2013). Such a strategy takes advantage of the fact that the electronic properties of the chromophore can be shaped by changes in the symmetry and size of the conjugated π-electron system in Chls, as caused by a reduction of the basic tetrapyrrole π-electron system to chlorin and further oxygenation of the peripheral substituents by molecular oxygen. Apparently, the structural effects dominate the environmental effects, in particular in terms of the S_1 state energy of isolated pigments. Indeed, solvent-induced shifts of the electronic transition energies in Chls amount merely to several nanometers (see the inset in Fig. 2). However, as mentioned above, the influence of the environment on Chl spectral characteristics might be greater than commonly assumed.

4. Functioning and fine-tuning of chlorophylls

The mechanisms of the functioning of Chls in photosynthesis have been investigated extensively in all aspects since the very moment of the discovery of their crucial involvement in this multi-step complex process. The majority of Chl molecules acts primarily as absorbers of photons in the photosynthetic light-harvesting antennae while only a few are involved in charge separation in the RCs. Often the number of antenna-bound Chl molecules exceeds the number of pigment molecules bound as the redox cofactors in the RCs associated with these antennae by almost three orders of magnitude. The large number of pigment molecules bound per single antenna complex serves to increase the effective cross-section of the antennae for photon absorption. When a photon is captured, the excitation energy migrates among the antenna pigments, in a practically loss-free manner, and then the excitation is transferred either directly to RC or via another LHC, again in practically loss-free steps. The high quantum and energetic efficiency of intra- and inter-antenna energy transfer are achieved via oligomerization of Chl molecules, which form supercomplexes of excitonically coupled chromophores. Smaller ensembles of Chls, such as dimers (special pair), are present in RCs, where they facilitate ultrafast electron transfer along the chain of coupled electron carriers. In short, the electron transfer is initiated on the special pair of Chls, which, after receiving the excitation from the antenna, becomes a strong electron donor (a reductant). The electron moves then from the special pair to the primary (accessory Chl) and secondary electron acceptors—Pheo and plastoquinones in PSII reaction center. The electron transfer along the chain of these cofactors results in a stable charge separation across the photosynthetic membrane, which drives further chemical reactions. In this way, the light energy is efficiently converted into the energy of chemical bonds (in ATP, NADPH and O_2). Here, with the primary charge separation in RC, the role of Chls as photoactive and redox cofactors in oxygenic photosynthesis is ended.

What is fascinating about Chls, and about Chl *a* in particular, is the fact that the electronic properties of a single chemical entity can be manipulated and tuned so broadly that it can be active in either energy or electron transfer in the photosystems, depending on the structure of the particular Chl-protein complex. Many aspects and mechanisms that underlie this gross and fine tuning of Chls are yet to be revealed. It is understood that the fine-tuning of Chl properties is achieved via a network of weak intermolecular interactions that are experienced by these cofactors within the photosynthetic structures. The effects of the tunings involve both inter-pigment (Chl-Chl and carotenoid (Crt)-Chl) and pigment-protein interactions.

4.1 The energy, quantum yield and lifetime of the S_1 state

The electronic properties of Chls are prone to modulation and fine-tuning to suit the specific tasks to be performed. For light energy absorption and excitation transfer in the antennae or charge separation in RCs, most relevant is tuning of the S_1 state of the pigment (or its assemblies). The initial level of this tuning is achieved via structural modifications, giving rise to a series of naturally occurring Chls *a-f* (Niedzwiedzki & Blankenship, 2010). Due to the modification at the C-2 position, i.e., stretching the π-electron system along the *Y* molecular axis, which lowers the S_1 state energy, the spectral limits of oxygenic photosynthesis (and the RC photochemistry) in organisms producing Chl *f* are pushed ~50 nm beyond the "red limit" of 700 nm as set by Chl *a* (Nürnberg et al., 2018). Conversely, when such a modification is made along the *X* axis, as in Chl *b*, it results in the narrowing of the Soret-Q energy gap so that more photons can be absorbed in the 400–600 nm region.

Structural factors also strongly affect the properties of the excited states of Chls. Thus, a drastic shortening of the S_1 lifetime (in solution) occurs when going from porphyrin (~11 ns) to Chl *a* (5.7 ns) and Chl *b* (3.7 ns), which seems to be related to the changes in molecular symmetry and the energy gap law. However, this does not quite agree with the concomitant trend seen in the value of the S_1 state quantum yield, which increases from very low in porphyrin (0.03) to above 0.20 in Chls (Karcz et al., 2014). Another important tuning factor is the centrally bound metal ion. The Chl π-electron system appears to be affected more strongly by the central metal ion than that of unsubstituted porphyrins, owing to a higher electron density localized on the core nitrogens in Chls and hence more effective interactions with the metal center (Sundholm, 1999). In the case of transition metal ions, the lowering of chelator symmetry may play a role too, because when the metal ion valence orbitals have an even (g) symmetry (due to the inversion operator), their interactions with the π-electron system of odd symmetry (u) may be weakened (Kania et al., 2014). Among the many divalent metal ions potentially bound in the central cavity of the tetrapyrrole, the influence of the central Mg^{2+} ion on the chromophore π-electron system appears to be the mildest, owing to the special character of the Mg^{2+}-macrocycle bonding. Basically, the properties of free bases are only moderately different from those of the Mg complexes. In contrast, the binding of non-spherical metal ions, such as the transition metal ions (Cu^{2+}, Co^{2+}, Fe^{2+}, Ni^{2+} etc.), results in a drastic deactivation of the S_1 excited state and complete quenching of fluorescence emission, in large parts due to a significant enhancement of internal conversion rates. The strongest effect seems to be generated by Ni^{2+}, which causes a shortening of the S_1 state lifetime to below 100 fs in Ni-substituted BChl *a* (Fiedor et al., 2000; Musewald et al., 1999). In complexes with heavy metal ions, such as Pd^{2+} and Pt^{2+}, S_1 relaxation is additionally affected by the strong spin-orbit coupling due to the heavy atom effect that favors the intersystem crossing (ISC) to the triplet state (Drzewiecka-Matuszek et al., 2005; Kotkowiak et al., 2017; Musewald et al., 1998).

Another level of Chl fine-tuning in the photosynthetic apparatus is realized through several types of non-covalent intermolecular interactions with the surrounding amino acid residues and pigment molecules. The functioning of the components of the photosynthetic apparatus in practically all instances involves oligomeric forms of Chls, whose type and size is strictly controlled by the protein matrix. In the photosynthetic ensembles, the pigment molecules stay in close contact with each other and their photophysical properties undergo specific changes. A large contribution to this fine-tuning of Chls comes from the π-π and excitonic interactions within the oligomers, which lead to significant shifts in their electronic properties (Rosenbach-Belkin, Fisher, & Scherz, 1991; van Amerongen, Valkunas, & van Grondelle, 2000). In principle, the photophysical features of Chls bound within the photosynthetic complexes in many aspects differ from those of isolated pigments in organic solvents. The most pronounced changes in the excited state properties of the pigment concern the S_1 state lifetimes which are shortened to below 1 ns. At the same time, the ISC probability in Chls and the yields of harmful ROS generation are significantly reduced. Also a lowering of the S_1 state energy occurs, which is important for creating excitation energy sinks and excitation funneling through down-hill energy transfer. The lowering of the S_1 energy is somewhat less obvious for Chls (e.g., P680, P700), but it is significant in the case of BChls (van Amerongen et al., 2000), hence favoring the absorption of light in the near-IR region of the spectrum and enabling the photosynthetic bacteria to carry out photosynthetic processes in deeper water reservoirs, under the screen of Chl-containing species.

Recent studies indicate that the role of weak interactions in the tuning of photosynthetic chromophores may have been considerably underestimated. The electronic absorption bands of Chl molecules in high vacuum are blue shifted with respect to the spectra in solution or in Chl-binding proteins, which points to the strong effect of the environment on the electronic levels of the pigment (Milne et al., 2015). Hence, the solvation effects, pigment-pigment interactions and the control exerted by the protein matrix over the photophysical properties of Chl via diverse weak interactions are much stronger than previously assumed. Another mechanism of Chl tuning, especially in LHCs, involves the ring deformations (Zucchelli, Brogioli, Casazza, Garlaschi, & Jennings, 2007). A study on the water-soluble Chl-binding protein from *Brassicaceae* (WSCP) sheds new light on how this fine-tuning in Chl-protein complexes is realized on the molecular level. A structural analysis of the native type II WSCPs reveals a significant protein-induced Chl-ring distortion, which is correlated with a considerable red shift in the chromophore S_1 state energy. Such a tuning mechanism may be the key to the robustness and high efficiency of photosynthetic LHCs (Bednarczyk et al., 2016).

5. Biosynthesis and biodegradation of chlorophylls

5.1 Biosynthesis

The biosynthetic pathway of Chls involves multiple enzymatic steps and several of the initial ones are shared with other tetrapyrroles, i.e., phycobilins and hemes. The first key precursor of the tetrapyrroles is δ-aminolevulinic acid (ALA). In animals and fungi the biosynthesis of ALA proceeds along the so called C-4 pathway, while in Chl biosynthesis it is the C-5 pathway which begins with a reduction of glutamic acid (Glu) to glutamate 1-semialdehyde, followed by its transamination to ALA (Beale, 2006). The condensation of two ALA molecules, catalyzed by ALA dehydratase, yields a monopyrrole porphobilinogen, four molecules of which, in a reaction catalyzed by PBG deaminase, form a linear tetrapyrrole 1-hydroxymethylbilane. The macrocycle closure, mediated by uroporphyrinogen III synthase, yields uroporphyrinogen III, which is then enzymatically converted to coproporphyrinogen III. The next intermediate, protoporphyrinogen (protogen) IX, formed in an oxidative decarboxylation catalyzed by coproporphyrinogen oxidase, is then converted to PPIX in the presence of protogen IX oxidase. At this branching point, two different divalent metal ions can be inserted into the macrocycle binding pocket; it is either the Fe^{2+} ion, inserted by Fe chelatase to yield heme, or the Mg^{2+} ion, inserted by Mg chelatase in the first committed step in Chl biosynthesis (Yaronskaya & Grimm, 2006). Chemically, the insertion of Mg^{2+} is a difficult reaction with high energy demand (Reid & Hunter, 2004). The next step involves the formation of the isocyclic pentenone ring (ring E), which yields [8-vinyl]-PChlide (MgDVP). The isocyclic ring, which is unique and characteristic of Chls, can be formed via two unrelated mechanisms catalyzed by aerobic or anaerobic cyclase, depending on the source of the oxygen atom for the carbonyl group. The coexistence of two enzymes that catalyze a single reaction seems to be a relic of the evolutionary transition from anaerobic to aerobic metabolism. MgDVP is reduced by 8-vinyl reductase to PChlide, which has been detected at five different levels of the Chl biosynthetic pathway (MgDVP, MgPMe, PChlide *a*, Chlide *a* and 8-vinyl Chl *a*) (Chen, 2014). PChlide is reduced on ring D by PChlide oxidoreductase (POR) to Chlide *a*. Two types of POR exist, a light-dependent POR (LPOR), which occurs only in angiosperms, and a light-independent POR (DPOR), which is found in most other photosynthetic organisms, i.e., in cyanobacteria and phototrophic bacteria, gymnosperms, algae, mosses and ferns. In the light-dependent conversion of PChlide to Chlide, which uses NADPH as the reductant, PChlide serves as a photochemically active cofactor in the PChlide-LPOR enzymatic complex (Gabruk & Myśliwa-Kurdziel, 2015). The light-independent conversion requires ferredoxin for substrate reduction (Reinbothe et al., 2010). The esterification of the C-17^3 propionic acid side chain with an activated (di-phosphorylated) phytol by Chl synthase constitutes the last step in Chl *a* biosynthesis. In rare cases, the propionic moiety can be esterified with geranylgeranyl pyrophosphate (PP) or a shorter chain farnesyl-PP. The isoprenoid side chains are derived from isopentenyl-pyrophosphate (IPP) or dimethylallyl-pyrophosphate (DMAPP), which are principally synthesized via the cytoplasmic mevalonic acid or the plastidial methylerythritol phosphate pathway (Lange, Rujan, Martin, & Croteau, 2000; Rohmer, 1999). IPP and DMPP condense to geranyl-PP (two C_5 units), farnesyl-PP (three C_5 units) and geranylgeranyl-PP (four C_5 units), the intermediate which is then reduced to phytyl-PP (Masuda & Fujita, 2008).

Chl *a* (or Chlide *a*) seems to be the precursor of all other types of Chls in those organisms that carry out oxygenic photosynthesis. In the Chl *b* pathway, the C-7 methyl in Chl *a* is oxidized to a formyl group in a reaction catalyzed by Chl *a* oxygenase (CAO). Chl *b* can be re-reduced to Chl *a* through 7-hydroxymethyl-Chl *a* when there is a need to optimize the Chl *a*/Chl *b* ratio to adapt to light conditions. This backward reaction is shared with the Chl *b* biodegradation pathway. The biosynthesis of Chl *f* starts from Chl *a* and resembles that of Chl *b* (further information can be found in chapters "Biosynthesis of chlorophylls and bacteriochlorophylls in green bacteria" and "Chlorophylls *d* and *f*: Synthesis, occurrence, light-harvesting, and pigment organization in chlorophyll-binding protein complexes" of this volume by Thweatt et al., and Chen, respectively). Chl *d*, in which the C-3 vinyl group is transformed to a formyl group is synthesized by a P-450 type enzyme (Allakhverdiev et al., 2016). Chl *e* was identified as early as 1966, but its biosynthetic pathway and structure remain unclear. In all the types of Chls (Scheme 1), both the C-13^1 carbonyl oxygen and the peripheral formyl oxygen in Chl *b*, Chl *d* and Chl *f*, are derived from molecular oxygen (Garg, Loughlin, Willows, & Chen, 2017; Porra, Schäfer, Cmiel, Katheder, & Scheer, 1993).

5.2 Biodegradation

The turnover of Chls in the biosphere occurs on a massive scale, amounting to over 1×10^9 tons of these pigments biosynthesized and biodegraded per annum. For a very long time the fate of Chls released from the photosystems during the normal life activity and senescence of plants remained a biological mystery. After the characterization of series of intermediate Chl catabolites and the enzymes involved in pigment biodegradation, the process is understood in more detail. It is now seen as a sequence of well controlled steps in the secure withdrawal of Chl molecules (see the chapter "Chlorophyll breakdown—Regulation, biochemistry and phyllobilins as its products" of this volume by Hörtensteiner et al.) (Hörtensteiner & Kraütler, 2011;

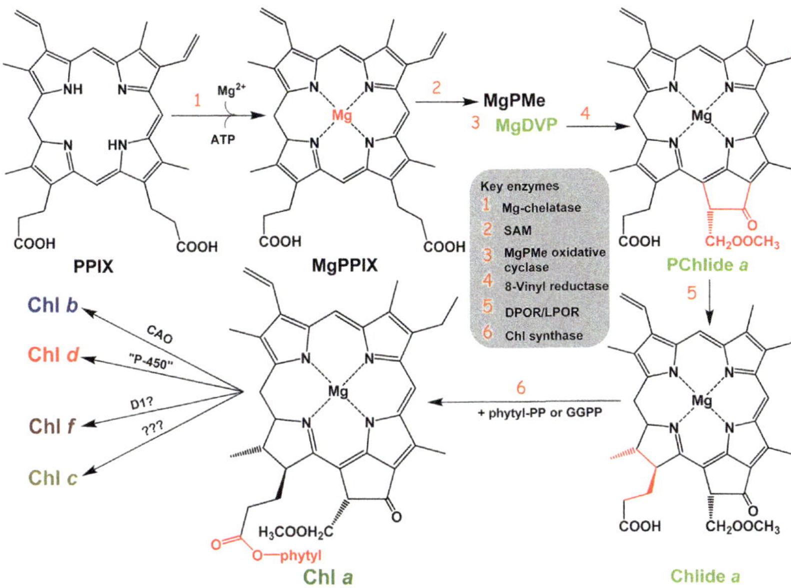

Scheme 1 The major steps chemical formulae of the key intermediates of chlorophyll biosynthetic pathways. The numbers above the arrows indicate the key enzymes (listed in the box) involved in the biosynthetic pathway of chlorophyll. *Abbreviations*: *Chlide a*, chlorophyllide *a*; *DPOR*, light-independent protochlorophyllide oxidoreductase; *GGPP*, geranylgeranyl pyrophosphate; *LPOR*, light-dependent protochlorophyllide oxidoreductase, *MgDVP*, divinyl protochlorophyllide; *MgPMe*, Mg-protoporphyrin IX monomethyl ester; *MgPPIX*, Mg-protoporphyrin IX, *PChlide a*, protochlorophyllide *a*; *phytyl-PP*, phytyl pyrophosphate; *PPIX*, protoporphyrin IX,; *SAM*, *S*-Adenosyl-L-methionine:magnesium-protoporphyrin IX *O*-methyltransferase.

Kuai, Chen, & Hörtensteiner, 2018), which, once released from their apoproteins, become highly photocytotoxic due to the photodynamic effect (Fiedor et al., 1993). Chl *a* takes the central position in the biodegradation of Chls in higher plants, and Chl *b*, in order to be degraded, needs to be converted to Chl *a* (Scheme 2). First, it is reduced by Chl *b* reductase (CBR) to 7-hydroxymethyl Chl *a*, which is then reduced to Chl *a* by hydroxymethyl Chl *a* reductase (HCAR). The breakdown of Chl *a* begins with the removal of the central Mg^{2+}, followed by the hydrolytic removal of the phytyl residue, yielding Pheide *a*, catalyzed by non-yellowing (NYE) proteins (Kuai et al., 2018) and pheophytinase (PPH) (Schelbert et al., 2009), respectively. The Pheide *a* macrocycle is opened up by Pheide *a* oxygenase (PAO) to yield the red Chl catabolite (RCC), reduced then to a primary fluorescent Chl catabolite (pFCC) via RCC reductase (RCCR). The pFCC is modified by the hydrolase TIC55 (translocon at the inner

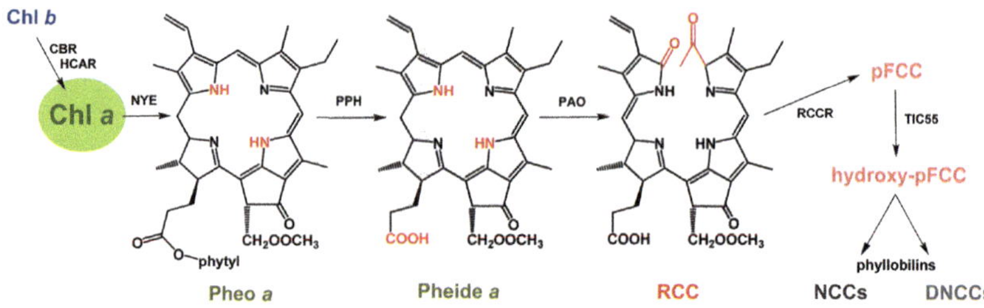

Scheme 2 The major steps and chemical formulae of the key intermediates of chlorophyll biodegradation pathways in higher plants. The numbers above the arrows indicate the key enzymes (listed in the box) involved in the biodegradation reactions of chlorophyll. *Abbreviations*: *CBR*, chlorophyll *b* reductase; *DNCCs*, dioxobilin-type non-fluorescent chlorophyll catabolites; *HCAR*, 7-OH chlorophyll *a* reductase; *hydroxyl-pFCC*, C3^3-hydroxylated primary fluorescent Chl catabolite; *NCCs*, non-fluorescent chlorophyll catabolites; *NYC*, non-yellowing protein; *PAO*, pheophorbide *a* oxygenase; *pFCC*, primary fluorescent chlorophyll catabolite; *Pheide a*, pheophorbide *a*; *Pheo a*, pheophytin *a*; *PPH*, pheophytinase; *RCC*, red chlorophyll catabolite; *RCCR*, red chlorophyll catabolite reductase; *TIC55*, translocon at the inner chloroplast envelope 55.

chloroplast envelope 55) to 3^2-hydroxylated pFCC (hydroxy-pFCC or secondary FCC). The further catabolites of Chl, termed phyllobilins, which are formed from hydroxy-pFCC via several modifications at various sites, are categorized into non-fluorescent Chl catabolites (NCCs) and dioxobilin-type NCCs (DNCCs) (Barry, 2009; Kuai et al., 2018). They are the terminal catabolites of Chl stored in vacuoles (Hörtensteiner & Kraütler, 2011).

Chlorophyllase (Chlase), the first plant enzyme discovered more than a century ago by Willstätter and Stoll, was considered for a long time to be an integral part of the Chl biodegradation pathway (Willstätter & Stoll, 1913). This highly hydrophobic hydrolase catalyzes, in a stereospecific manner, the hydrolysis (and transesterification) of the ester bond at the C-17^3 side chain (Fiedor, Rosenbach-Belkin, Sai, & Scherz, 1996; Fiedor, Rosenbach-Belkin, & Scherz, 1992). There are many plant species in which Chlase is constitutively expressed and substantial levels of the enzyme are present in the leaves during the entire vegetation season (Brown, Houghton, & Hendry, 1991; Fiedor et al., 1992; Hendry, Houghton, & Brown, 1987). However, Chlase being located outside the plastids and catalyzing the Chlide release upon cell damage, is now proposed to rather function in a plant defense system against pathogens (Kariola, Brader, Li, & Palva, 2005) and herbivores (Hu et al., 2015). Chlase seems to be regulated at the level of gene induction and expression; ethylene induces de novo synthesis of the enzyme, both at the level of mRNA and protein (Azoulay-Shemer et al., 2008; Jacob-Wilk, Holland, Goldschmidt, Riov, & Eyal, 1999; Trebitsh, Goldschmidt, & Riov, 1993). In *Arabidopsis*, two genes, *AtCLH1* and *AtCLH2*, have been found to encode Chlases. The level of the *AtCLH*1 mRNA can be enhanced by several stimuli: jasmonic acid (Tsuchiya et al., 1999), wounding, or pathogen attack (Kariola et al., 2005), and it decreases during senescing (Lin & Wu, 2004).

Recently, many different Chl catabolic genes (CCGs), which undergo regulation during the processes of senescence and Chl breakdown, were discovered. Three important phytohormones, jasmonic acid, ethylene and abscisic acid, seem to be the major regulation factors of Chl breakdown and they act at the level of regulation of several key CCGs. Abscisic acid promotes the transcription of *NYE1–3, NYC 1, PAO, PPH* and *RCCR*, involved in the degreening during seed maturation and abiotic stress-induced leaf senescence. The regulatory property of ethylene leads to the transcription of *NYE1* and *NYC1*, being an important part of the degreening during fruit maturation and natural leaf senescence, whereas jasmonic acid influences the expression of *NYC1, NYE1/2* and *PAO*. Some CCGs, in particular *NYE1, NYC1* and *PPH*, were shown to be regulated by light (Kuai et al., 2018). Intriguingly, also the expression of the *AtCLH1* and *AtCLH2* gene products is light-controlled, during both regular Chl turnover and senescence. The light effect on *AtCLH1* is mediated by a complicated interplay of photoreceptors, while up-regulation of *AtCLH2* is mediated by cryptochromes and modulated by phototropin 1 and phytochrome A (Banaś, Łabuz, Sztatelman, Gabryś, & Fiedor, 2011).

The enzymes involved in Chl biosynthesis/biodegradation seem to be highly stereoselective toward their substrates. In particular, the LPOR-catalyzed reduction of PChlide, the phytylation, the last step of Chl biosynthesis, and phytyl removal by Chlase, were shown to be stereospecific and occur only with the non-prime diastereoisomers of the pigments as the substrates. Thus, the reactions yield stereochemically pure non-prime products (13^2-*S* diastereoisomers) (Fiedor et al., 1992; Helfrich, Schoch, Lempert, Cmiel, & Rüdiger, 1994; Helfrich, Schoch, Schäfer, Ryberg, & Rüdiger, 1996). The available high resolution crystal structures of various Chl-protein complexes show that strict stereoselectivity is also retained during their assembly, In conclusion, a high degree of stereoselectivity is generally seen at all levels of Chl-protein interactions.

6. Photosensitizing versus photoprotective properties of chlorophylls

Chls show very strong photosensitizing abilities owing to intrinsically high quantum yields of triplet states in these molecules, reaching up to 70% in organic solvents, followed by highly efficient energy transfer to molecular oxygen (Bowers & Porter, 1967; Fiedor et al., 1993). It appears then to be a paradox that nature has selected their S_1 states to be involved in the conversion of photon energy into biochemically useful energy. These singlet-singlet energy transfers must compete both with a very efficient $S_1 \rightarrow T_1$ transition and the sensitization of O_2, the very (by)product of photosynthesis. To this end, the photophysical processes involved in photosynthetic light energy conversion proceed with very high rates and efficiencies in order to minimize the risk of the formation of highly harmful triplets. This is secured by interplay of many weak pigment-pigment and pigment-protein interactions, which functionally stabilize optimal structures of the pigment-protein complexes, as outlined above.

Such high yields of triplet state formation in Chls are not unusual, being comparable to other aromatic systems, in which ISC is a privileged S_1 relaxation pathway, even in the absence of heavy elements in these molecules. Apparently, not the heavy atom effect, but some other mechanisms are responsible for strong spin-orbit coupling in such compounds and they are still being investigated. In porphyrins and related macrocyclic compounds, vibronic coupling is the major mechanism for the enhancement of spin-orbit interactions (Etinski et al., 2016; Gouterman, Schwarz, & Smith, 1973; Perun, Tachten, & Marian, 2008). Thus, photosynthetic energy conversion inevitably always competes with the formation of Chl triplet states and the generation of reactive oxygen species, which, especially under high light stress, may damage the photosynthetic apparatus. Because photoprotection is essential for maintaining high efficiency of photosynthesis and the robustness of the photosynthetic apparatus, many different sophisticated photoprotective mechanisms have evolved to secure a harmless flow of excitation energy through the photosynthetic apparatus and Chls seem to be in the frontline of photoprotection (Kotkowiak et al., 2017). Recent investigations show that the binding of Mg^{2+} in Chls activates several intramolecular photoprotective mechanisms which act on various molecular levels. One mechanism, which operates on the level of spin state mixing, lowers the efficiency of ISC and elevates the quantum yield of S_1, while another one reduces the efficiency of energy transfer from T_1 to O_2. The next level of photoprotection is due to the axial ligands, whose binding affects the relaxation pathway of T_1 in Chl, once it has been populated. The axial ligands pose a significant steric hindrance, which reduces the rate of effective collisions with O_2 molecules. These intrinsic photoprotective mechanisms in Chls, embedded in their molecular design, are highly effective. They both suppress the formation of Chl triplet states and the efficiency of energy transfer to O_2, each by 20–25%, and elevate the probability of S_1 population, which is critical for photosynthetic light energy conversion. Such built-in photoprotective effects must have created a large evolutionary advantage for the Mg complexes during their evolution as the principal photoactive cofactors of photosynthetic proteins. The involvement of the Mg^{2+} in the activation of these intrinsic photoprotective mechanisms underlies its choice as the central metal ion in Chls (Kotkowiak et al., 2017).

Chls are involved in another photoprotective mechanism against the harmful effects of over-excitation of the photosystems. Depending on the specific Chl-protein complex, this photoprotection by Chls also takes place at several levels. In virtually all types of photosynthetic complexes, a way to avoid ROS formation is efficient energy transfer from Chls in the triplet state to Crts bound at specific sites proximal to Chls. This excess energy is then safely, in a non-radiative manner, dissipated to its surroundings while the Crt molecules return to their ground state (Young & Britton, 1993). In the antenna complexes of PSII, the excess excitation energy is partly dissipated by the emission of fluorescence from Chl *a*, while in PSI, especially in its trimeric form in cyanobacteria, the "red emitters," i.e., Chl *a* molecules with far-red shifted emission, seem to play the role of such a "safety valve" (Karapetyan et al., 2014; Ruban et al., 2007). In the RC of PSII, the special Chl *a* molecule, Chl_Z, participates in cyclic electron flow which is thought to be another "safely valve." In addition, oxidized Chl_Z acts as the excitation energy trap near the P680 site (Barber, 2016).

7. Applications of chlorophylls

Because the advantageous photophysical features of Chls meet the criteria of excellent photosensitizers (see above), their most important applications are related to photodynamic therapy (PDT) (Perez-Galvez, Viera, & Roca, 2017; Solymosi & Mysliwa-Kurdziel, 2017; Staroń, Boroń, Karcz,

Szczygieł, & Fiedor, 2015). The intense electronic transitions of Chls, which span the spectral range from ~630 to 800 nm (in the case of BChls), fit the so called "phototherapeutic window" well (Fig. 2). In addition, they show low dark toxicity and relatively rapid clearance from the body, and their chemistry is well established, facilitating structural modifications. Since a long time, Chls have attracted attention as photodynamic agents, in spite of their chemical and photochemical lability and relatively high costs of isolation. The use of Chls in PDT requires modifications to improve their stability and bioavailability, while preserving their favorable features, and this can be done enzymatically (Fiedor et al., 1996). Most of these alterations are made rather to the peripheral groups which do not conjugate to the π-electron system. The C-3 substituents, such as vinyl in Chl *a* or acetyl in BChl *a*, are also modified, but because of their coupling to the π-electron system these modifications affect the spectral characteristics of the pigments. Aiming to obtain more active PDT agents, several modifications were also made at the isocyclic ring, yielding water-soluble derivatives with red-shifted Q_Y absorption maxima. The physico-chemical properties and pharmacokinetics of Chl-based photosensitizers are also enhanced via the replacement of the central Mg^{2+} ion with various divalent metal ions (Staroń et al., 2015). Heavy central metal ions, such as Pd^{2+} or Pt^{2+}, introduce strong spin-orbit coupling, which favors ISC as the relaxation pathway (Drzewiecka-Matuszek et al., 2005; Hartwich et al., 1998; Kotkowiak et al., 2017). Due to the heavy metal atom effect the quantum yield of the T_1 state reaches 1, enhancing the photodynamic activity. In addition, the chelation of transition metal ions elevates the photostability of the pigment (Fiedor, Fiedor, Kammhuber, Scherz, & Scheer, 2002; Pilch et al., 2013). So far, several photodynamic drugs containing Chl-derived pigments as the active compound have been admitted for use in clinics. Among these, two derivatives of Chl *a*, chlorin e_6 (N-aspartyl-functionalized chlorophyllin) and Photochlor® (derivatized pyroPheide), as well as Tookad®, a palladium complex of modified bacterioPheide, are the most prominent examples (Perez-Galvez et al., 2017; Solymosi & Mysliwa-Kurdziel, 2017; Staroń et al., 2015).

Other applications of Chls are related to their most obvious feature—a strong green color, which naturally makes them good candidates as food colorants. Long since, a water-soluble Cu-chlorophyllin has been used as a color-enhancing additive, which is a blend of several Cu-substituted derivatives obtained by alkalic treatment of a raw Chl *a*/Chl *b* mixture. Also partly-separated or raw Chl *a* and Chl *b* have been accepted for use as food colorants. In the EU, the Chl-derived food additives are known as E140/E141 (Solymosi & Mysliwa-Kurdziel, 2017).

Inevitably, due to the consumption of plant-derived foods, Chls interact with animal and human organisms, and their daily uptake with a balanced diet may reach as much as 200 mg (Perez-Galvez et al., 2017). Intriguingly, the first steps of Chl catabolism in animal's body under acidic conditions release two toxic xenobiotics: a highly phototoxic Pheide and phytol. The latter is oxidatively converted into phytanic acid, which accumulates in a number of metabolic disorders. It is toxic at high levels in the body and therefore needs to be metabolized (Wanders, Komen, & Ferdinandusse, 2011). The accumulation of phototoxic Pheide poses an even more severe risk to the body due to the photodynamic effect, and indeed this phototoxin is strictly avoided, owing to the activity of specialized xenobiotic vectorial transport system (Jonker et al., 2002). Pheide is recognized and actively eliminated from the body through a specialized transporter known as the breast cancer resistance protein (BCRP or ABCG2), which is universally found in the animal world (Szafraniec, Szczygieł, Urbanska, & Fiedor, 2014). Interestingly, the transporter is able to distinguish between the Mg-containing pigment and its Zn-substituted analog (Szczygieł et al., 2008).

Acknowledgments

This work was supported by a grant from the National Science Centre, Poland to L.F. (No. UMO-2012/07/B/NZ1/02876) and by an internal grant No. K/ZDS/008227.

References

Agostiano, A., Catucci, L., Colafemmina, G., & Scheer, H. (2002). Role of functional groups and surfactant charge in regulating chlorophyll aggregation in micellar solutions. *The Journal of Physical Chemistry. B*, *106*, 1446–1454.

Airs, R. L., Temperton, B., Sambles, C., Farnham, G., Skill, S. C., & Llewellyn, C. A. (2014). Chlorophyll f and chlorophyll *d* are produced in the cyanobacterium *Chlorogloeopsis fritschii* when cultured under natural light and near-far radiation. *FEBS Letters*, *588*, 3770–3777.

Allakhverdiev, S. I., Kreslavski, V. D., Zharmukhamedov, S. K., Voloshin, R. A., Korol'kova, D. V., Tomo, T., et al. (2016). Chlorophylls d and f and their role in primary photosynthetic processes of cyanobacteria. *Biochemistry (Moscow)*, *81*, 201–212.

Azoulay-Shemer, T., Harpaz-Saad, S., Belausov, E., Lovat, N., Krokhin, O., Spicer, V., et al. (2008). *Citrus* chlorophyllase dynamics at ethylene-induced fruit color-break: A study of chlorophyllase expression, posttranslational processing kinetics, and in situ intracellular localization. *Plant Physiology*, *148*, 108–118.

Balaban, T. S. (2005). Relevance of the diastereotopic ligation of magnesium atoms of chlorophylls in the major light-harvesting complex II (LHC II) of green plants. *Photosynthesis Research*, *86*, 251–262.

Balaban, T. S., Fromme, P., Holzwarth, A. R., Krauss, N., & Prokhorenko, V. I. (2002). Relevance of the diastereotopic ligation of magnesium atoms of chlorophylls in photosystem I. *Biochimica et Biophysica Acta, 1556*, 197–207.

Banaś, A. K., Łabuz, J., Sztatelman, O., Gabryś, H., & Fiedor, L. (2011). Expression of enzymes involved in chlorophyll catabolism in *Arabidopsis* is light controlled. *Plant Physiology, 157*, 1497–1504.

Barber, J. (2016). Photosystem II: The water splitting enzyme of photosynthesis and the origin of oxygen in our atmosphere. *Quarterly Reviews of Biophysics, 49*, 1–21.

Barry, C. S. (2009). The stay-green revolution: Recent progress in deciphering the mechanisms of chlorophyll degradation in higher plants. *Plant Science, 176*, 325–333.

Beale, S. I. (2006). Biosynthesis of 5-aminolevulinic acid. In B. Grimm, R. J. Porra, W. Rüdiger, & H. Scheer (Eds.), *Chlorophylls and bacteriochlorophylls: Vol. 25* (pp. 147–158). Dordrecht: Springer.

Bechaieb, R., Fredj, A. B., Akacha, A. B., & Gerard, H. (2016). Interactions of copper(II) and zinc(II) with chlorophyll: Insights from density functional theory studies. *New Journal of Chemistry, 40*, 4543–4549.

Bednarczyk, D., Dym, O., Prabahar, V., Peleg, Y., Pike, D. H., & Noy, D. (2016). Fine tuning of chlorophyll spectra by protein-induced ring deformation. *Angewandte Chemie, International Edition, 55*, 6901–6905.

Ben-Fredj, A., Ben Lakhdar, Z., & Ruiz-Lopez, M. F. (2009). Six-coordination in chlorophylls: The fundamental role of dispersion energy. *Chemical Physics Letters, 472*, 243–247.

Bowers, P. G., & Porter, G. (1967). Quantum yields of triplet formation in solutions of chlorophyll. *Proceedings of the Royal Society of London. Series A: Mathematical and Physical Sciences, 296*, 435–441.

Brown, S. B., Houghton, J. D., & Hendry, G. A. (1991). Chlorophyll breakdown. In H. Scheer (Ed.), *Chlorophylls* (pp. 465–489). Boca Raton: CRC Press.

Cai, Z.-L., Crossley, M. J., Reimers, J. R., Kobayashi, R., & Amos, R. D. (2006). Density functional theory for charge transfer: The nature of the N-bands of porphyrins and chlorophylls revealed through CAM-B3LYP, CASPT2, and SAC-CI calculations. *The Journal of Physical Chemistry. B, 110*, 15624–15632.

Callahan, P. M., & Cotton, T. M. (1987). Assignment of bacteriochlorophyll a ligation state from absorption and resonance Raman spectra. *Journal of the American Chemical Society, 109*, 7001–7007.

Chen, M. (2014). Chlorophyll modifications and their spectral extension in oxygenic photosynthesis. *Annual Review of Biochemistry, 83*, 317–340.

Chen, M., & Blankenship, R. E. (2011). Expanding the solar spectrum used by photosynthesis. *Trends in Plant Science, 16*, 427–431.

Chen, M., Li, Y., Birch, D., & Willows, R. D. (2012). A cyanobacterium that contains chlorophyll *f*—A red-absorbing photopigment. *FEBS Letters, 586*, 3249–3254.

Chen, M., Schliep, M., Willows, R. D., Cai, Z. L., Neilan, B. A., & Scheer, H. (2010). A red-shifted chlorophyll. *Science, 329*, 1318–1319.

Christ, B., & Hörtensteiner, S. (2014). Mechanism and significance of chlorophyll breakdown. *Journal of Plant Growth Regulation, 33*, 4–20.

Clarke, R. H., & Frank, H. A. (1977). Investigation of the effect of metal substitution on the triplet state of chlorophyll by optically detected zero-field magnetic resonance. *Chemical Physics Letters, 51*, 13–17.

Dahlbom, M. G., & Reimers, J. R. (2005). Successes and failures of time-dependent density functional theory for the low-lying excited states of chlorophylls. *Molecular Physics, 103*, 1057–1065.

Davis, C. M., Parkes-Loach, P. S., Cook, C. K., Meadows, K. A., Bandilla, M., Scheer, H., et al. (1996). Comparison of the structural requirements for bacteriochlorophyll binding in the core light-harvesting complexes of *Rhodospirillum rubrum* and *Rhodobacter sphaeroides* using reconstitution methodology with bacteriochlorophyll analogs. *Biochemistry*, *35*, 3072–3084.

Drzewiecka-Matuszek, A., Skalna, A., Karocki, A., Stochel, G., & Fiedor, L. (2005). Effects of heavy central metal on the ground and excited states of chlorophyll. *Journal of Biological Inorganic Chemistry*, *10*, 453–462.

Etinski, M., Petkovic, M., & Ristic, M. M. (2016). A quantum-chemical study of the chlorophyll phosphorescence spectrum: Electron-vibrational coupling and coordination effects. *Chemical Physics Letters*, *647*, 139–144.

Fiedor, L. (2006). Hexacoordination of bacteriochlorophyll in photosynthetic antenna LH1. *Biochemistry*, *45*, 1910–1918.

Fiedor, J., Fiedor, L., Kammhuber, N., Scherz, A., & Scheer, H. (2002). Photodynamics of the bacteriochlorophyll-carotenoid system. 1. Influence of central metal, solvent and β-carotene on photobleaching of bacteriochlorophyll derivatives. *Photochemistry and Photobiology*, *76*, 145–152.

Fiedor, L., Gorman, A. A., Hamblett, I., Rosenbach-Belkin, V., Salomon, Y., Scherz, A., et al. (1993). A pulsed laser and pulse radiolysis study of amphiphilic chlorophyll derivatives with PDT activity toward malignant melanoma. *Photochemistry and Photobiology*, *58*, 506–511.

Fiedor, L., Kania, A., Myśliwa-Kurdziel, B., & Stochel, G. (2008). Understanding chlorophylls: Central magnesium and phytyl as structural determinants. *Biochimica et Biophysica Acta*, *1777*, 1491–1500.

Fiedor, L., Rosenbach-Belkin, V., Sai, M., & Scherz, A. (1996). Preparation of tetrapyrrole-amino acid covalent complexes. *Plant Physiology and Biochemistry*, *34*, 393–398.

Fiedor, L., Rosenbach-Belkin, V., & Scherz, A. (1992). The stereospecific interaction between chlorophylls and chlorophyllase. Possible implication for chlorophyll biosynthesis and degradation. *The Journal of Biological Chemistry*, *267*, 22043–22047.

Fiedor, L., Scheer, H., Hunter, C. N., Tschirschwitz, F., Voigt, B., Ehlert, J., et al. (2000). Introduction of a 60 fs deactivation channel in the photosynthetic antenna LH1 by Ni-bacteriopheophytin a. *Chemical Physics Letters*, *319*, 145–152.

Fiedor, L., Stąsiek, M., Myśliwa-Kurdziel, B., & Strzałka, K. (2003). Phytol as one of the determinants of chlorophyll interactions in solution. *Photosynthesis Research*, *78*, 47–57.

Gabruk, M., & Myśliwa-Kurdziel, B. (2015). Light-dependent protochlorophyllide oxidoreductase: Phylogeny, regulation and catalytic properties. *Biochemistry*, *54*, 5255–5262.

Garg, H., Loughlin, P. C., Willows, R. D., & Chen, M. (2017). The $C2^1$-formyl group in chlorophyll *f* originates from molecular oxygen. *The Journal of Biological Chemistry*, *292*, 19279–19289.

Gouterman, M., Schwarz, F. P., & Smith, P. D. (1973). Porphyrins. XXVII. Spin-orbit coupling and luminescence of group IV complexes. *The Journal of Chemical Physics*, *59*, 676–690.

Gouterman, M., Wagniere, G. H., & Snyder, L. C. (1972). Spectra of porphyrins. Part II. Four orbital model. *Journal of Molecular Spectroscopy*, *11*, 108–127.

Gudowska-Nowak, E., Newton, M. D., & Fajer, J. (1990). Conformational and environmental effects on bacteriochlorophyll optical spectra: Correlations of calculated spectra with structural results. *The Journal of Physical Chemistry*, *94*, 5795–5801.

Hanson, L. K. (1991). Molecular orbital theory of monomer pigments. In H. Scheer (Ed.), *Chlorophylls* (pp. 993–1012). Boca Raton: CRC Press.

Hartwich, G., Fiedor, L., Simonin, I., Cmiel, E., Schäfer, W., Noy, D., et al. (1998). Metal-substituted bacteriochlorophylls. 1. Preparation and influence of metal and coordination on spectra. *Journal of the American Chemical Society*, *120*, 3675–3683.

Heimdal, J., Jensen, K. P., Devarajan, A., & Ryde, U. (2007). The role of axial ligands for the structure and function of chlorophylls. *Journal of Biological Inorganic Chemistry*, *12*, 49–61.

Helfrich, M., Schoch, S., Lempert, U., Cmiel, E., & Rüdiger, W. (1994). Chlorophyll synthetase cannot synthesize chlorophyll *a'*. *European Journal of Biochemistry, 219*, 267–275.
Helfrich, M., Schoch, S., Schäfer, W., Ryberg, M., & Rüdiger, W. (1996). Absolute configuration of protochlorophyllide *a* and substrate specificity of NADPH-protochlorophyllide oxidoreductase. *Journal of the American Chemical Society, 118*, 2606–2611.
Hendry, G. A. F., Houghton, J. D., & Brown, S. B. (1987). The degradation of chlorophyll—A biological enigma. *The New Phytologist, 107*, 255–302.
Hoober, J. K., Eggink, L. L., & Chen, M. (2007). Chlorophylls, ligands and assembly of light-harvesting complexes in chloroplasts. *Photosynthesis Research, 94*, 387–400.
Horn, R., & Paulsen, H. (2002). Folding in vitro of light-harvesting chlorophyll *a/b* protein is coupled with pigment binding. *Journal of Molecular Biology, 318*, 547–556.
Hörtensteiner, S., & Kraütler, B. (2011). Chlorophyll breakdown in higher plants. *Biochimica et Biophysica Acta, 1807*, 977–988.
Hu, X., Makita, S., Schelbert, S., Sano, S., Ochiai, M., Tsuchiya, T., et al. (2015). Reexamination of chlorophyllase function implies its involvement in defense against chewing herbivores. *Plant Physiology, 167*, 660–670.
Hu, Q., Miyashita, H., Iwasaki, I., Kurano, N., Miyachi, S., Iwaki, M., et al. (1998). A photosystem I reaction center driven by chlorophyll *d* in oxygenic photosynthesis. *Proceedings of the National Academy of Sciences of the United States of America, 95*, 13319–13323.
Jacob-Wilk, D., Holland, D., Goldschmidt, E. E., Riov, J., & Eyal, Y. (1999). Chlorophyll breakdown by chlorophyllase: Isolation and functional expression of the Chlase1 gene from ethylene-treated citrus fruit and its regulation during development. *The Plant Journal, 20*, 653–661.
Jonker, J. W., Buitelaar, M., Wagenaar, E., van der Valk, M. A., Scheffer, G. L., Scheper, R. J., et al. (2002). The breast cancer resistance protein protects against a major chlorophyll-derived dietary phototoxin and protoporphyria. *Proceedings of the National Academy of Sciences of the United States of America, 99*, 15649–15654.
Jordan, P., Fromme, P., Witt, H. T., Klukas, O., Saenger, W., & Krauß, N. (2001). Three-dimensional structure of cyanobacterial photosystem I at 2.5 A resolution. *Nature, 411*, 909–917.
Kania, A., & Fiedor, L. (2006). Steric control of bacteriochlorophyll ligation. *Journal of the American Chemical Society, 128*, 454–458.
Kania, A., Pilch, M., Rutkowska-Zbik, D., Susz, A., Heriyanto, Stochel, G., et al. (2014). High-pressure and theoretical studies reveal significant differences in the electronic structure and bonding of magnesium, zinc and nickel ions in metalloporphyrinoids. *Inorganic Chemistry, 53*, 8473–8484.
Karapetyan, N. V., Bolychevtseva, Y. V., Yurina, N. P., Terekhova, I. V., Shubin, V. V., & Brecht, M. (2014). Long-wavelength chlorophylls in photosystem I of cyanobacteria: Origin, localization, and functions. *Biochemistry (Moscow), 79*, 213–220.
Karcz, D., Boroń, B., Matwijczuk, A., Furso, J., Staroń, J., Ratuszna, A., et al. (2014). Lessons from chlorophylls: Modifications of porphyrinoids towards optimized solar energy conversion. *Molecules: A Journal of Synthetic Chemistry and Natural Product Chemistry, 19*, 15938–15954.
Kariola, T., Brader, G., Li, J., & Palva, E. T. P. (2005). Chlorophyllase 1, a damage control enzyme, affects the balance between defense pathways in plants. *Plant Cell, 17*, 282–294.
Kotkowiak, M., Dudkowiak, A., & Fiedor, L. (2017). Intrinsic photoprotective mechanisms in chlorophylls. *Angewandte Chemie, International Edition, 56*, 10457–10461.
Kuai, B., Chen, J., & Hörtensteiner, S. (2018). The biochemistry and molecular biology of chlorophyll breakdown. *Journal of Experimental Botany, 69*, 751–767.
Küpper, H., Setlik, I., & Spiller, M. (2002). Heavy metal-induced inhibition of photosynthesis: Targets of in vivo heavy metal chlorophyll formation. *Journal of Phycology, 38*, 429–441.

Lange, B. M., Rujan, T., Martin, W., & Croteau, R. (2000). Isoprenoid biosynthesis: The evolution of two ancient and distinct pathways across genomes. *Proceedings of the National Academy of Sciences of the United States of America, 97*, 13172–13177.

Larkum, A. W. D., & Kühl, M. (2005). Chlorophyll d: The puzzle resolved. *Trends in Plant Science, 10*, 355–357.

Li, Y., Scales, N., Blankenship, R. E., Willows, R. D., & Chen, M. (2012). Extinction coefficient for red-shifted chlorophylls: Chlorophyll *d* and chlorophyll *f*. *Biochimica et Biophysica Acta, 1817*, 1292–1298.

Lin, J.-F., & Wu, S.-H. (2004). Molecular events in senescing *Arabidopsis* leaves. *The Plant Journal, 39*, 612–628.

Linnanto, J., & Korppi-Tommola, J. (2000). Spectroscopic properties of Mg-chlorin, Mg-porphin, and chlorophylls a, b, c1, c2, c3 and d studied by semi-empirical and ab initio MO/CI methods. *Physical Chemistry Chemical Physics, 2*, 4962–4970.

Linnanto, J., & Korppi-Tommola, J. (2004). Semiempirical PM5 molecular orbital study on chlorophylls and bacteriochlorophylls: Comparison of semiempirical, ab initio, and density functional results. *Journal of Computational Chemistry, 25*, 123–137.

Linnanto, J., & Korppi-Tommola, J. (2006). Quantum chemical simulation of excited states of chlorophylls, bacteriochlorophylls and their complexes. *Physical Chemistry Chemical Physics, 8*, 663–687.

Liu, Z., Yan, H., Wang, K., Kuang, T., Jiping, Z., Gui, L., et al. (2004). Crystal structure of spinach major light-harvesting complex at 2.72 A resolution. *Nature, 428*, 287–292.

Masuda, T., & Fujita, Y. (2008). Regulation and evolution of chlorophyll metabolism. *Photochemical & Photobiological Sciences*, 7, 1131–1149.

Mazaki, H., Watanabe, T., Takahashi, T., Struck, A., & Scheer, H. (1992). Pheophytinization of eight chlorophyll derivatives in aqueous acetone. *Bulletin of the Chemical Society of Japan, 65*, 3212–3214.

Milne, B. F., Toker, Y., Rubio, A., & Nielsen, S. B. (2015). Unraveling the intrinsic color of chlorophyll. *Angewandte Chemie, International Edition, 54*, 2170–2173.

Miyashita, H., Ohkubo, S., Komatsu, H., Sorimachi, Y., Fukayama, D., Fujinuma, D., et al. (2014). Discovery of chlorophyll d in *Acaryochloris marina* and chlorophyll *f* in a unicellular cyanobacterium, strain KC1, isolated from Lake Biwa. *Journal of Physical Chemistry & Biophysics, 4*, 149.

Musewald, C., Hartwich, G., Lossau, H., Gilch, P., Pöllinger-Dammer, F., Scheer, H., et al. (1999). Ultrafast photophysics and photochemistry of [Ni]-bacteriochlorophyll a. *The Journal of Physical Chemistry, 103*, 7055–7060.

Musewald, C., Hartwich, G., Pöllinger-Dammer, F., Lossau, H., Scheer, H., & Michel-Beyerle, M. E. (1998). Time-resolved spectral investigation of bacteriochlorophyll *a* and its transmetalated derivatives [Zn]-bacteriochlorophyll *a* and [Pd]-bacteriochlorophyll *a*. *The Journal of Physical Chemistry B, 102*, 8336–8342.

Myśliwa-Kurdziel, B., Kruk, J., & Strzałka, K. (2004). Fluorescence lifetimes and spectral properties of protochlorophyllide in organic solvents and their relations to the respective parameters in vivo. *Photochemistry and Photobiology, 79*, 62–67.

Niedzwiedzki, D. M., & Blankenship, R. E. (2010). Singlet and triplet excited state properties of natural chlorophylls and bacteriochlorophylls. *Photosynthesis Research, 106*, 227–238.

Nürnberg, D. J., Morton, J., Santabarbara, S., Telfer, A., Joliot, P., Antonaru, L. A., et al. (2018). Photochemistry beyond the red limit in chlorophyll *f*-containing photosystems. *Science, 360*, 1210–1213.

Oba, T., & Tamiaki, H. (2002). Which side of the π-macrocycle plane of (bacterio)chlorophylls is favored for binding of the fifth ligand? *Photosynthesis Research, 74*, 1–10.

Oba, T., & Tamiaki, H. (2014). Asymmetry of chlorophylls in photosynthetic proteins: From the viewpoint of coordination chemistry. *Journal of Porphyrins and Phthalocyanines, 18*, 919–932.

O'Malley, P. J. (2000). The effect of oxidation and reduction of chlorophyll *a* on its geometry, vibrational and spin density properties as revealed by hybrid density functional methods. *Journal of the American Chemical Society*, *122*, 7798–7801.

Orzeł, Ł., Kania, A., Rutkowska-Zbik, D., Susz, A., Stochel, G., & Fiedor, L. (2010). Structural and electronic effects in the metalation of porphyrinoids. Theory and experiment. *Inorganic Chemistry*, *49*, 7362–7371.

Parusel, A. B. J., & Grimme, S. (2000). A theoretical study of the excited states of chlorophyll *a* and pheophytin *a*. *The Journal of Physical Chemistry B*, *104*, 5395–5398.

Paulsen, H., Finkenzeller, B., & Kühlein, N. (1993). Pigments induce folding of light-harvesting chlorophyll *a*/b-binding protein. *European Biophysics Journal*, *215*, 809–816.

Perez-Galvez, A., Viera, I., & Roca, M. (2017). Chemistry in the bioactivity of chlorophylls: An overview. *Current Medicinal Chemistry*, *24*, 4515–4536.

Perun, S., Tachten, J., & Marian, C. M. (2008). Singlet and triplet excited states and intersystem crossing in free-base porphyrin: TDDFT and DFT/MRCI study. *Chemphyschem*, *9*, 282–292.

Pilch, M., Dudkowiak, A., Jurzyk, B., Łukasiewicz, J., Susz, A., Stochel, G., et al. (2013). Molecular symmetry determines the mechanism of a very efficient ultrafast excitation-to-heat conversion in Ni-substituted chlorophylls. *Biochimica et Biophysica Acta*, *1827*, 30–37.

Porra, R. J., Schäfer, W., Cmiel, E., Katheder, I., & Scheer, H. (1993). Derivation of the formyl-group oxygen of chlorophyll *b* from molecular oxygen in greening leaves of a higher plant (*Zea mays*). *FEBS Letters*, *323*, 31–34.

Porra, R. J., Thompson, W. A., & Kreidmann, P. E. (1989). Determination of accurate extinction coefficients and simultaneous equations for assaying chlorophylls a and b extracted with four different solvents: Verification of the concentration of chlorophyll standards by atomic absorption spectroscopy. *Biochimica et Biophysica Acta*, *975*, 384–394.

Reid, J. D., & Hunter, C. N. (2004). Magnesium-dependent ATPase activity and cooperativity of magnesium chelatase from *Synechocystis* sp. PCC6803. *The Journal of Biological Chemistry*, *279*, 26893–26899.

Reinbothe, C., El Bakkouri, M., Buhr, F., Muraki, N., Nomata, J., Kurisu, G., et al. (2010). Chlorophyll biosynthesis: Spotlight on protochlorophyllide reduction. *Trends in Plant Science*, *15*, 614–624.

Rohmer, M. (1999). The discovery of a mevalonate-independent pathway for isoprenoid biosynthesis in bacteria, algae and higher plants. *Natural Product Reports*, *16*, 565–574.

Rosenbach-Belkin, V., Fisher, J. R. E., & Scherz, A. (1991). Effect of nonexitonic interactions among the paired molecules on the Qy transition of bacteriochlorophyll dimers. Applications to the primary electron donors P-860 and P-960 in bacterial reaction centers. *Journal of the American Chemical Society*, *113*, 676–678.

Ruban, A. V., Berera, R., Ilioaia, C., van Stokkum, I. H. M., Kennis, J. T. M., Pascal, A. A., et al. (2007). Identification of a mechanism of photoprotective energy dissipation in higher plants. *Nature*, *450*, 575–578.

Rutkowska-Zbik, D., Witko, M., & Fiedor, L. (2013). Ligation of water to magnesium chelates of biological importance. *Journal of Molecular Modeling*, *19*, 4661–4667.

Saito, K., Suzuki, T., & Ishikita, H. (2018). Absorption-energy calculations of chlorophyll *a* and b with an explicit solvent model. *Journal of Photochemistry and Photobiology A: Chemistry*, *358*, 422–431.

Scheer, H. (1991). Structure and occurrence of chlorophylls. In H. Scheer (Ed.), *Chlorophylls* (pp. 3–30). Boca Raton: CRC Press.

Scheer, H. (2006). An overview of chlorophylls and bacteriochlorophylls: Biochemistry, biophysics, functions and applications. In B. Grimm, R. J. Porra, W. Rüdiger, & H. Scheer (Eds.), *Chlorophylls and bacteriochlorophylls: Vol. 25* (pp. 1–26). Dordrecht: Springer.

Schelbert, S., Aubbry, S., Burla, B., Agne, B., Kessler, F., Krupinska, K., et al. (2009). Pheophytin pheophorbide hydrolase (pheophytinase) is involved in chlorophyll breakdown during leaf senescence in *Arabidopsis*. *Plant Cell, 21*, 767–785.

Scherer, P. O. J., & Fischer, S. F. (2009). Photophysics of chlorophylls and carotenoids. In *Theoretical molecular biophysics* (2nd ed., pp. 321–333)). Berlin Heidelberg: Springer-Verlag GmbH.

Schliep, M., Cavigliasso, G., Quinnell, R. G., Stranger, R., & Larkum, A. W. D. (2013). Formyl group modification of chlorophyll *a*: A major evolutionary mechanism in oxygenic photosynthesis. *Plant, Cell & Environment, 36*, 521–527.

Schunck, E., & Marchlewski, L. (1894). Zur chemie des chlorophylls. *Liebigs Annalen der Chemie, 278*, 329–345.

Solymosi, K., & Mysliwa-Kurdziel, B. (2017). Chlorophylls and their derivatives used in food industry and medicine. *Mini-Reviews in Medicinal Chemistry, 17*, 1194–1222.

Staroń, J., Boroń, B., Karcz, D., Szczygieł, M., & Fiedor, L. (2015). Recent progress in chemical modifications of chlorophylls and bacteriochlorophylls for the applications in photodynamic therapy. *Current Medicinal Chemistry, 22*, 3054–3074.

Sundholm, D. (1999). Density functional theory calculations of the visible spectrum of chlorophyll *a*. *Chemical Physics Letters, 302*, 480–484.

Szafraniec, M. J., Szczygieł, M., Urbanska, K., & Fiedor, L. (2014). Determinants of the activity and substrate recognition of breast cancer resistance protein (ABCG2). *Drug Metabolism Reviews, 46*, 459–474.

Szczygieł, M., Urbańska, K., Jurecka, P., Stawoska, I., Stochel, G., & Fiedor, L. (2008). Central metal determines pharmacokinetics of chlorophyll-derived xenobiotics. *Journal of Medicinal Chemistry, 51*, 4412–4418.

Trebitsh, T., Goldschmidt, E. E., & Riov, J. (1993). Ethylene induces de novo synthesis of chlorophyllase, a chlorophyll degrading enzyme, in Citrus fruit peel. *Proceedings of the National Academy of Sciences of the United States of America, 90*, 9441–9445.

Tsuchiya, T., Ohta, H., Okawa, K., Iwamatsu, A., Shimada, H., Masuda, T., et al. (1999). Cloning of chlorophyllase, the key enzyme in chlorophyll degradation: Finding of a lipase motif and the induction by methyl jasmonate. *Proceedings of the National Academy of Sciences of the United States of America, 96*, 15362–15367.

Umena, Y., Kawakami, K., Shen, J.-R., & Kamiya, N. (2011). Crystal structure of oxygen-evolving photosystem II at a resolution of 1.9 A. *Nature, 473*, 55–60.

van Amerongen, H., Valkunas, L., & van Grondelle, R. (2000). *Photosynthetic excitons. Vol. 105*. Singapore: World Scientific.

Wanders, R. J. A., Komen, J., & Ferdinandusse, S. (2011). Phytanic acid metabolism in health and disease. *Biochimica et Biophysica Acta, 1811*, 498–507.

Watanabe, T., Kobayashi, M., Hongu, A., Nakazato, M., Hiyama, T., & Murata, N. (1985). Evidence that a chlorophyll *a*' dimer constitutes the photochemical reaction centre 1 (P700) in photosynthetic apparatus. *FEBS Letters, 191*, 252–256.

Willstätter, R., & Stoll, A. (1913). *Untersuchungen über chlorophyll*. Berlin: Verlag von Julius Springer.

Yaronskaya, E., & Grimm, B. (2006). The pathway from 5-aminolevulinic acid to protochlorophyllide and protoheme. In B. Grimm, R. J. Porra, W. Rüdiger, & H. Scheer (Eds.), *Chlorophylls and bacteriochlorophylls: Vol. 25* (pp. 173–188). Dordrecht: Springer.

Young, A., & Britton, G. (Eds.), (1993). *Carotenoids in photosynthesis*. London: Chapmann & Hall.

Zucchelli, G., Brogioli, D., Casazza, A. P., Garlaschi, F. M., & Jennings, R. C. (2007). Chlorophyll ring deformation modulates Qy electronic energy in chlorophyll-protein complexes and generates spectral forms. *Biophysical Journal, 93*, 2240–2254.

CHAPTER TWO

Biosynthesis of chlorophylls and bacteriochlorophylls in green bacteria

Jennifer L. Thweatt[a], Daniel P. Canniffe[b], Donald A. Bryant[a,c,*]
[a]Department of Biochemistry and Molecular Biology, The Pennsylvania State University, University Park, PA, United States
[b]Institute of Integrative Biology, University of Liverpool, Liverpool, United Kingdom
[c]Department of Chemistry and Biochemistry, Montana State University, Bozeman, MT, United States
*Corresponding author: e-mail address: dab14@psu.edu

Contents

Abstract

Green bacteria include chlorophototrophic members of the phyla *Chlorobi*, *Chloroflexi*, and *Acidobacteria* and are defined by their use of chlorosomes for light-harvesting. Despite their shared use of chlorosomes as light-harvesting antenna and their chlorophototrophic metabolic mode, these organisms exhibit a surprising diversity in their ecological and physiological characteristics. Chlorosomes contain thousands of unique chlorin molecules known as bacteriochlorophyll (BChl) *c*, *d*, *e*, or *f* that are arranged in supramolecular aggregates. Additionally, all green bacteria can synthesize BChl *a*, the green members of the phyla *Chlorobi* and *Acidobacteria* can synthesize chlorophyll (Chl) *a*, and *Chloracidobacterium thermophilum* uniquely synthesizes Zn-BChl *a*′. This review describes the biosynthetic pathways leading to the production of Chls and BChls in green bacteria. Special attention is paid to how the physiological diversity of green bacteria affects the particular combinations of biosynthetic enzymes that are used in these distantly related, chlorosome-producing bacteria.

Advances in Botanical Research, Volume 90
ISSN 0065-2296
https://doi.org/10.1016/bs.abr.2019.03.002

Abbreviations

3-HPP	3-hydroxypropionate
3V	C3 vinyl
5-ALA	5-aminolevulinic acid
8V	C8 vinyl
8VR	8 vinyl reductase
A.	*Arabidopsis*
Ac	acetyl
ATP	adenosine triphosphate
BChl	bacteriochlorophyll
BChlide	bacteriochlorophyllide
BPhe	bacteriopheophytin
BPheide	bacteriopheophorbide
C.	*Chloroploca*
Ca.	*Candidatus*
Cab.	*Chloroacidobacterium*
CAO	chlorophyllide *a* oxygenase
Cba.	*Chlorobaculum*
Cfl.	*Chloroflexus*
Chl.	*Chlorobium*
Chl	chlorophyll
Chlide	chlorophyllide
Chp.	*Chloroherpeton*
Coprogen III	coproporphyrinogen III
COR	chlorophyllide oxidoreductase
cryo-EM	cryogenic electron microscopy
DPOR	dark operative protochlorophyllide reductase
E.	*Escherichia*
Et	ethyl
F	farnesyl
FAD	flavin adenine dinucleotide
FAP	filamentous anoxygenic phototrophs
FMO	Fenna-Matthews-Olsen protein
G.	*Gemmatimonas*
GG	geranylgeranyl
GSB	green sulfur bacteria
HYSCORE	hyperfine sublevel correlation
LH	light harvesting complex
MgCH	magnesium chelatase
Mg-PME	Mg-Proto IX 13-monomethyl ester
NAD	nicotinamide adenine dinucleotide
NADP	nicotinamide adenine dinucleotide phosphate
Osc.	*Oscillochloris*
P	phytyl
PChlide	protochlorophyllide
PD	Δ2,6 phytadienyl
Phe	pheophytin

Pheide	pheophorbide
PP	diphosphate
Proto IX	protoporphyrin IX
Protogen IX	protoporphyrinogen IX
Ptc.	*Prostecochloris*
Rba.	*Rhodobacter*
RC	reaction center
sp.	species
spp.	species (plural)
SAM	*S*-adenosyl-L-methionine
T.	Thermochlorobacter
TCA	tricarboxylic acid
V.	*Viridilinea*
WT	wild type

1. Introduction

Green bacteria are an eclectic group of chlorophototrophic (i.e., chlorophyll (Chl) and/or bacteriochlorophyll (BChl)-dependent) bacteria, which are defined by their ability to make large light-harvesting antenna complexes known as chlorosomes (Bryant et al., 2012; Bryant & Canniffe, 2018; Bryant & Frigaard, 2006; Frigaard & Bryant, 2004, 2006; Oostergetel, van Amerongen, & Boekema, 2010; Orf & Blankenship, 2013; Saer & Blankenship, 2017; Thiel, Tank, & Bryant, 2018). Chlorosomes contain self-assembled supramolecular aggregates of BChl *c*, *d*, *e*, or *f*, which are chlorin molecules uniquely found in chlorosomes (Bryant & Canniffe, 2018; Gomez Maqueo Chew & Bryant, 2007a). Green bacteria currently include organisms from three different bacterial phyla: *Chlorobi*, *Chloroflexi*, and *Acidobacteria* (Bryant et al., 2007, 2012; Thiel et al., 2018) (see Fig. 1A). In addition to the green bacteria, chlorophototrophic bacteria are also found in the phyla *Proteobacteria* (commonly called purple bacteria), *Gemmatimonadetes*, *Cyanobacteria*, and *Firmicutes* for a total of seven bacterial phyla which contain chlorophototrophs. When compared to the other major bacterial phyla, green bacteria are phylogenetically distant from each other and in some cases are more closely related to other chlorophototrophic bacteria than to each other (Fig. 1A).

The term "green bacteria" was initially used in a non-specific fashion to differentiate green-colored chlorophototrophic bacteria from purple chlorophototrophic bacteria. However, this common name eventually became

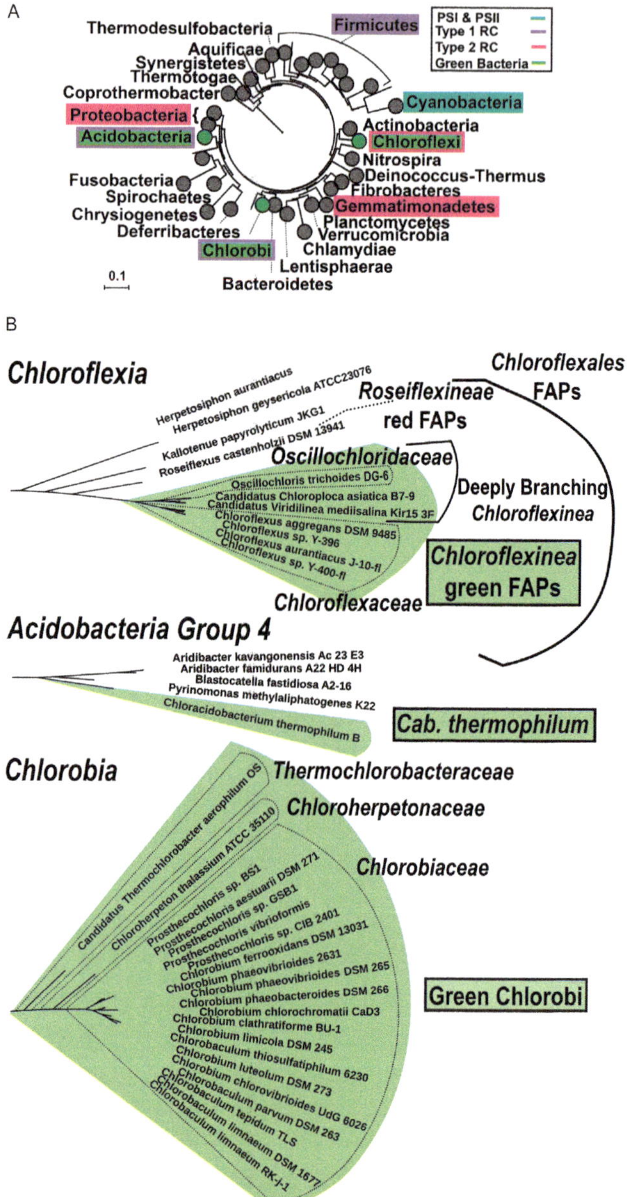

Fig. 1 Phylogeny of Green Bacteria. (A) Phylogenetic tree modified from SILVA living tree project 16S rRNA gene tree (release 128) to show the major phyla of Bacteria (Munoz et al., 2011). Boxed phyla contain at least one chlorophototrophic member. Phyla with type-1 RCs are boxed or highlighted in purple while phyla with type-2 RCs are boxed or highlighted in magenta; cyanobacteria which contain PSI and PSII are highlighted in turquoise. Phyla containing at least one member of the green bacteria (indicating the use of chlorosomes for light harvesting) are additionally highlighted in green. (B) Phylogenic tree of the class level branches which contain green bacteria. Green bacterial members within each class are highlighted in green. The 16S rRNA genes from cultured type strains available in the RDP database in addition to those from organisms used in Fig. 4 were aligned with the SINA aligner, and the tree was computed with FastTree (Cole et al., 2014; Price, Dehal, & Arkin, 2009; Pruesse, Peplies, & Glöckner, 2012). Phylogentic trees were modified using iTol (Letunic & Bork, 2016).

synonymous with green sulfur bacteria (GSB). GSB contain chlorosomes and are chlorophototrophic members of the phylum *Chlorobi*, families *Chlorobiaceae and Chloroherpetonaceae*, whose first isolates were green-colored (Imhoff, 2017; Imhoff & Thiel, 2010; Liu et al., 2012; Nadson, 1906). When *Chloroflexus* (*Cfl.*) *aurantiacus* was initially isolated and shown to contain chlorosomes, it and other chlorosome-containing members of the phylum *Chloroflexi*, such as *Oscillochloris* (*Osc.*) *trichoides*, were included within the green bacteria (Bryant et al., 2012; Gorlenko & Korotkov, 1979; Keppen, Baulina, & Kondratieva, 1994; Pierson & Castenholz, 1974a, 1974b). Chlorosome-containing members of the *Chloroflexi* are now referred to as green *f*ilamentous *a*noxygenic *p*hototrophs, or green *FAP*s, which were formerly and commonly known as green non-sulfur bacteria or green gliding bacteria (Bryant & Frigaard, 2006; Bryant et al., 2012; Garrity et al., 2001). One of the most recent additions to the green bacteria is *Chloracidobacterium* (*Cab.*) *thermophilum* (Bryant et al., 2007). *Cab. thermophilum* produces chlorosomes and is currently the only known chlorophototrophic species of the phylum *Acidobacteria* (Garcia Costas et al., 2011, 2012; Tank & Bryant, 2015a, 2015b; Tank, Garcia Costas, & Bryant, 2019; Tank, Thiel, Ward, & Bryant, 2017; Thiel et al., 2018). Despite the common trait that they use chlorosomes for light-harvesting, members of these three phyla are otherwise markedly different in their physiological properties and photosynthetic apparatus. We will briefly describe each group of green bacteria and their photosynthetic apparatus before discussing the biosynthesis of their unique complements of (B)Chls.

The phylum *Chlorobi* is represented by two classes, the non-chlorophototrophic *Ignavibacteria* and the chlorophototrophic *Chlorobia* (Iino et al., 2010). *Ignavibacterium album* and *Melioribacter roseus* are the best described members of the *Ignavibacteria* (Iino et al., 2010; Kadnikov et al., 2013; Liu et al., 2012; Podosokorskaya et al., 2013). The class *Chlorobia* contains all chlorophototrophic members of the phylum *Chlorobi*: GSB as well as "*Candidatus* (*Ca.*) Thermochlorobacter aerophilum" (Liu, Klatt, et al., 2012) (Fig. 1B). GSB are obligately anaerobic, anoxygenic Chl-dependent chlorophotolithoautotrophic organisms that include all members of the family *Chlorobiaceae* as well as *Chp. thalassium* (and possibly other members of the family *Chloroherpetonaceae*) (Liu, Frigaard, et al., 2012; Liu, Klatt, et al., 2012). These organisms fix carbon dioxide *via* the reverse TCA cycle, fix dinitrogen with molybdenum-containing nitrogenase, and use reduced sulfur sources, molecular hydrogen, or ferrous iron as their electron sources (Frigaard & Dahl, 2009; Heising, Richter, Ludwig, & Schink, 1999;

Wahlund & Madigan, 1993; Wahlund & Tabita, 1997). Despite their name, GSB may be either green-colored or brown-colored, depending on the type of BChl in their chlorosomes (see below). "*Ca.* T. aerophilum" is unique within the *Chlorobia.* It is an aerobic, chlorophotoheterotophic organism that contains chlorosomes, lacks the ability to fix CO_2 and N_2, and does not oxidize sulfur compounds as electron sources (Liu, Frigaard, et al., 2012; Tank et al., 2017). Because of these significant physiological and metabolic differences, a separate family, *Thermochlorobacteriaceae*, has been proposed for this organism and its close relatives (Liu, Klatt, et al., 2012). All members of the *Chlorobia* have homodimeric, type-1 reaction centers (RCs), and employ the BChl *a*-binding Fenna-Matthews-Olson (FMO) protein as the conduit for energy transfer from the baseplate BChl *a*-binding protein, CsmA, to the core antenna of the RC (Bryant & Canniffe, 2018; Bryant et al., 2012; Saer & Blankenship, 2017).

The phylum *Chloroflexi* includes many non-chlorophototrophic members in addition to FAPs (Thiel et al., 2018) (Fig. 1B). Unlike the *Chlorobia*, FAPs lack FMO and have type-2 RCs (Cardona, 2015; Thiel et al., 2018). Red FAPs include members of the chlorosome-less genera *Roseiflexus* and *Heliothrix*, and the recently discovered "*Ca.* Roseilinea gracile" (Tank et al., 2017; Thiel et al., 2018). The green FAPs contain chlorosomes and include members of the genera *Chloroflexus, Oscillochloris, Chloronema,* and "*Ca.* Chloroploca asiatica", "*Ca.* Viridilinea mediisalina" and "*Ca.* Chloranaerofilum" (Bryant et al., 2012; Grouzdev, Rysina, Bryantseva, Gorlenko, & Gaisin, 2018; Tank et al., 2017; Thiel et al., 2018). Under oxic conditions *Chloroflexus* spp. grow chemoheterotrophically, but under anoxic conditions most are Chl-dependent photoheterotrophs and a few are photoautotrophs. *Chloroflexus* spp. do not fix N_2; can use organic carbon sources, H_2, or H_2S as electron donors; and can fix CO_2 using the 3-hydroxypropionate pathway (3-HPP) (Fuchs, 2011; Klatt, Bryant, & Ward, 2007; Klatt et al., 2013). *Oscillochloris* spp. and the *Candidatus* strains listed above are part of a deeply branching subclade of the suborder *Chloroflexineae*, but "*Ca.* Chloroploca" and "*Ca.* Viridilinea" likely differ from *Oscillochloris* at the family level (Grouzdev et al., 2018). *Osc. trichoides* is the best characterized member of this suborder. *Oscillochloris* spp. are unique among FAPs because they are strict anaerobes that can fix N_2; moreover, they grow chlorophotoautotrophically by fixing CO_2 *via* the Calvin-Benson-Basham cycle with H_2S (or H_2) as the electron donor or chlorophotoheterotrophically using acetate or pyruvate (Ivanovsky et al., 1999; Keppen et al., 1994). "*Ca. C. asiatica*" and "*Ca.* V. mediisalina" have been studied *via* enrichment cultures and recent draft genomes. Like *Osc. trichoides* both of these strains have been shown to grow as anaerobic chlorophototrophs, however; like *Chloroflexus* spp. they have genes for the 3-HPP CO_2 fixation pathway (Grouzdev et al., 2018). Green FAPs can actually be brown, green, orange or reddish in color depending on growth conditions, in particular the oxygen concentration that controls their pigment composition (Thiel et al., 2018).

The phylum *Acidobacteria* comprises many physiologically diverse, non-phototrophic organisms but only a single chlorophototrophic species, *Cab. thermophilum,* has been described to date (Bryant et al., 2007; Tank & Bryant, 2015a, 2015b; Tank et al., 2017; Thiel et al., 2018) (Fig. 1B). *Cab. thermophilum* is an anoxygenic, chlorophotoheterotrophic organism that grows optimally under microoxic conditions. It does not fix N_2 and lacks the genes encoding key enzymes of all known pathways for autotrophic CO_2 fixation (Garcia Costas, Liu, et al., 2012; Thiel et al., 2018). Furthermore, *Cab. thermophilum* cannot synthesize lysine, branched-chain amino acids, or vitamin B_{12} (cobalamin); however, it can degrade branched-chain amino acids, and it takes up and assimilates all amino acids except aspartic acid and glutamic acid (Tank & Bryant, 2015a, 2015b). Like the *Chlorobia, Cab. thermophilum* has type-1 homodimeric RCs (Tsukatani, Romberger, Golbeck, & Bryant, 2012). Homodimeric type-1 RCs are otherwise only found in strictly anaerobic members of the phyla *Firmicutes* (heliobacteria) and *Chlorobi* (with the exception of "*Ca.* T. aerophilum" as noted above). An additional similarity to the *Chlorobia* is the presence of the BChl *a*-binding FMO antenna protein (Tsukatani, Wen, Blankenship, & Bryant, 2010; Wen et al., 2011).

1.1 Light-harvesting in green bacteria

As noted above, the defining property of green bacteria is the presence of chlorosomes, light-harvesting organelles that are tightly appressed to the cytoplasmic surface of the cell membrane (Cohen-Bazire, Pfennig, & Kunisawa, 1964; Staehelin, Golecki, & Drews, 1980; Staehelin, Golecki, Fuller, & Drews, 1978). Chlorosomes are supramolecular structures assembled from self-aggregated BChl *c*, *d*, *e*, or *f* molecules within a protein-stabilized, phospholipid monolayer envelope (Frigaard & Bryant, 2006; Tsukatani et al., 2016) (Fig. 2). Chlorosomes can contain up to 250,000 BChl molecules and a single cell can contain up to ~250 chlorosomes (Frigaard & Bryant, 2006; Montaño et al., 2003). Chlorosomes additionally contain large amounts

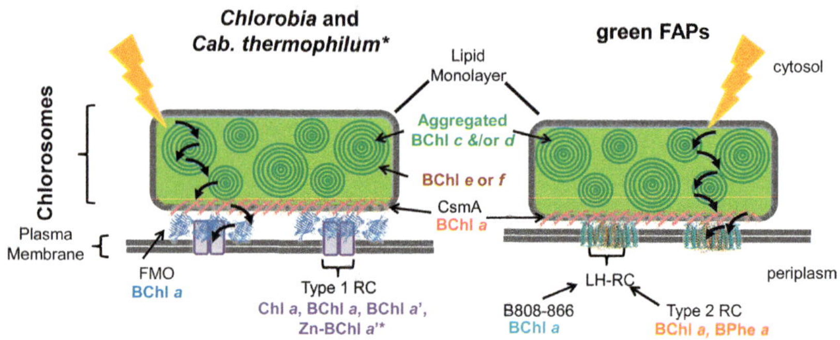

Fig. 2 Localization of Chlorophylls and Bacteriochlorophylls in Green Bacteria. Diagram of light-harvesting apparatus of green bacteria indicating localization of (B)Chls and path of excitation energy. The "*" indicates a molecule found only in *Cab. thermophilum*. The crystal structure of FMO (PDB: 3ENI) from *Cba. tepidum* was used to represent the FMO proteins of *Chlorobia* and *Cab. thermophilum*. The structure of CsmA (PDB: 2K37) from *Cba. tepidum* was used to represent the CsmA baseplate arrays of *Chlorobia*, *Cab. thermophilum*, and green FAPs. The cryo-EM structure of RC-LH core complex (PDB:5YQ7) of *Roseiflexus castenholzii* with the cytochrome subunit removed from view was used to represent the LH-RC of green FAPs.

of carotenoids and quinones, and the envelope can contain up to ~10 different proteins (Frigaard & Bryant, 2006); however, mutational studies in *Chlorobaculum* (*Cba.*) *tepidum* have shown that only one chlorosome protein, CsmA, is essential (Bryant et al., 2012; Bryant, Vassilieva, Frigaard, & Li, 2002; Frigaard, Li, Milks, & Bryant, 2004; Garcia Costas et al., 2011; Li & Bryant, 2009; Li, Frigaard, & Bryant, 2013).The extremely high pigment:protein ratio of chlorosomes makes them a remarkably efficient light-harvesting antenna from a cost-benefit perspective, especially when compared to predominantly proteinaceous antenna complexes (e.g., the phycobilisome; Bryant & Canniffe, 2018; Frigaard & Bryant, 2006). This efficiency, as well as the enormous number of BChls (up to 30% of the cell carbon), allows green bacteria to live in extremely low light intensities that exclude other chlorophototrophs, which have antenna complexes with many fewer pigments per RC (Bryant & Canniffe, 2018; Frigaard & Bryant, 2006). After light absorption by BChl aggregates within the chlorosome, excitation energy is rapidly transferred to the BChl *a* molecules associated with the baseplate (Bryant & Canniffe, 2018; Pšenčík, Butcher, & Tuma, 2014; Pšenčík et al., 2009; Saer & Blankenship, 2017). The baseplate forms the interface between the chlorosome and cytoplasmic membrane and is made up of an array of CsmA dimers (Bryant & Canniffe, 2018; Li, Frigaard, & Bryant, 2006; Nielsen et al., 2016). Each CsmA subunit binds one BChl *a* molecule, and baseplate arrays of CsmA occur in the chlorosomes of all green bacteria (Bryant et al., 2012; Garcia Costas et al., 2011; Li et al., 2006; Nielsen et al., 2016; Pšenčík et al., 2014, 2009).

In *Chlorobia* and *Cab. thermophilum*, which have type-1 RCs, the baseplate transfers excitation energy to the FMO protein, a BChl *a*-binding, peripheral antenna complex that also binds to RCs (Huang, Wen, Blankenship, & Gross, 2012; Tsukatani et al., 2010; Wen et al., 2011; Wen, Zhang, Gross, & Blankenship, 2009). FMO functions as a homo-trimer, and each subunit binds 7 BChl *a* molecules and 1 intersubunit BChl *a* molecule; thus, each FMO trimer binds 24 BChl *a* molecules, and several trimers are probably associated with each RC (Busch, Mueh, Madjet, & Renger, 2011; Fenna & Matthews, 1975; Larson et al., 2011; Tronrud, Wen, Gay, & Blankenship, 2009). The FMO protein transfers excitation energy to the BChl *a* molecules associated with the core antenna of the type-1 RC (Busch et al., 2011; Milder, Brueggemann, van Grondelle, & Herek, 2010). The GSB RCs mostly contain BChl *a*, but a few (~4) Chl *a* molecules are also bound; the special pair, known as P840, consists of two molecules of BChl *a*', while Chl *a* is the primary electron acceptor. An additional 14 BChl *a* and a total of 4 Chl *a* molecules, two forming A_0 and two accessory pigments, have been reported in the RC (Ohashi et al., 2010; Permentier et al., 2000). This relatively low pigment content of the RC has recently been called into question. The crystal structure of the heliobacterial RC had a much larger than expected number of pigments (~60 BChl *g* and 8^1-OH-Chl *a* molecules), and this suggests that Chl numbers in other homodimeric type-1 RCs should be reinvestigated (Gisriel et al., 2017). In *Cab. thermophilum* the type-1 RCs contain BChl *a*, Zn-BChl *a*', and Chl *a* with an estimated ratio of 12.8:2:8, respectively (Tsukatani et al., 2012). The primary electron acceptor in this RC was recently identified as Chl *a*, and thus, all type-1 RCs utilize a chlorin, either Chl *a* or 8^1-OH-Chl *a*, as the electron acceptor (Zill et al., 2018). Very recent results from HYSCORE analysis of $P840^+$ indicate that Zn-BChl *a*' is the special pair pigment in this RC (Charles et al., 2019).

Unlike GSB and *Cab. thermophilum*, green FAPs lack the FMO protein and have type-2 RCs. The chlorosome baseplate complex instead transfers excitation energy to a B808–866 complex, which is similar to the LH1 complex of purple bacteria (Novoderezhkin et al., 1998; Saer & Blankenship, 2017; Xin, Lin, Montaño, & Blankenship, 2005). The B808–866 complex is a ring-shaped, integral membrane protein complex that binds ~45 BChl *a* molecules (Collins, Wen, & Blankenship, 2012; Majumder et al., 2016; Xin et al., 2018), and as in purple bacteria, excitation energy is transferred from

B808–866 to the type-2 RC. The heterodimeric RC comprises the PufL and PufM subunits that bind three BChl *a* and three bacteriopheophytin (BPhe) *a* molecules (Pierson & Thornber, 1983; Xin et al., 2018). The special pair, known as P870, is made up of two BChl *a* molecules that act as the primary electron donor, and BPhe *a* is the primary acceptor (Blankenship, 1994; Bruce, Fuller, & Blankenship, 1982; Woodbury, Becker, Middendorf, & Parson, 1985). A "voyeur" BChl *a* molecule is present on the active branch of the electron transport chain as in bacterial reaction centers, but the inactive branch has two BPhe *a* molecules (Xin et al., 2018).

1.2 Chlorophylls and bacteriochlorophylls of green bacteria

Like heliobacteria and cyanobacteria that also have type-1 RCs, *Chlorobia* and *Cab. thermophilum* synthesize derivatives of Chl *a*. The C–C bond at the C17–C18 position of the D ring is a double bond in porphyrins, but in chlorins, including Chl *a*, this is a single bond (Fig. 3). In the RCs of both GSB and *Cab. thermophilum*, the C17^3 carboxylate moiety of Chl *a* is esterified with Δ2,6-phytadienol, denoted here as Chl a_{PD} to differentiate it from Chl *a* esterified with phytol (Chl a_P), which is the form of Chl *a* commonly found in cyanobacteria, algae and plants. Because Chl a_{PD} is only found in small amounts in the RC, it is a relatively minor pigment in these organisms. For example, it comprises only ~0.3% of the cellular (B)Chl content in the model GSB *Cba. tepdium,* but is somewhat more abundant in *Cab. thermophilum* (Frigaard & Bryant, 2004; Tsukatani et al., 2012).

BChl *a* derivatives occur in all green bacteria, most purple chlorophototrophic bacteria and the recently discovered aerobic anoxygenic chlorophotoheterotroph, *Gemmatimonas phototrophica* (Bryant et al., 2012; Garcia Costas et al., 2012; Imhoff, 2017; Thiel et al., 2018; Zeng, Feng, Medova, Dean, & Koblizek, 2014; Zeng & Koblížek, 2017). BChl *a* is a bacteriochlorin, meaning that, in addition to the reduced D ring present in chlorins, the C7=C8 double bond of the B ring is also reduced to a single bond (Fig. 3). The only other difference between BChl *a* and Chl *a* occurs at the C3 position, where BChl *a* has an acetyl group but Chl *a* has a vinyl group. In green bacteria BChl *a* is esterified with phytol at the C17^3 position, and is denoted here as BChl a_P. Additionally, BChl a_P' and Zn-BChl a_P' are found in the RCs of GSB and *Cab. thermophilum*, respectively (Ohashi et al., 2010; Permentier et al., 2000; Tsukatani et al., 2012). BChl *a*' differs from BChl *a* in the stereochemistry at the C13^2 position while Zn-BChl *a*

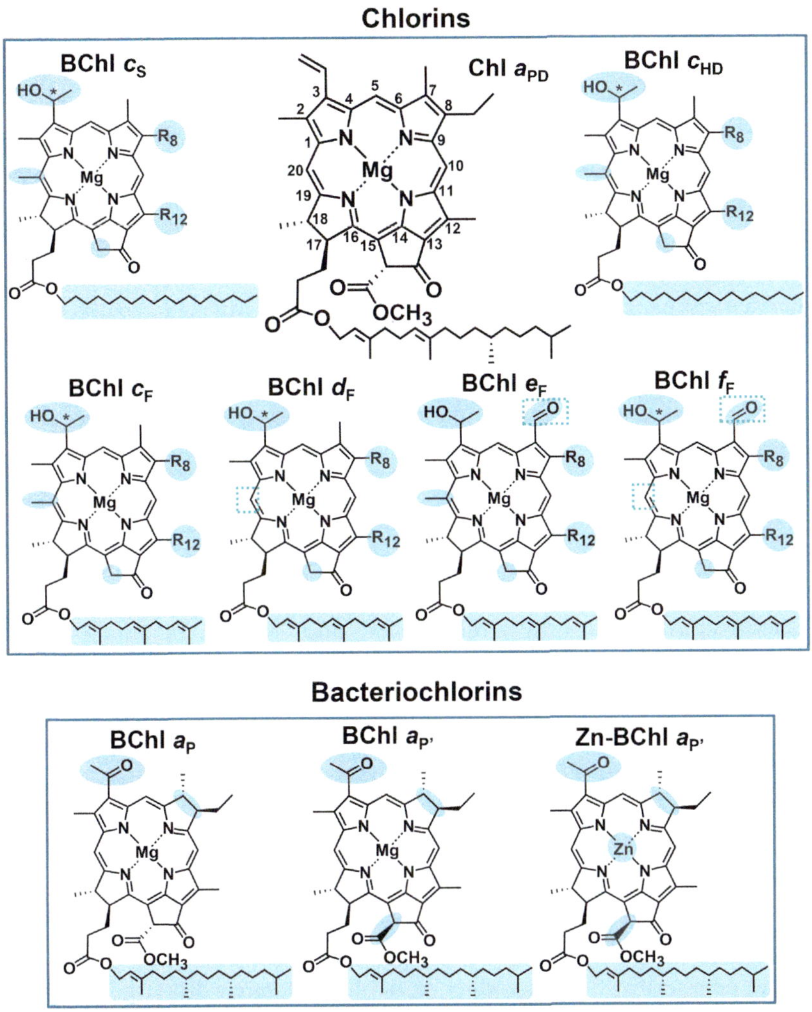

Fig. 3 Structure of Chlorophylls and Bacteriochlorophylls from Green Bacteria. Structures of (B)Chls found in green bacteria. Numbering listed on Chl a_{PD} molecule is used throughout review. Blue shading indicates structural differences from Chl a_{PD}. Dashed boxes indicate differences in BChl *d*, *e*, and *f* from BChl *c*.

derivatives differ by the substitution of a Zn^{++} ion for the canonical Mg^{++} ion in the macrocycle (Fig. 3). BPhe *a*, which differs from BChl *a* by the absence of any metal ion in the macrocycle, is found in type-2 RCs of purple bacteria, FAPs, and *G. phototrophica*.

Despite their names, BChls *c*, *d*, *e* and *f* are not bacteriochlorins. These molecules are chlorins that are commonly referred to as "*Chlorobium* Chls" and are only found in green bacteria. They have unique structural features that allow them to form the supramolecular structures found within chlorosomes. Unlike all other (B)Chls, they lack the $C13^2$ methylcarboxyl group. Removal of this group decreases steric hindrance and enhances the ability of the molecules to self-assemble. All "*Chlorobium* Chls" also have a hydroxyethyl group at the C3 position, while most other (B)Chls have either a vinyl group or an acetyl group (Fig. 3). Within chlorosomes or in BChl aggregates *in vitro*, the hydroxyl group at $C3^1$ forms a coordination bond with the central Mg of a neighboring BChl molecule that is crucial for self-aggregation. The $C3^1$ carbon atom is a chiral center, and the hydroxyethyl group occurs with both *R* and *S* stereochemistry at this position, which results in structural heterogeneity among the BChl molecules in chlorosomes. Additional heterogeneity exists at the $C8^2$, $C12^1$, and $C17^3$ positions of BChl *c*, *d*, *e*, and *f*. As a result of the structural heterogeneity of these molecules, it is important to note that the terms BChl *c*, *d*, *e*, or *f*, describe a family of related homolog molecules and thus do not refer to molecules with a defined structure. In *Chlorobia*, *Cab. thermophilum*, *Osc. trichoides*, (and probably "*Ca.* *C. asiatica*" and "*Ca.* V. mediisalina") but not *Chloroflexus* spp., the $C8^2$ and $C12^1$ positions are differentially methylated. The exact nature and extent of these modifications are dependent on the organism and the growth conditions. The C8 side chains of these BChls may be ethyl, n-propyl, isobutyl, or neopentyl groups, while the C12 sidechain may be a methyl or an ethyl moiety. The degree of methylation at $C8^2$ and $C12^1$ affects the absorbance properties of the BChl aggregates in chlorosomes, with higher degrees of methylation causing spectral broadening and correlating with greater amounts of BChl in cells (Gomez Maqueo Chew, Frigaard, & Bryant, 2007). The $C17^3$ carboxyl group of BChls *c*, *d*, *e*, or *f* can be esterified with a variety of alcohols in green bacteria. The esterifying alcohol for these BChls in GSB is predominantly farnesol, but smaller amounts of other alcohols, including geranylgeraniol, phytol and various straight-chain alcohols, often occur (Airs, Borrego, Garcia-Gil, & Keely, 2001; Glaeser, Bañeras, Rütters, & Overmann, 2002; Gomez Maqueo Chew, Frigaard, & Bryant, 2008). In *Cab. thermophilum* the $C17^3$ carboxyl group is mostly esterified with a mixture of straight-chain alcohols between 15 and 18 carbons in length, predominantly octadecanol; smaller amounts of farnesol and geranylgeraniol also occur (Garcia Costas, Tsukatani, et al., 2012). In green FAPs, the $C17^3$ carboxyl moiety of BChl *c* is also mostly esterified with stearol and small amounts of geranylgeraniol. The chain length of the esterifying alcohols has been shown to affect the spacing of aggregated BChls in chlorosomes and may be modified in response to changes in growth conditions (Pšenčík et al., 2010). The remarkable heterogeneity of BChl homologs in green bacteria may be a mechanism to tune the absorption properties of the BChls in chlorosomes, where no proteins are available to perform this function.

BChl *c*, *d*, *e*, and *f* are structurally similar, but the subtle differences among them produce major differences in the absorption spectra and coloration of the organisms that produce them. Green bacteria naturally synthesize either BChl *c*, *d*, or, *e*, and mutational studies have produced organisms that synthesize BChl *f*, which has not yet been shown to occur naturally (Harada, Mizoguchi, Tsukatani, Noguchi, & Tamiaki, 2012; Orf et al., 2013; Vogl, Tank, Orf, Blankenship, & Bryant, 2012). Organisms that synthesize BChl *c* or *d* are green-colored, while organisms synthesizing BChl *e* or *f* are brown-colored (Thiel et al., 2018). BChl *c* is the most common *Chlorobium* Chl and is found in most GSB, most if not all green FAPs, and in *Cab. thermophilum*. BChl *d* is similar in structure to BChl *c* but lacks the methyl group on the methine bridge at C20 (Fig. 3). BChl *d* is found in GSB that occur closer to the surface of stratified environments (Maresca et al., 2004; van Gemerden & Mas, 1995), and is found in some strains of *Cba. parvum*, "*Ca.* T. aerophilum," and some *Chloronema* spp. (Borrego, Garcia-Gil, Cristina, Vila, & Abella, 1998; Dubinina & Gorlenko, 1975; Liu, Klatt, et al., 2012; Smith & Goff, 1985; Tank et al., 2017). BChl *e*, which differs from BChl *c* by replacement of the C7 methyl group with a formyl group, is only made by brown-colored GSB. Finally, BChl *f*, like BChl *d*, lacks a methyl group at C20, but has a formyl group at C7 like BChl *e*.

2. Biosynthesis

The biosynthetic pathways for Chl *a* and BChl *a* were initially studied in plants, algae, cyanobacteria, and purple bacteria. Homologs of the genes encoding enzymes for BChl *a* synthesis in purple bacteria were subsequently found in the genomes of green bacteria (Bryant et al., 2012; Eisen et al., 2002; Garcia Costas, Liu, et al., 2012; Grouzdev et al., 2015; Tang et al., 2011). The unique enzymatic steps in the biosynthetic pathways of BChl *c*, *d*, *e*, and *f* were characterized in the model GSB *Cba. tepidum* and *Cba. limnaeum*, and bioinformatic analyses and biochemical studies have confirmed that essentially the same pathways are present in other green bacteria as well. To date, finished or nearly complete permanent draft genomes are available for 17 *Chlorobia*, 7 green FAPs, and *Cab. thermophilum* (Fig. 4).

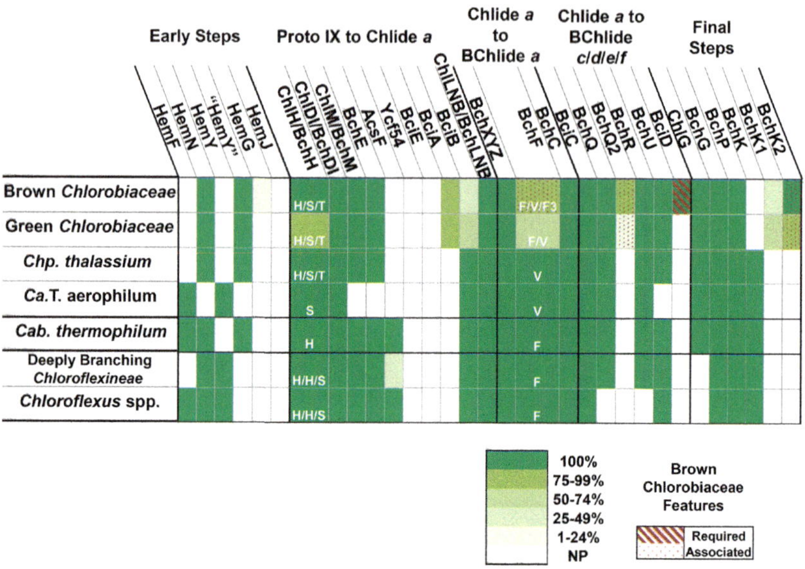

Fig. 4 Distribution of Biosynthetic Enzymes for (B)Chl Biosynthesis in Green Bacteria. Color-coded array indicating the percentage of organisms in each category whose genomes encode the indicated biosynthetic enzyme. The term "HemY" refers to enzymes described as HemY-related in the text. In the ChlH/BchH column text indicates which paralogs are present in each group. In the green-colored *Chlorobiaceae*, *Chl. chlorochromatii* is missing ChlH/BchH but has BchS and BchT. In "Ca. T. aerophilum," there is no full-length member of the BchH/BchS/BchT family. Instead, there are two genes that separately encode the N- and C-terminal domains of what appears to be a single BchS-type Mg-chelatase. The BchF column text indicates which paralog(s) are present. The brown-colored *Chlorobiaceae*, *Prosthecochloris* sp. BS1 miss the BchV paralogs, but contains BchF and BchF3. In the green-colored *Chlorobiaceae*, *Ptc. bathyomarina* and *Chl. phaeovibriodes* have a BchF3 paralog instead of BchV and *Chl. limicola* has 2 paralogs of BchV in addition to BchF. In the BchQ2 column it should be noted that BchQ and BchQ2 do not form clearly delineated clades, so for this figure a second copy of BchQ has been considered BchQ2. In the brown-colored *Chlorobiaceae*, *Prosthecochloris* sp. BS1 has only a single copy of BchQ, however the gene encoding BchQ is directly upstream of the gene cluster which encodes BciD, which is located where other brown-colored GSB encode BchQ2. This likely represents a loss of BchQ, as is true of BchV for this organism. *Chl. phaeobacteroides* possesses a truncated copy of BchQ which is encoded distantly from the gene cluster that includes *bciD*, *bchF3*, and *bchQ2*. In green-colored *Chlorobiaceae*, *Chl. limicola* encodes a second copy of BchQ, as is true for BchV in this organism. Genomes analyzed for this figure are as follows: Brown-colored *Chlorobiaceae* (6): *Cba. limnaeum* DSM 1677^T, *Cba. limnaeum* Rk-j-1, *Prosthecochloris* sp. BS-1, *Chl. clathratiforme* DSM5477^T, *Chl. phaeobacteroides* DSM 266, *Prosthecochloris* sp. CIB2401; Green-colored *Chlorobiaceae* (9): *Cba. tepidum* TLS, *Chl. limicola* DSM245, *Chl. chlorochromatii* CaD3, *Chl. phaeovibriodes* DSM 265, *Ptc. bathyomarina* GSB1, *Chl. luteolum* DSM273, *Chl. ferrooxidans* DSM 13031, *Ptc. aestuarii* DSM 271, *Cba. parvum* DSM 263; *Chp. thalassium* ATCC 35110 (1); "*Ca.* T. aerophilum"* (1); *Cab. thermophilum* B (1); Deep Branching *Chloroflexineae* (3); *Osc. trichoides* DG-6, "*Ca.* Chloroploca asiatica" B7–9*, "*Ca.* Viridilinea mediisalina" Kir15-3F*; *Chloroflexus* spp. (4): *Cfl. aurantiacus* J-10-fl, *Cfl.* sp. Y-400-fl, *Cfl.* sp. Y-396-1, *Cfl.* aggregans DSM 9485. * in this list indicate organisms whose draft genomes were assembled from metagenomic data. NP, not present.

The availability of complete genomes for many green bacteria has made it possible to compare their biosynthetic capabilities. In the following sections we will discuss the biosynthetic pathways of (B)Chls in green bacteria.

2.1 Early steps

The biosynthesis of all hemes, Chls, BChls, cobalamin, and other tetrapyrroles proceeds *via* the same basic pathway from 5-aminolevulinate to uroporphyrinogen III. The pathways for biosynthesis of Chls and BChls, and the porphyrin-dependent heme pathway, then proceed *via* the same enzymatic steps from uroporphyrinogen III to protoporphyrin IX (Proto IX), while the pathways for cobalamin and other tetrapyrroles, and the siroheme-dependent heme pathway, diverge at uroporphyrinogen III, and the coproporphyrin-dependent heme pathway diverges at coproporphyrinogen (Dailey et al., 2017). An in-depth discussion of these reactions can be found elsewhere in this volume (see also chapter "Porphyrin and heme synthesis" by Layer et al., Vol. 91 and chapter "Evolution of tetrapyrrole pathway in eukaryotic phototrophs" of this volume by Cihlář et al.), so they will only be briefly considered here to address distribution or other features important to green bacteria. A recent review of heme biosynthesis in prokaryotes has renamed several genes and gene products in this pathway (Dailey et al., 2017). Here we will use the old gene designations followed by the new one in parentheses.

Two steps in the pathway from 5-aminolevulinic acid to Proto IX can be catalyzed by alternative enzymes (Fig. 5). The first of these reactions is the oxidative decarboxylation of two of the four propionate groups of coproporphyrinogen III to produce the vinyl groups of protoporphyrinogen IX. In some aerobic organisms these reactions are performed by coproporphyrinogen decarboxylase, HemF (CgdC), which is an oxygen-dependent enzyme. In anaerobic organisms, an unrelated oxygen-independent, radical-*S*-adenosyl-L-methione (SAM) enzyme, coproporphyrinogen dehydrogenase, HemN (CgdH), forms protoporphyrinogen IX (Layer et al., 2005). Genes encoding both HemF (CgdC) and HemN (CgdH) are found in the genomes of *Chloroflexus* spp., *Cab. thermophilum*, and *Ca.* T. aerophilum that can grow aerobically. In contrast, the strictly anaerobic GSB, *Osc. trichoides*, "*Ca. C. asiatica*" and "*Ca.* V. mediisalina" only contain genes encoding HemN (CgdH) (Fig. 4). Two copies of *hemN (cgdH)* are present in the genomes of all green bacteria examined for this review except *Ca.* T. aerophilum (Fig. 4), and the second paralog has been variably called *hemN2*

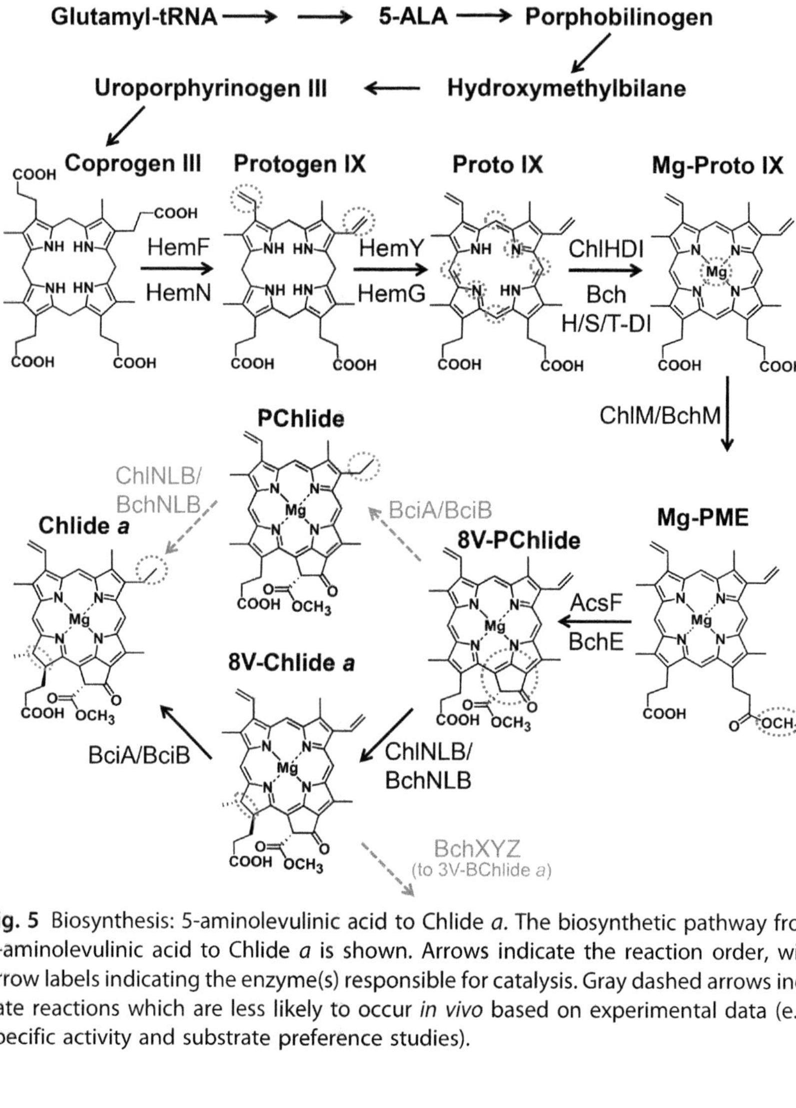

Fig. 5 Biosynthesis: 5-aminolevulinic acid to Chlide *a*. The biosynthetic pathway from 5-aminolevulinic acid to Chlide *a* is shown. Arrows indicate the reaction order, with arrow labels indicating the enzyme(s) responsible for catalysis. Gray dashed arrows indicate reactions which are less likely to occur *in vivo* based on experimental data (e.g., specific activity and substrate preference studies).

or *hemZ* (Eisen et al., 2002; Garcia Costas, Liu, et al., 2012; Tang et al., 2011). Recently, homologs of HemN (CgdH) in other bacteria have been characterized as heme-binding proteins without dehydrogenase activity and renamed HemW (Abicht, Martinez, Layer, Jahn, & Solioz, 2012; Haskamp et al., 2018). Comparing the sequences of green bacterial HemN homologs to biochemically characterized HemW, it is apparent that *Ca.* T. aerophilum lacks *hemN* and that all green bacterial genomes have one copy of *hemW*.

The next reaction in the pathway, the six-electron oxidative conversion of protoporphyrinogen IX to Proto IX, is also performed by different isofunctional enzymes (Fig. 5). In some aerobic organisms an oxygen-dependent protoporphyrinogen IX oxidase, HemY (PgoX), performs this oxidation (Dailey & Dailey, 1996; Dailey et al., 2017). HemY (PgoX) uses a single FAD molecule to perform three sequential, two-electron oxidations using three molecules of O_2 (the electron acceptor) and producing Proto IX and three molecules of H_2O_2 (Dailey et al., 2017). Recently it has been shown that some homologs of HemY (PgoX) actually prefer coproporphrinogen III as a substrate. These enzymes have been designated coproporphyrinogen oxidases (CgoX), and they function in the coproporphyrin-dependent heme biosynthetic pathway that does not include Proto IX as an intermediate (Dailey et al., 2017). While these oxidases were originally described as oxygen-dependent enzymes, HemY from *Plasmodium falciparum* functions as a protoporphyrinogen oxidase under anoxic conditions in the presence of FAD, NAD^+, or $NADP^+$ (Nagaraj, Arumugam, Prasad, Rangarajan, & Padmanaban, 2010). In some anaerobic organisms the oxygen-independent enzyme protoporphryinogen dehydrogenase, HemG (PgdH1), is used. This enzyme carries out the same six-electron oxidation as HemY, but can use various electron acceptors including ubiquinone and menaquinone *in vitro*. It has been suggested that HemG (PgdH1) activity may be coupled to the electron transport chain *in vivo* (Boynton, Daugherty, Dailey, & Dailey, 2009; Möbius et al., 2010; Sasarman et al., 1993). A third enzyme, known as HemJ (PgdH2; note that PgdH1 and PgdH2 are not paralogs and do not belong to the same protein family), has recently been identified that can perform this reaction in cyanobacteria; however, no homologs of HemJ are encoded in any green bacterial genomes (Boynton et al., 2011; Dailey et al., 2017; Kato, Tanaka, Sano, Tanaka, & Hosaka, 2010; Skotnicová et al., 2018).

Genes encoding HemY (PgoX) are found in the microaerophilic chlorophototrophs, *Chloroflexus* spp. and in the aerobic *Ca.* T. aerophilum. Searches of the genomes of GSB for genes related to protoporphyrinogen IX oxidation have produced conflicting results. One study failed to identify any known gene(s) for protoporphyrinogen IX oxidation in GSB (Kato et al., 2010). However, more recent analyses show the presence of *hemY* (*pgoX*)-related genes in several GSB, while *hemG* (*pgdH1*) was only found in *Prosthecochloris* (*Ptc.*) sp. BS-1 (Dailey et al., 2017; Kobayashi, Masuda, Tajima, Wada, & Sato, 2014). Surprisingly, the microaerophile *Cab. thermophilum* has a *hemY* homolog that is more closely related to those found in GSB, while the anaerobes *Osc. trichoides*, "*Ca. C. asiatica*" and "*Ca.*

V. mediisalina" have a *hemY* gene closely related to those of *Chloroflexus* spp. and *Ca.* T. aerophilum (Fig. 4) (J. L. Thweatt & D. A. Bryant, unpublished data). Considering that GSB are strict anaerobes, it is likely that their *hemY*-related genes encode an oxygen-independent enzyme for protoporphyrinogen oxidation, or that another as-yet unidentified enzyme performs this reaction. Further research on the putative HemY enzymes of green bacteria is needed to understand how protoporphyrinogen IX oxidation occurs in these organisms.

2.2 Protoporphyrin IX to chlorophyllide *a*

2.2.1 Magnesium chelation

The biosynthesis of all (B)Chls in green bacteria proceeds *via* the same pathway from Proto IX to a branch point at chlorophyllide (Chlide) *a*. The first committed step of (B)Chl biosynthesis is the insertion of the Mg^{++} ion into the macrocycle. This reaction requires an ATP-dependent enzyme, Proto IX magnesium (Mg) chelatase (MgCH). MgCH uses ~15 ATP molecules, Mg^{++}, and Proto IX to produce Mg-Proto IX (Reid & Hunter, 2004; more details in chapter "The Mg branch chlorophyll synthesis: Biosynthesis of chlorophyll *a* from protoporphyrin IX" of this volume by Willows) (Fig. 5). This enzyme is made up of three subunits, denoted ChlH/BchH, ChlD/BchD, and ChlI/BchI in all chlorophototrophic bacteria; conventionally, the Chl or Bch prefix is used depending on whether the gene/protein in question is associated with a Chl or BChl biosynthetic pathway, respectively (Willows, Gibson, Kanangara, Hunter, & von Wettstein, 1996). ChlH/BchH is the large subunit of MgCH, which binds the Proto IX and Mg^{++} substrates (Sirijovski et al., 2008). ChlD/BchD and ChlI/BchI are smaller subunits with similar structures, which make homohexamers or homoheptamers, contain AAA+ domains, and associate with ChlH/BchH (Fodje et al., 2001; Lundqvist et al., 2010; Reid, Siebert, Bullough, & Hunter, 2003). Although both subunits have AAA+ domains, the ATPase activity resides in the ChlI/BchI subunit (Jensen, Gibson, & Hunter, 1999).

Multiple copies of the *chlH/bchH* genes are present in most green bacteria, with exceptions in *Cab. thermophilum* and *Ca.* T. aerophilium (Fig. 4) (Bryant & Liu, 2013; Bryant et al., 2012; Eisen et al., 2002; Tang et al., 2011). Because all green bacteria must make multiple types of (B)Chl molecules (see above), it has been hypothesized that organisms may have different types of ChlH/BchHDI to channel Mg-Proto IX toward different end products and to allow differential feedback regulation of the first committed step of (B)Chl biosynthesis (Gomez Maqueo Chew, Frigaard, &

Bryant, 2009). Mutational studies and *in vitro* gene expression of *bchH* paralogs from *Cba. tepidum* seem to support this hypothesis. Three *bchH* genes, denoted *bchH*, *bchS*, and *bchT*, are present in this GSB. Single mutants of all three paralogs were shown to have MgCH activity producing all three (B) Chls found in GSB; however, a mutant lacking BchS showed a significant decrease in BChl *c* content (Gomez Maqueo Chew et al., 2009). Double mutants lacking *bchH* and *bchT* (functional BchS) or *bchS* and *bchT* (functional BchH) also made all three (B)Chls; however, attempts to make a *bchHbchS* double mutant (functional BchT) failed. The *bchS bchT* double mutant made similarly low amounts of BChl *c* as the *bchS* single mutant (Gomez Maqueo Chew et al., 2009). *In vitro* studies with BchH, BchS, or BchT from *Cba. tepidum* produced in *E. coli* and reconstituted with BchD and BchI showed that BchHDI, BchSDI, and BchTDI all had MgCH activity. The specific activity of BchSDI was two orders of magnitude higher than that of BchHDI and five orders of magnitude higher than that of BchTDI (Johnson & Schmidt-Dannert, 2008). These data agree well with the *in vivo* studies and the substrate channeling hypothesis. Additionally, in *Chloroflexus* spp., which synthesize only two types of (B)Chl but experience both oxic and anoxic conditions, there are also three paralogs of *bchH*: two *bchH*-like paralogs and one *bchS*-like paralog. It has been suggested that the *bchH*-like paralogs might be differentially regulated on the basis of oxygen levels (Tang et al., 2011).

There is only one copy of the *bchH* gene in *Cab. thermophilum*, which synthesizes Chl *a*, BChl *a*, Zn-BChl *a*', and BChl *c* and probably experiences variations in oxygen levels (Bryant & Liu, 2013; Garcia Costas, Liu, et al., 2012; Garcia Costas, Tsukatani, et al., 2012). At present the nature of the chelatase, if any, which inserts Zn^{++} into the macrocycle for the biosynthesis of Zn-BChl *a*' in *Cab. thermophilum* is unknown. *In vitro* studies have shown that Proto IX ferrochelatase of *Rba. sphaeroides*, which catalyzes Fe^{++} insertion into Proto IX during heme biosynthesis, can insert various divalent metal cations into the macrocycle, including Zn^{++} (Jones & Jones, 1970). However; it is unknown if this enzyme performs this function *in vivo* in *Cab. thermophilum*. The mechanism of Zn^{++} insertion into the macrocycle is an open question in both purple bacteria and *Cab. thermophilum* (Jaschke, Hardjasa, Digby, Hunter, & Beatty, 2011). It is possible that this reaction occurs spontaneously in the latter organism, which only requires very small amounts of Zn-BChl *a*' in RCs. Finally, there are two genes in "*Ca. T. aerophilum*" which together encode a single complete BchS protein (Fig. 4).

2.2.2 C13 propionate methylation

The insertion of Mg into Proto IX is followed by the conversion of Mg-Proto IX into Mg-Proto IX 13-monomethyl ester. This reaction involves methylation of the C13 propionate group of Mg-ProtoIX and is catalyzed by Mg-ProtoIX methyltransferase, ChlM/BchM (Fig. 5). ChlM/BchM is a SAM-dependent methyltransferase and is present in most chlorophototrophic bacteria (Bollivar, Jiang, Bauer, & Beale, 1994; Gibson & Hunter, 1994). Unlike the previous enzyme in the pathway, only one copy of *chlM/bchM* is present in the genomes of green bacteria and thus must be used to synthesize all (B)Chl molecules (Bryant & Liu, 2013). ChlHDI/BchHDI and ChlM/BchM have been shown to interact and possibly form a multi-enzyme complex (Sawicki & Willows, 2010). Studies in cyanobacteria and purple bacteria established that ChlH/BchH affects the activity of ChlM/BchM (Hinchigeri, Hundle, & Richards, 1997; McLean & Hunter, 2009; Shepherd, McLean, & Hunter, 2005). In green bacteria, experiments comparing the effect of BchH, BchS, and BchT from *Cba. tepidum* on BchM showed that BchS and BchT increased BchM activity while BchH decreased activity (Johnson & Schmidt-Dannert, 2008). These experiments further support the idea that the BchH subunits of green bacteria may be involved in substrate channeling and regulation of (B)Chl biosynthesis. The draft genome of *Ca.* T. aerophilum is notably lacking *bchM* and also lacks any known genes for isocyclic *E*-ring formation (see below; Fig. 4). Possible explanations are that these genes are in a currently unsequenced region of the genome, that a different methyltransferase is used in lieu of BchM, or that the E-ring is formed *via* an uncharacterized reaction which does not require the formation of the methylester.

2.2.3 Isocyclic E-ring formation

Mg-Proto IX 13-monomethyl ester is converted to 8-vinyl-protochlorophyllide (8V-PChlide) by one of several enzymes classified as oxidative isocylic ring cyclases, also known as Mg-Proto IX monomethylester oxidative cyclase. Like the oxidative reactions discussed above, this reaction is performed by different enzymes in aerobic and anaerobic organisms, although many chlorophototrophs contain genes encoding more than one enzyme of this reaction. Each enzyme performs a multi-step cyclization and oxidation which converts the C13 propionate methyl ester group into the isocyclic E ring found in all (B)Chls (Chen, Canniffe, & Hunter, 2017; Gough, Petersen, & Duus, 2000; Piñta, Picaud, Reiss-Husson, & Astier, 2002) (Fig. 5). In organisms inhabiting oxic environments the dominant cyclase enzyme is an oxygen-dependent di-iron oxygenase, AcsF (also variously known as CHL27, CRD1, CycI) (Piñta et al., 2002). The oxygen atom in the keto group of the E ring is derived from O_2 (Walker, Mansfield, Smith, & Castelfranco, 1989). Plant and cyanobacterial AcsF enzymes require an additional subunit, Ycf54, for activity (Albus et al., 2012; Hollingshead et al., 2012), while AcsF paralogs found in purple chlorophototrophic members of the α-proteobacteria require BciE (Chen et al., 2017). AcsF enzymes found in other chlorophototrophic purple bacteria do not require Ycf54 or BciE. For example, AcsF from *Rubrivivax gelatinosus* encodes a functional cyclase requiring no additional subunit when produced in *E. coli* (Chen et al., 2018). Ycf54 and BciE are not found together in the genome of any chlorophototroph studied to date (Chen et al., 2017). The *acsF* gene is not found in the strictly anaerobic GSB, *Osc. trichoides, or* "*Ca. C. asiatica*" but is present in "*Ca.* V. mediisalina," *Chloroflexus* spp. and *Cab. thermophilum* (Bryant et al., 2012); however, *ycf54* and *bciE* are absent from green bacteria containing *acsF* (Fig. 4).

In anaerobic organisms the oxidative cyclase reaction is performed by an oxygen-independent and oxygen-sensitive enzyme, BchE. BchE is a radical-SAM enzyme, which contains an oxygen-labile [4Fe-4S] cluster and requires a cobalamin cofactor for catalysis (Gough et al., 2000). As with HemN (CgdH) discussed above, this oxidation occurs under anoxic conditions *via* radical chemistry. The keto group on the isocyclic E ring derives its oxygen atom from H_2O instead of O_2 (Porra et al., 1996). BchE is a member of the BchE/*P*-methyltransferase family, which also includes C8 and C12 methyltransferases involved in the BChl *c, d, e,* and *f* biosynthetic pathway of some organisms (see below). BchE is present in FAPs and *Cab. thermophilum* that also contain *acsF* (Fig. 4). In *Cab. thermophilum bchE* and *acsF* transcript abundances are oxygen-dependent and have opposite expression patterns as would be expected if BchE functions under microoxic and anoxic conditions while AcsF functions under oxic conditions (Liu & Bryant, 2012). The green FAP *Cfl. aurantiacus* primarily relies on BchE for the production of BChls; however, AcsF was detected in a chlorosome preparation under anoxic conditions, and unlike *bchE*, the expression of *acsF* does not change with O_2 tension in this organism (Tang, Wen, Li, & Blankenship, 2009). These observations led to the hypothesis that the *Cfl. aurantiacus* AcsF may have an alternative function; for example, this di-iron protein could play a role in electron transfer or iron transport under anoxic conditions. The strictly anaerobic green bacteria, GSB, *Osc. trichoides,* and "*Ca. C. asiatica*" only contain the *bchE* gene encoding the oxygen-independent enzyme and lack *acsF*

(Liu & Bryant, 2012). As mentioned above, the draft genome of "*Ca.* T. aerophilum" did not contain *bchM, bchE,* or *acsF* (Fig. 4). This raises the interesting prospect that this organism may form the isocyclic ring by a very different enzymatic mechanism than other organisms (Liu, Klatt, et al., 2012). Furthermore, chlorophototrophic stramenopile algae also appear to lack *acsF* and *bchE*, providing additional evidence for the existence of a missing isocyclic ring cyclase enzyme, which may be shared with "*Ca.* T. aerophilum" (Matsuo & Inagaki, 2018).

2.2.4 Reduction of C17=C18 double bond

The reduction of the C17=C18 double bond of the D ring results in the conversion of 8V-PChlide to 8V-Chlide *a* (or PChlide to Chlide *a*, see following section). In green bacteria the enzyme responsible for this reaction is the dark-operative (i.e., light-independent) PChlide oxidoreductase (DPOR) (Bröcker et al., 2008; Reinbothe et al., 2010) (Fig. 5). DPOR is an ATP-dependent oxidoreductase made up of three subunits: BchN, BchB, and BchL. The enzyme consists of a homodimer of BchL, which binds an intrasubunit [4Fe-4S] cluster, and a heterotetramer of BchB and BchN which binds two [4Fe-4S] clusters; the electron donor for this reaction is reduced ferredoxin (Bröcker, Virus, et al., 2008; Nomata, Swem, Bauer, & Fujita, 2005). BchLBN is structurally similar to nitrogenase NifHDK (Bröcker et al., 2010; Muraki et al., 2010; Sarma et al., 2008). While BchLBN was initially described as a PChlide reductase that reduced PChlide to Chlide *a*, it was later demonstrated that 8V-PChlide can also be reduced by DPOR (see below) (Bröcker et al., 2008).

2.2.5 Reduction of C8 vinyl group

Mutagenesis studies in purple bacteria described strains disrupted at the *bchJ* locus that were unable to grow phototrophically and excreted a pigment that was later identified as 8V-PChlide (Bollivar, Suzuki, Beatty, Dobrowolski, & Bauer, 1994; Zsebo & Hearst, 1984; Gomez Maqueo Chew & Bryant, 2007a, 2007b). Although these mutants could still synthesize BChl *a* carrying an ethyl group at C8, BchJ was nevertheless designated as an 8V-reductase (8VR) or at least a subunit of 8VR. This interpretation naturally led to the assumption that the substrate for 8VR was 8V-PChlide.

Arabidopsis thaliana mutants of the AT5G18660 locus accumulate 8V-Chl *a* and 8V-Chl *b* (Nagata, Tanaka, Satoh, & Tanaka, 2005; Nakanishi et al., 2005). The recombinant protein encoded by this gene, which is unrelated to BchJ, is able to reduce the 8-vinyl group of 8V-Chlide *a* to an ethyl group to form Chlide *a*, confirming that this gene product defines a first functional class of 8VR. Subsequently, an ortholog of this gene from *Cba. tepidum*, *bciA*, was also shown to encode an 8VR. Heterologously produced BciA can reduce the 8-vinyl group of 8V-PChlide to form PChlide *a* in the presence of NADPH (Gomez Maqueo Chew & Bryant, 2007b).

The genomes of many cyanobacteria that synthesize Chl *a* do not contain orthologs of *bciA,* indicating that another class of 8VR must exist in these organisms. Mutants in open reading frame slr1923 of *Synechocystis* sp. PCC 6803 are unable to grow under high light and accumulate 8V-Chl *a* (Islam et al., 2008; Ito, Yokono, Tanaka, & Tanaka, 2008). The ortholog of slr1923 from the GSB *Chloroherpeton* (*Chp.*) *thalassium* (Ctha_1208) could complement the previously described *ΔbciA* mutant of *Cba. tepidum,* thereby demonstrating the activity of a second class of 8VR, BciB (Liu & Bryant, 2011b). *In vitro* characterization of this enzyme showed that BciB was able to convert 8V-PChlide to PChlide in the presence of either dithionite or ferredoxin, NADPH, and ferredoxin:$NADP^+$ oxidoreductase. BciB binds one FAD cofactor and two [4Fe-4S] clusters, which are likely involved in the catalytic mechanism (Saunders, Golbeck, & Bryant, 2013).

All green bacteria have at least one gene encoding BciA or BciB, while some have multiple paralogs for one or the other, or even genes for both enzymes (Fig. 4) (Liu & Bryant, 2011b). In organisms where the $C8^2$ position of BChl *c*, *d*, *e*, or *f* is differentially methylated, the reduction of the C8 vinyl group is a prerequisite reaction that must occur before the further modifications of the side chain by the BchQ methyltransferase. In mutants lacking 8VR activity, the $C8^2$ position cannot be methylated, and this defect causes a decrease in the half bandwidth of the Q_y absorbance band compared to wild-type. This defect also leads to lower total BChl *c* content in cells, a blue-shift in the absorbance maximum of BChl *c* aggregates, and slower growth under low irradiance conditions (Gomez Maqueo Chew & Bryant, 2007b; Liu & Bryant, 2011b). Thus, in low-light environments 8VR activity provides an important competitive advantage for organisms that can potentially methylate the C8 position. The ecophysiological importance of this selective advantage may explain why some green bacteria encode multiple, redundant 8VRs.

All tested higher plant enzymes demonstrate a preference for 8V-Chlide *a* as a substrate, as seen in the *A. thaliana* ortholog of BciA, and some are unable to reduce 8V-PChlide entirely (Wang et al., 2013). Similarly, BciA from *Rba. sphaeroides* was shown to act on 8V-Chlide *a*, and could only reduce 8V-PChlide when the DPOR enzyme was absent (Canniffe, Chidgey, & Hunter, 2014). The same study also indicated that the BciB enzyme from *Synechocystis* sp. PCC6803 prefers 8V-Chlide *a*, but this enzyme is also able to reduce 8V-PChlide. Under pigment-accumulating conditions, only a small amount of reduced PChlide could be detected in the WT (Canniffe et al., 2014). These studies indicate that the C8-vinyl reduction most likely occurs after the reduction of the C17=C18 bond, but as is common in the pigment biosynthesis pathways, this enzyme has demonstrated substrate flexibility, which could allow 8-vinyl reduction to occur under conditions that cause precursor pigments to accumulate.

Further analyses of *bchJ* mutants have been made. The deletion of this gene from the GSB *Cba. tepidum* and the purple bacterium *Rba. sphaeroides* did not perturb synthesis of (B)Chls, while disruption of *bciA* led to the accumulation of 8V-(B)Chl species (Canniffe, Jackson, Hollingshead, Dickman, & Hunter, 2013; Gomez Maqueo Chew & Bryant, 2007b). A lack of direct evidence remains for the activity of BchJ in the reduction of the C8 vinyl group. It has been suggested that BchJ may act as a substrate carrier/chaperone in the (B)Chl biosynthesis pathway in anoxygenic chlorophototrophic bacteria, and, thus, might play a similar porphyrin-binding role to that of GUN4 in plants and cyanobacteria (Sawicki & Willows, 2010).

A third enzyme with 8VR activity was hinted at in *bciA* mutants of *Rba. sphaeroides* and *Cba. tepidum*, both of which lack *bciB* genes but still make BChl *a* with an 8-ethyl moiety (Canniffe et al., 2013; Mizoguchi, Harada, & Tamiaki, 2012). The finding that a *bciA* mutant of *Cba. tepidum* can make 8-vinyl derivatives of Chl *a* and BChl *c* but reduced BChl *a* conflicted with an earlier characterization of *bciA* mutants in *Cba. tepidum* that found 8-vinyl derivatives of all three (B)Chls (Gomez Maqueo Chew & Bryant, 2007b; Mizoguchi et al., 2012). Additionally, the red FAPs in the genus *Roseiflexus* have BChl *a* in their photosynthetic apparatus, but lack orthologs of *bciA* or *bciB* in their genomes. These observations strongly suggested that a third enzyme class capable of 8-vinyl group reduction must exist (see discussion of Chlide reductase (COR) below).

2.3 Chlorophyllide *a* to bacteriochlorophyllide *a*

As in purple bacteria, the conversion of Chlide *a* to BChlide *a* proceeds *via* the action of three enzymes in all green bacteria (Bollivar, Suzuki, et al., 1994; Harada et al., 2014; Liu & Bryant, 2012; Suzuki, Bollivar, & Bauer, 1997). These enzymes catalyze the reduction of the C7=C8 double bond

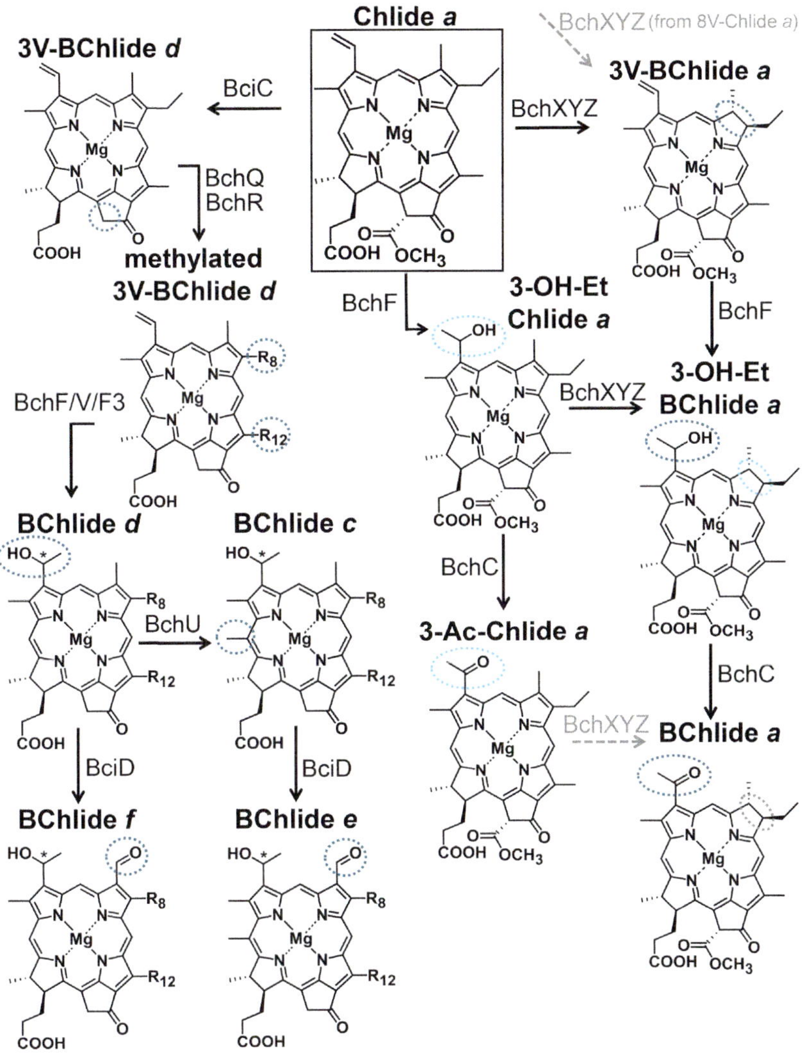

Fig. 6 Biosynthesis: Chlide *a* to BChlide *a* and BChlide *c, d, e,* or *f*. The biosynthetic pathways from Chlide *a* to BChlide *a* and BChlide *c, d, e,* or *f* are shown. Arrows indicate the reaction order, with arrow labels indicating the enzyme(s) responsible for catalysis. Gray dashed arrows indicate reactions which are less likely to occur *in vivo* based on experimental data (e.g., specific activity). R8: ethyl, *n*-propyl, isobutyl, or neopentyl. R12: ethyl or methyl.

(BchXYZ), hydration of the C3-vinyl group to form a hydroxylethyl group (BchF), and subsequent dehydrogenation of the C3-hydroxyethyl group to form an acetyl group (BchC) (Fig. 6). Biosynthetic pathways commonly list the reactions in this order which was established based on substrate accumulation studies in various studies of purple bacteria combined with limited *in vitro* characterization (Lange et al., 2015). However, recent *in vitro* characterization of all three enzymes has shown that they exhibit expanded substrate reactivity; this suggests that multiple possible pathway orders may exist for the three terminal steps of BChlide *a* biosynthesis (Harada et al., 2014, 2015; Kiesel et al., 2015; Lange et al., 2015; Teramura, Harada, & Tamiaki, 2018).

2.3.1 Reduction of the C7=C8 double bond

The first committed step of BChlide *a* biosynthesis is the reduction of the C7=C8 double bond on ring B of the chlorin substrate producing a bacteriochlorin. The enzyme responsible for this reaction is Chlide *a* oxidoreductase (COR), encoded by *bchX*, *bchY*, and *bchZ* (Fig. 6). BchX, BchY, and BchZ are paralogs of the BchL, BchN, and BchB subunits of DPOR and nitrogenase as described above. DPOR and COR are structurally and functionally similar enzymes which catalyze the reduction of a double bond on the macrocycle of a Chlide molecule in an ATP-dependent manner (Kiesel et al., 2015; Nomata, Mizoguchi, Tamiaki, & Fujita, 2006; Wätzlich et al., 2009). Importantly, duplication of the DPOR-encoding genes leading to the evolution of COR, and thus the synthesis of bacteriochlorins, permitted the use of near-infrared light for anoxygenic photosynthesis.

Initial *in vitro* characterization of COR showed that it could convert Chlide *a* into 3V-BChlide *a* (Nomata et al., 2006). Subsequently, substrate-specificity assays of purified COR enzyme have shown that it is also active with other chlorins, including Zn-containing analogs of 3-hydroxyethyl-Chlide *a* and 3-acetyl-Chlide *a* (Kiesel et al., 2015). The specific activity of COR using the 3-hydroxy-ethyl-Chlide *a* analog as the substrate was shown to be equal to that with Childe *a* as the substrate. In contrast, the specific activity with the 3-acetyl-Chlide *a* analog was only 30% of that observed with Chlide *a*. Based on these data, it seems equally likely that Chlide *a* or 3-hydroxyethyl-Chlide *a* may act as a substrate *in vivo*, while 3-acetyl-Chlide *a* may also be used but does not appear to be the preferred substrate for this enzyme.

Surprisingly, it was recently demonstrated that COR from the BChl *a*-producing purple chlorophototrophic bacterium *Rba. capsulatus* was also able to reduce the 8V group of Chlide *a* (Tsukatani, Yamamoto, Harada, et al., 2013). This form of the enzyme (COR*a*) was able to catalyze the formation of 8-ethyl bacteriochlorin with both 8-vinyl and 8-ethyl Chlide substrates. The COR enzymes from the BChl *b*- or BChl *g*-producing strains,

Blastochloris viridis or *Heliobacterium modesticaldum*, respectively (COR*b*) were not able to act on 8-ethyl Chlide *a*, but when 8V-Chlide *a* was provided as substrate, these enzymes formed a product carrying an ethylidene group at C8 (known as 3V-BChlide *b* or BChlide *g*) (Tsukatani, Yamamoto, Harada, et al., 2013; Tsukatani, Yamamoto, Mizoguchi, Fujita, & Tamiaki, 2013). Replacement of the COR*a*-encoding genes of *Rba. sphaeroides* with the COR*b* ortholog from *Blc. viridis* led to the switch from the biosynthesis of BChl *a* to BChl *b* in this strain when the native 8VR, BciA, was absent (Canniffe & Hunter, 2014). These studies delineate a third class of 8VR enzyme providing an explanation for the anomalous results mentioned previously. It is hypothesized that organisms using BChl *a* employ COR*a* to reduce the 8V group of any Chlide molecule that has bypassed the conventional 8VR. This mechanism also accounts for absence of *bciA* or *bciB* in *Roseiflexus* spp., and clarifies why *bciA* mutants of *Cba. tepidum* and *Rba. sphaeroides* are able to synthesize BChl *a* (Canniffe et al., 2013; Mizoguchi et al., 2012). It is interesting to note that while COR enzymes may have 8VR activity, they require ATP for catalysis and are therefore energetically more costly than the BciA and BciB enzymes. Considering the very large numbers of (B)Chls produced by most green bacteria, this energetic penalty of the COR enzymes suggests why they are probably not the preferred route of 8V reduction whenever BciA or BciB are available. This penalty might also provide an explanation for why GSB have laterally acquired an enzyme (BciA) that is quite similar to the enzymes found in higher plants (Gomez Maqueo Chew & Bryant, 2007b). The 8VR activity of COR provides further evidence that DPOR enzymes use 8V-PChlide instead of PChlide *a* as the substrate in many circumstances. In fact, in *Roseiflexus* spp. that lack *bciA* and *bciB* genes (see above), BChl *a* biosynthesis almost certainly proceeds from 8V-PChlide to 8V-Childe *a* catalyzed by DPOR, then from 8V-Chlide *a* to 3-vinyl-BChlide *a*, *via* a Chlide *a* intermediate, catalyzed by COR.

2.3.2 Hydration of the C3 vinyl group

BChl *a* has an acetyl group at the C3 position, while BChl *c*, *d*, *e*, and *f* have a hydroxyethyl group. The formation of each of these requires the action of a 3-vinyl hydratase enzyme that can hydrate the vinyl group to form a hydroxyethyl moiety. In the case of BChl *a*, this sidechain is further modified by a stereoselective, NAD^+-dependent dehydrogenase that results in the formation of the acetyl group.

The C3 hydratase in BChl *a* biosynthesis was first identified in purple bacteria, in which the encoding gene was denoted *bchF* (Burke, Alberti, & Hearst, 1993). A single homolog of *bchF* is found in green FAPs and *Cab. thermophilum*. For the most part green-colored GSB, such as *Cba. tepidum,* have two paralogs of *bchF,* denoted *bchF* and *bchV*, and brown-colored GSB contain a third paralog denoted *bchF3* (Bryant & Liu, 2013). In green FAPs and *Cab. thermophilum,* BchF must perform C3 hydratase activity required for the synthesis of both BChl *a* and BChl *c*. Initial attempts to inactivate *bchF* in *Cba. tepidum* were unsuccessful, which led to the interpretation that *bchF* was required for BChl *a* biosynthesis making it an essential gene for GSB, while BchV was specific to BChl *c* biosynthesis and did not function in BChl *a* biosynthesis (Frigaard, Gomez Maqueo Chew, Li, Maresca, & Bryant, 2003; Gomez Maqueo Chew, Frigaard, & Bryant, 2004; Liu & Bryant, 2012). Further discussion of the roles of BchF and BchV during the synthesis of BChlides *c, d, e* and *f* from Chlide *a* in green bacteria is found below.

2.3.3 3-hydroxyethyl dehydrogenase

Following formation of the C3 hydroxyethyl moiety, 3-hydroxyethyl BChlide *a* dehydrogenase, BchC, catalyzes the oxidation of the hydroxyethyl group at C3 to yield an acetyl group. Similar to *bchF*, *bchC* was first shown to encode the 3-hydroxyethyl dehydrogenase in purple bacteria (McGlynn & Neil Hunter, 1993). In *in vitro* experiments the substrate specificity of heterologously produced BchC from *Cba. tepidum* has been examined. The BchC dehydrogenase exhibited activity with a demetallated analog of 3-hydroxyethyl BChlide *a* and a Zn-containing analog of 3-hydroxyethyl Chlide *a* when NAD^+, but not $NADP^+$, as the hydride acceptor (Lange et al., 2015). The specific activity with the Zn-containing analog of 3-hydroxyethyl Chlide *a* was about 20 times higher than that with the demetallated analog of 3-hydroxyethyl BChlide *a*. The replacement of Mg by Zn is tolerated by many enzymes in (B)Chl biosynthesis; however, the complete absence of a metal in the macrocycle is often not tolerated (Harada, Saga, Yaeda, Oh-Oka, & Tamiaki, 2005; Kiesel et al., 2015; Teramura, Harada, Mizoguchi, Yamamoto, & Tamiaki, 2016; Teramura, Harada, & Tamiaki, 2016, 2019; Teramura et al., 2018). With this in mind it is difficult to say if BchC shows a preference for 3-hydroxyethyl Chlide *a* or 3-hydroxyethyl BChlide *a,* because the analogs used may not have allowed for a valid comparison. It is clear, however, that BchC has significant activity with 3-hydroxyethyl Chlide *a* in addition to 3-hydroxyethyl BChlide *a*.

As a result of the expanded substrate reactivities of BchC, BchF, and COR, an alternative reaction sequence has been proposed in which Chlide *a* is first acted upon by BchF to form 3-hydroxyethyl Chlide *a*. This step is either followed by conversion of the 3-hydroxyethyl group to an acetyl group by BchC or by reduction of the C7=C8 double bond by COR (BchXYZ) (Fig. 6).

2.4 Chlorophyllide *a* to bacteriochlorophyllide *c*, *d*, *e*, and *f*

2.4.1 Demethoxycarbonylation of the C13^2 methylcarboxyl group

The first committed step of the biosynthetic pathway leading from Chlide *a* to BChls *c*, *d*, *e*, and *f* is the removal of the C13^2 methylcarboxyl group from Chlide *a* to produce 3-vinyl-BChlide *d* (Fig. 6). A similar reaction occurs in Chl degradation, and two different enzymes are known to catalyze the removal of this group to produce pheophorbide (Pheide) *a* in plants, algae, and some cyanobacteria. The enzyme pheophorbidase catalyzes a methylesterase reaction, which is followed by spontaneous decarboxylation. An unknown enzyme in *Chlamydomonas reinhardtii* has been reported to have demethyoxycarbonylase activity, which completes the removal of the C13^2 methylcarboxyl group in a single step (Doi & Shioi, 2002; Hörtensteiner & Kräutler, 2011). A candidate gene for this activity in green bacteria was identified by phylogenetic profiling, and the corresponding deletion mutant in *Cba. tepidum* produced Chl *a* and BChl *a* but no BChl *c*; the strain also accumulated (B)Pheides that retained the intact C13^2 methylcarboxyl group (Liu & Bryant, 2011a). These data showed that the gene was essential for removal of the C13^2 methylcarboxyl group, and the gene was designated *bciC*. BciC is not homologous to known pheophorbidases or methylesterases, and so the mechanism of the reaction remained unclear. Recently, the *bciC* gene from *Cba. tepidum* was heterologously expressed in *E. coli,* and the encoded protein catalyzed the removal of the C13^2 methylcarboxyl group of Chlide *a* and 8V-Chlide *a,* with a preference for Chlide *a.* BciC could not catalyze this reaction when Pheide *a*, PChlide *a,* or 8V-PChlide *a* were tested as substrates (Teramura, Harada, Mizoguchi, et al., 2016). Further studies of BciC showed that the enzyme can use various Chlides modified at the C3 or C7 position, but at a reduced rate compared to Chlide *a,* and could also act on BChlide *a* and 3V-BChlide *a* at greatly reduced rates. However, no activity was observed with any porphyrins tested or any Chlide variant with C13^2 *S* stereochemistry (Teramura et al., 2019). The substrate specificity data supports Chlide *a* as the currently accepted branching point compound. BciC was not inhibited by addition of methanol to the reaction, which suggests that it does not function *via* the methylesterase mechanism seen in pheophorbidase enzymes, but instead it apparently acts as a demethoxycarbonylase. Homologs of the *bciC* gene are present in the genomes of all green bacteria, but are not found in any other organism.

2.4.2 Methylation at C8 and C12

Following the removal of the methylcarboxyl group 3-vinyl-BChlide *d* may be methylated at the $C8^2$ and $C12^1$ positions to form methylated-3-vinyl-BChlide *d*, which describes a mixture of methylated species (Fig. 6). The methylations at these two positions are performed by independent methytransferases; each is a radical-SAM enzyme that derives its methyl-donating substrate from the cofactor SAM (Gomez Maqueo Chew et al., 2007; Huster & Smith, 1990; Kenner, Rimmer, Smith, & Unsworth, 1978). The $C8^2$-methyltransferase is encoded by *bchQ* while the $C12^1$-methytransferase is encoded by *bchR*; both are distant homologs of BchE, the oxygen-independent cyclase discussed above (Gomez Maqueo Chew et al., 2007). The ethyl group at the $C8^2$ position may be methylated one to three times, sequentially producing n-propyl, isobutyl, or neopentyl groups. The methyl side chain at C12 is only methylated once to produce an ethyl moiety (Airs et al., 2001; Bobe, Pfennig, Swanson, & Smith, 1990; Borrego, Gerola, Miller, & Cox, 1999; Glaeser et al., 2002; Gomez Maqueo Chew et al., 2007).

BchQ and BchR are found in *Chlorobia*, *Cab. thermophilum*, *Osc. trichoides*, "*Ca. C. asiatica*" and "*Ca.* V. mediisalina" but notably are absent in *Chloroflexus* spp. (Bryant et al., 2012). Brown-colored GSB have a second paralog of BchQ, denoted BchQ2. The *bchQ2* gene is located just upstream from a conserved gene cluster found in all brown-colored GSB. This cluster contains *bchF3*, a $C3^1$ hydratase, and *bciD*, which is required for BChl *e* or *f* biosynthesis (see below) (Thweatt, Ferlez, Golbeck, & Bryant, 2017). While it is likely that *bchQ2* is associated with BChl *e* biosynthesis, its exact function remains unclear. Brown-colored GSB tend to live in extremely low-light environments, which favor BChls highly methylated at the $C8^2$ position (Gomez Maqueo Chew et al., 2007). It is possible that *bchQ2* is responsible for increased methylation activity in these organisms. Because brown-colored GSB can live at lower light intensities than other GSB, it is possible that increased methylation of BChl *e* by BchQ2 provides an evolutionary advantage by expanding the absorption cross-section of these organisms.

2.4.3 Hydration of the C3^1 position

The BChls found in chlorosomes (BChl *c*, *d*, *e* or *f*) carry a hydroxyethyl group at the C3 position, which is formed by the action of C3^1 hydratase on the C3 vinyl moiety of a chlorin substrate. Green bacteria may have one, two, or three paralogs of the gene encoding the C3^1 vinyl hydratase; in green FAPs and *Cab. thermophilum,* in which only a single paralog occurs, BchF is responsible for the C3^1 hydration reaction in both BChl *a* and BChl *c*/*d* biosynthesis. However, in *Chlorobia* the situation is more complicated.

As mentioned above, an initial report led to the hypothesis that BchF was specific to BChl *a* biosynthesis and was thus an essential enzyme for members of the *Chlorobia*, and that BchV was specific to BChl *c* biosynthesis (Gomez Maqueo Chew et al., 2004). However, individual mutants in the genes encoding these enzymes have now been successfully constructed in *Cba. tepidum*, and their roles have been characterized more thoroughly (Harada et al., 2015). The *bchF* mutant made significantly less BChl *a,* increased levels of 3-vinyl-BChl *a*, and less BChl *c* than the wild type. The BChl *c* homologs in this mutant were primarily *S*-epimers, and when compared to the wild type, a higher proportion of these homologs were more highly methylated at the C8^2 position. The *bchV* mutant contained less BChl *c* than the wild type, very little of which had the *S*-configuration at C3^1; ethyl and propyl groups at C8 were present, but unlike the wild type, almost no homologs carried isobutyl side chains. The cells contained more BChl *a* and 3-vinyl-BChl *c* as a proportion of total (B)Chl. These results indicate that BchF is the main C3 vinyl hydratase responsible for BChl *a* biosynthesis in *Cba. tepidum,* and it is also important for the biosynthesis of BChl *c* homologs with ethyl or propyl groups at C8 in this organism. Additionally, these mutants show that the BchV hydratase must also have some C3 hydratase activity in the BChl *a* biosynthetic pathway of GSB; that BchV tolerates more highly methylated substrates than BchF and contributes a greater ratio of *S*-epimers. In contrast, BchF preferentially forms *R*-epimers, which in turn provide the substrate for the BchC dehydratase for biosynthesis of BChlide *a*.

Studies using lysates of *E. coli* producing BchF and BchV from *Cba. tepidum* showed that each enzyme has C3 hydratase activity with a variety of substrates. Hydroxyethyl groups could be formed with Chlide *a*, 8V-Chlide *a*, 3V-[8-ethyl,12-methyl]-BChlide *d*, Zn-3V-[8-ethyl, 12-methyl]-BPheide *d*, Zn-3V-[8-ethyl/8-propyl/8-isobutyl,12-ethyl]-BPheide *d*, Zn-3V-[8-ethyl, 12-methyl]-BPheide *c*,and Zn-3V-[8-ethyl/8-propyl,12-ethyl]-BPheide *c*. Only BchF had activity with 3V-BChlide *a*, and neither enzyme could use Zn-3V-[8-isobutyl, 12-ethyl]-BPheide *c* (Harada et al., 2015; Teramura, Harada, & Tamiaki, 2016; Teramura et al., 2018). The single BchF paralogs from *Cfl. aurantiacus* and *Cab. thermophilum*, and BchF, BchV and BchF3 from *Cba. limnaeum* were all able to use Chlide *a*, Zn-3V-[8-ethyl, 12-methyl]-BPheide *d* and 3V-BChlide *a* as substrates (Teramura et al., 2018).

On the basis of these results, it is likely that all three BchF homologs participate in BChl *a* biosynthesis in *Cba. limnaeum*; however, their exact functions *in vivo* remain unclear (Teramura et al., 2018). The role of BchF3 in brown-colored GSB may be related to the fact that these organisms often live in very low-light environments in which BChls with higher degrees of C8 and C12 methylation are favored. However, further studies are needed to determine if BchF3 is important for the production of the C3 hydroxyethyl group in BChl *e* molecules carrying higher degrees of methylation and the resulting spectral broadening caused by C8 and C12 methylation.

It is interesting to note that brown-colored GSB, but not green-colored GSB, possess an additional copy of *bchQ*, which could also be related to the prevalence of high degrees of C8 and C12 methylation (see above). The apparent lack of *in vitro* hydratase activity observed for recombinant BchV from *Cba. tepidum* when using 3V-BChlide *a* as the substrate suggests that either the activity was too low to be detected or that BchV contributes to BChl *a* biosynthesis by hydration of chlorins instead of bacteriochlorins *in vivo*. The ability of C3 hydratase enzymes to use a variety of substrates, in combination with the expanded COR reactivity discussed above, suggests that the order of reactions within the BChlide *a* pathway is currently ambiguous. However, the experiments with recombinant BchF, BchV, and BchF3 discussed in this section were performed with crude lysates, and therefore did not yield specific activities for the enzyme and substrate combinations tested. Based on *in vitro* hydratase activity with both 3V-BChlide *c* and 3V-BChlide *d* analogs, Teramura, Harada, and Tamiaki (2016); Teramura et al. (2018) suggest a possible alternative reaction order in the biosynthetic pathway, in which methylation of the C20 position comes before or both before and after $C3^1$ hydration of BChlide *c* homologs with C8 ethyl or C8 propyl groups. However, this is in conflict with the crystal structure of the C20 methyltransferase, BchU, for which BChlide *d* was modeled into the active site, indicating that hydrogen bonding between the $C3^1$ hydroxyl group and the enzyme was likely to be important in substrate binding and specificity (see below; Wada et al., 2006). Further investigation of the substrate specificities of C3 hydratases is needed to elucidate the preferred substrates for each enzyme.

2.4.4 Methylation of the C20 methine bridge

In green bacteria that contain BChl *c* or BChl *e*, the C20 methyltransferase BchU is required to methylate the C20 methine bridge of BChlide *d* to form BChlide *c*. BchU is a SAM-dependent methyltransferase, but it is not a homolog of BchR and BchQ nor is it a radical-SAM methyltransferase. Crystal structures of BchU show that the enzyme is a homodimer and that the co-product *S*-adenosylhomocysteine binds near the C-terminal active site (Wada et al., 2006). The *bchU* gene was first identified in *Cba. tepidum*. When this gene was inactivated, the mutant produced BChl *d* instead of *c* (Maresca et al., 2004). Growth rate and competition studies showed that in high-light conditions *Cba. tepidum* containing BChl *c* or *d* grew at the same rate, but in low-light conditions *Cba. tepidum* containing BChl *c* had a growth advantage; these observations agree with other comparisons of BChl *c* or *d*-containing GSB (Bobe et al., 1990; Broch-Due & Ormerod, 1978; Maresca et al., 2004; Saga, Oh-oka, Hayashi, & Tamiaki, 2003). BchU from *Cba. tepidum* produced by heterologous expression in *E. coli* could catalyze the conversion of Zn-BChlide *d* to Zn-BChlide *c* (Harada et al., 2005).

Mutants of *bchU* in the model brown-colored GSB *Cba. limnaeum* resulted in strains that produce BChl *f*, an unnatural BChl (Tamiaki et al., 2011; Tsukatani, Harada, Mizoguchi, & Tamiaki, 2013; Vogl et al., 2012). BChl *f* is able to form supramolecular structures and assemble functional chlorosomes (Niedzwiedzki et al., 2014). The *bchU* mutants of *Cba. limnaeum* containing BChl *f* exhibited a significantly slower growth rate under low-light conditions than wild-type strains containing BChl *e* (Vogl et al., 2012). A comparison of the energy transfer characteristics of chlorosomes containing BChl *e* or *f* by time-resolved fluorescence spectroscopy showed that energy transfer from the BChl *f* aggregates to the baseplate BChl *a*-binding CsmA complexes was much less efficient than from BChl *e* aggregates (Orf et al., 2013). Based on these data BChl *f* containing organisms would require anoxic, high-light conditions. Such environments are rare and may be more favorable for other phototrophs. This finding could explain why BChl *f* has not been found in nature (Orf et al., 2013). Furthermore, the triplet excitonic state of BChl *f* is energetically more favorable for the production of singlet oxygen than BChl *c*, *d*, or *e* (Hartzler et al., 2014).

2.4.5 Formation of the C7 formyl group of BChlide e

Brown-colored GSB require an enzyme known as BChlide *c* $C7^1$ hydroxylase, which converts the C7 methyl group of BChlide *c* or *d* to the formyl group of BChlide *e* or *f*, respectively (Thweatt et al., 2017). Similar to the biosynthesis of Chl *b*, which also carries a formyl group at C7, this conversion is believed to occur *via* two sequential hydroxylation reactions followed by spontaneous dehydration. The formyl group of Chl *b* is formed *via* an oxygen-dependent enzyme known as Chlide *a* oxygenase (CAO) (Tanaka et al., 1998). However, GSB are obligate anaerobes and do not contain homologs of CAO. Brown-colored GSB use the radical-SAM enzyme BciD. BciD contains a single [4Fe-4S] cluster and catalyzes the hydroxylation reactions *via* radical chemistry (Thweatt et al., 2017). BciD from *Cba. limnaeum* heterologously produced in *E. coli* converted BChlide *c* or *d* into BChlide *e* or *f* in the presence of SAM and sodium dithionite. Additionally, small amounts of the putative intermediates 7^1-OH-BChlide *c* or *d* were identified (Thweatt et al., 2017). The oxygen source for this reaction must be water, but further research is required to determine the detailed mechanism of this reaction.

The *bciD* gene was first identified as a candidate gene related to BChl *e* and *f* biosynthesis *via* bioinformatic comparisons of the genomes of green-colored and brown-colored GSB, which identified a gene cluster unique to brown-colored GSB. This cluster includes (i) *cruB*, a gene required for isorenieratene biosynthesis (Maresca, Romberger, & Bryant, 2008), (ii) *bciD*, which encodes the $C7^1$ hydroxylase (Thweatt et al., 2017), (iii) *bchF3*, which encodes a $C3^1$ hydratase, and (iv) *bchQ2*, which encodes a paralog of the C8 methyltransferase. Mutation of *bciD* in *Cba. limnaeum* resulted in cells producing BChl *c* instead of BChl *e*, while a double mutant of *bchU* and *bciD* resulted in cells producing BChl *d* (Harada et al., 2013; Mizoguchi, Harada, Yamamoto, & Tamiaki, 2015; Thweatt et al., 2017).

2.5 The final steps in (B)Chl biosynthesis

The final steps in the biosynthetic pathways of (B)Chls involve the formation of the esterifying alcohol at the $C17^3$ position. After the formation of the (B)Chlide intermediate two types of enzymes are involve in forming the final product. A specific Chl or BChl synthase esterifies an alcohol moiety to the $C17^3$ position and in the case of phytyl or Δ2,6-phytadienol, geranylgeranyl reductase, ChlP/BchP, reduces the geranylgeranyl precursor to the appropriate moiety (Fig. 7). It remains uncertain whether reduction happens before or after esterification (or both), as *in vitro* assays have demonstrated that esterification can occur with diverse substrate alcohols at various stages of saturation (see below). However, the occurrence of BChl *a* carrying a geranylgeraniol moiety and BPhe *a* esterified with phytol in the purple bacterium *Rhodospirillum rubrum*, and the subsequent identification of

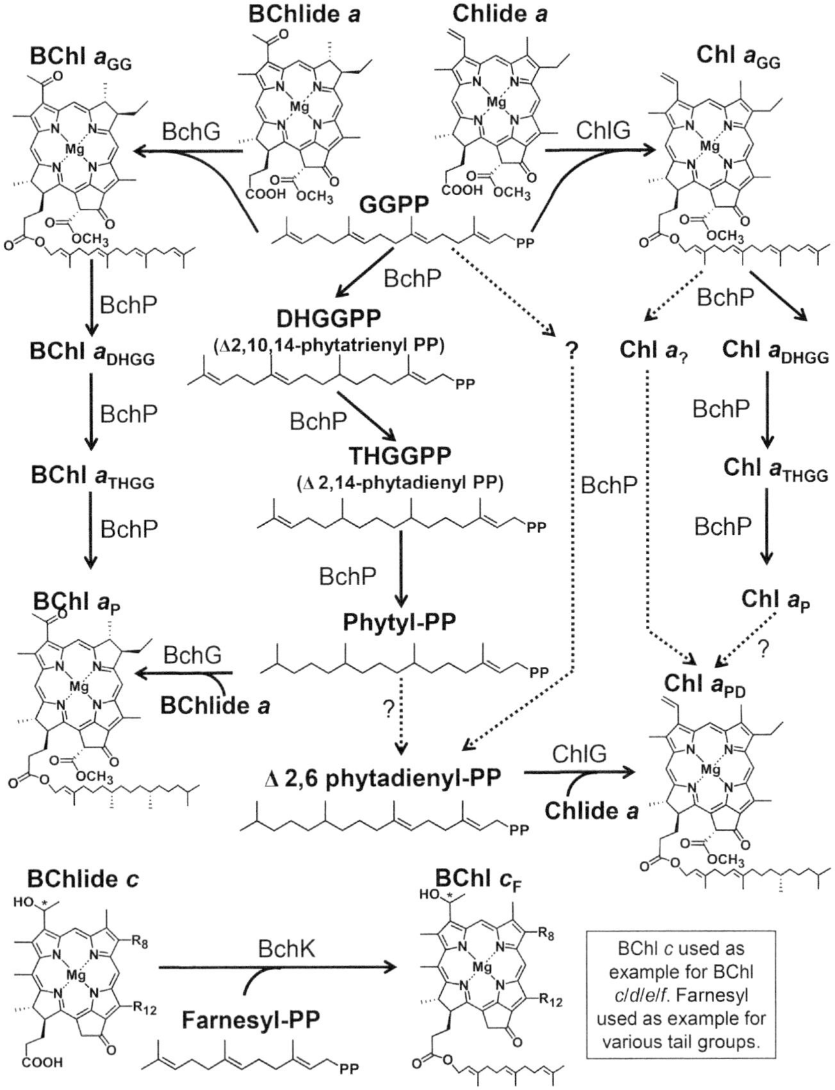

Fig. 7 Biosynthesis: Final Steps. The final steps in the biosynthetic pathways of Chl a_{PD}, Bchl a_P, and Bchl c_F are represented. Arrows indicate the reaction order, with arrow labels indicating the enzyme(s) responsible for catalysis. Dashed arrows represent possible reaction which have not be characterized. The "?" represents an unknown intermediate (in place of a molecule), or possible unknown dehydrogenase (in place of arrow label). GG, geranylgeranyl; P, phytyl; F, farnesyl; GGPP, geranylgeranyl diphosphate; DH, dihydro; TH, tetrahydro.

a mutation in its *bchP* gene resulting in the production of a mutated BchP which is only able to reduce alcohols attached to demetallated pigments, indicates that esterification occurs first in this organism (Addlesee & Hunter, 2002). Therefore, the following sections are ordered in this manner, although observations possibly contradicting this order will be mentioned.

2.5.1 Esterification of (B)Chlide

Green bacteria contain multiple (B)Chl synthase homologs known as Chl *a* synthase (ChlG), BChl *a* synthase (BchG), and BChl *c* synthase (BchK). Green FAPs have two types of BChl synthases, BchG and BchK that esterify BChl *a* and BChl *c* or *d*, respectively. *Chlorobia* and *Cab. thermophilum* additionally have a Chl *a* synthase, ChlG, giving these bacteria three types of (B)Chl synthase. Some GSB have multiple copies of the *bchK* gene, which encodes BChl *c*, *d*, *e*, or *f* synthase (Frigaard, Voigt, & Bryant, 2002; Thweatt et al., 2017). Phylogenetic analysis of these three types of synthases shows that the Chl/BChl synthase clade branches from other prenyl-transferases, and that BchG arose by a duplication within an existing ChlG/BchK clade. This implies that the chlorin synthases BchK and ChlG existed prior to the branching of BchG in evolutionary history (J.L. Thweatt and D.A. Bryant, unpublished data).

In Chl *a*-containing organisms, ChlG esterifies Chlide *a* at the $C17^3$ carboxylate moiety to form Chl *a* by using an isoprenoid alcohol-diphosphate substrate. The *chlG* genes of plants and cyanobacteria were first identified by homology to the *bchG* genes of purple bacteria and were later used to identify *chlG* in green bacterial genomes (Bollivar, Wang, Allen, & Bauer, 1994; Eisen et al., 2002; Frigaard et al., 2003; Gaubier, Wu, Laudié, Delseny, & Grellet, 1995; Kaneko et al., 1995). *In vitro* characterization of ChlG enzymes has shown that ChlG can esterify the $C17^3$ position of Chlide *a*, but not BChlide *a*, showing that ChlG is specific for the Chlide *a* substrate (Oster, Bauer, & Rüdiger, 1997). In experiments comparing different alcohol-diphosphate substrates, ChlG from cyanobacteria and plants, which produce Chl a_P, can use both geranylgeranyl-PP or phytyl-PP; some enzymes show a preference for geranylgeranyl-PP and others show a preference for phytyl-PP (Oster et al., 1997; Rüdiger, Benz, & Guthoff, 1980). No *in vitro* study of ChlG from green bacteria has yet been reported; however, some hints at substrate specificity are available from *bchP* and *bciC* mutants in *Cba. tepidum.* The *bchP* mutant was shown to produce Chl a_{GG}, while the *bciC* mutant showed over-production of Chl *a* derivatives resulting in mostly Chl a_P in addition to the normal Chl a_{PD}

(Harada et al., 2008; Liu & Bryant, 2011a). These *in vivo* data indicate that ChlG in *Cba. tepidum* can use geranylgeranyl-PP as a substrate. However, it is unclear if Chl a_P and Chl a_{PD} are the result of reductions that occur after esterification with geranylgeranyl-PP or esterification with reduced diphosphate counterparts.

BChl *a* synthase, BchG, was initially discovered and characterized in purple bacteria (Addlesee, Fiedor, & Hunter, 2000; Bollivar, Wang, et al., 1994; Oster et al., 1997). The activities of homologs from green bacteria were first confirmed by expression of two BchG homologs from *Cfl. aurantiacus* (Schoch, Oster, Mayer, Feick, & Rudiger, 1999). The first homolog (BchG) had synthase activity with BChlide *a* derivatives but not Chlide *a* or BChlide *c*, *d*, or *e* derivatives; the second, now known as BchK, exhibited activity with BChlide *c*, *d*, and *e* derivatives. The *bchG* gene from *Cba. tepidum* was identified *via* complementation of BChl *a* synthase activity in the purple bacterium *Rba. capsulatus* (Xiong, Fischer, Inoue, Nakahara, & Bauer, 2000). The *bchG* gene was subsequently identified in the sequenced genomes of other green bacteria as well. A recent study characterized the *in vitro* activity of heterologously produced BchG from *Cba. tepidum,* which also showed specificity for BChlide *a*. The enzyme is unable to esterify Chlide *a* or BChlide *c* (Saga, Hirota, Harada, & Tamiaki, 2015). As observed with ChlG, BchG enzymes are able to esterify BChlide *a* with a variety of diphosphate alcohols. BchG from *Cfl. aurantiacus* could use the diphosphate derivatives of geranylgeraniol, phytol, farnesol, and geraniol with a strong preference for geranylgeranyl-PP, while BchG of *Cba. tepidum* was shown to use geranylgeranyl-PP or farnesyl-PP. Unlike the studies with *Cfl. aurantiacus,* these experiments did not test phytyl-PP as a substrate or show that BchG from *Cba. tepidum* preferentially used a particular esterifying alcohol. The data confirm that BchG from *Cba. tepidum* is a BChl synthase specific for BChl *a,* but it has broad specificity for the esterifying alcohol *in vitro*.

Surprisingly, BChl *a* derivatives *in vivo* almost exclusively have phytyl tails in *Cba. tepidum,* except in the case of a *bchP* mutant that produced BChl a_{GG} instead (Gomez Maqueo Chew et al., 2008; Harada et al., 2008; Saga, Hirota, et al., 2015). This observation suggests that substrate channeling or sequestration may occur *in vivo* or the pools of other alcohol-diphosphate derivatives must be too small to support rapid esterification of BChlide *a*. However, it has been observed that BChl a_P is still exclusively produced by *Cba. tepidum* in the presence of exogenous alcohols, but BChl *c* contains various exogenously added alcohols (Larsen, Miller, & Cox, 1995; Steensgaard, Cox, & Miller, 1996). These findings, suggest that the concentration of alcohol-diphosphates alone does not account for the selective nature of BChl a_P production.

The BChl *c* (*d*/*e*/*f*) synthase BchK is the third type of (B)Chl synthase that occurs in green bacteria. As noted above BChl *c* synthase activity was demonstrated with a BchG paralog from *Cfl. aurantiacus*; however, the gene and gene product were not named at the time (Schoch et al., 1999). Null mutants of the *bchK* gene in *Cba. tepidum* were unable to synthesize BChl *c*, but could grow photoautotrophically and still synthesize BChl *a* and Chl *a* (Frigaard et al., 2002). As a result, it was concluded that *bchK* encodes BChl *c* synthase. It is important to note that in *bchU* mutants of *Cba. tepidum*, BchK must function as a BChl *d* synthase. All green bacteria have at least one copy of *bchK* in their genomes, while some have multiple copies. Phylogenetic analysis of BchK sequences from GSB showed two distinct clades of BchK sequences, which have been denoted BchK1 and BchK2 (Thweatt et al., 2017). Interestingly, all brown-colored GSB have at least one BchK2 homolog. Despite *in vitro* experiments showing BciD is sufficient to convert BChlide *c* or *d* to BChlide *e* or *f*, attempts to produce BChl *e* or *f* in *Cba. tepidum*, which naturally produced BChl *c* and only encodes BchK1, have failed. Together these observations led to the hypothesis that the enzymes in the BchK2 clade may be able to accommodate the C7 formyl group found on BChlide *e* and *f* while enzymes in the BchK1 clade cannot. Additional phylogentic analyses showed that *Chloroflexus* spp. form a separate clade of BchK enzymes, meaning BchK from *Cfl. aurantiacus* is not part of either the BchK1 or BchK2 clades of GSB (J. L. Thweatt and D. A. Bryant, unpublished data). Further studies are needed to determine the substrate specificity of GSB BchK enzymes with relationship to the BChlide substrate. Several *in vivo* studies feeding a wide variety of exogenous alcohols to *Cba. tepidum* suggest that the substrate specificity of BchK for the esterifying alcohol is extremely broad and is primarily driven by substrate availability (Larsen et al., 1995; Nishimori, Mizoguchi, Tamiaki, Kashimura, & Saga, 2011; Saga, Hayashi, Mizoguchi, & Tamiaki, 2014; Saga, Hayashi, Hirota, Harada, & Tamiaki, 2015, Saga, Yamashita, & Hirota, 2016; Steensgaard et al., 1996).

2.5.2 Reduction of the alcohol moiety

ChlP/BchP produces Δ2,6 phytadienol, or phytol moieties by sequential reductions of geranylgeraniol for biosynthesis of Chl a_{PD} and BChl a_P, respectively. As noted above, these reductions may occur either on the diphosphate derivative of geranylgeraniol or after the alcohol is esterified to the $C17^3$ position of (B)Chlide *a* (Gomez Maqueo Chew et al., 2008; Harada et al., 2008; Liu & Bryant, 2012). Elucidation of the structures of

BChl *a* in *Rhodopseudomonas palustris*, a purple bacterium that produces BChl *a* esterified with phytyl, tetrahydrogeranylgeranyl (THGG), dihydrogeranylgeranyl, and geranylgeranyl moieties in amounts amenable to purification, led to the proposal of a pathway for ordering the subsequent reductions of the tail moiety by BchP. In this scheme the double bonds of geranylgeranyl are sequentially reduced at C6–C7, C10–C11, and C14–C15 to produce Δ2,10,14 phytatrienyl, Δ2,14 phytadienyl, and phytyl sequentially (Harada et al., 2008; Mizoguchi, Harada, & Tamiaki, 2006). This pathway easily explains the production of BChl a_{P} in both purple and green bacteria. However, the above pathway does not account for the production of Chl a_{PD}, which contains Δ2,6 phytadienyl, whereas Δ2,14 phytadienyl was identified as the predicted intermediate in the sequential reductions leading from geranylgeranyl to phytyl. It is unclear if the BchP enzymes of Chl *a*-containing green bacteria follow a different pathway for sequential reduction of the double bonds in geranylgeraniol, or if the Δ2,6 phytadienol originates from a phytyl intermediate that is re-oxidized at the C6–C7 bond (Harada et al., 2008).

Green FAPs and *Cab. thermophilum* have single copies of *bchP*. However, multiple copies of *bchP* are present in genomes in several species of GSB (Canniffe, Thweatt, Gomez Maqueo Chew, Hunter, & Bryant, 2018). In *Cba. tepidum*, which has two *bchP* paralogs, null mutants showed that one gene, denoted *bchP*, was required for geranylgeranyl reductase activity *in vivo*, while the second gene, originally denoted *bchO*, had no phenotype affecting (B)Chl biosynthesis (Gomez Maqueo Chew et al., 2008; Harada et al., 2008). More recent studies have shown that the *bchO* mutant of *Cba. tepidum* cannot synthesize 1′,2′-dihydrochlorobactene (Canniffe et al., 2018). Moreover, *bchO* from *Cba. tepidum* and two paralogs of *bchO* from *Cba. limnaeum* are able to catalyze reduction of 1,2 double bonds in carotenoid substrates produced in *Rba. sphaeroides*. Thus, these former *bchO* genes have been renamed *cruI* to reflect their role in carotenoid biosynthesis. None of the other *bchO* paralogs tested in this way exhibited CruI activity or BchP activity when assayed in *Rba. sphaeroides*. Thus, the roles of these other *bchP* paralogs remain unknown at this time (Canniffe et al., 2018).

3. Concluding remarks

Over the past two decades, the biosynthetic pathways of Chls and BChls of green bacteria have been nearly fully elucidated. This was largely due to the availability of complete genomes of many organisms, and similarities to the pathways for the synthesis of these molecules in purple bacteria, cyanobacteria, and plants. Some remaining questions in these biosynthetic pathways are also unresolved in other chlorophototrophs while others relate largely to the unique properties of green bacteria. For example, the insertion of Zn^{++} in the biosynthesis of Zn-BChl *a'* in *Cab. thermophilum* and in acidophilic purple bacteria has not been explained. Another major area of uncertainty concerns the regulation of these pathways, particularly how green bacteria make and channel the appropriate amounts of different (B)Chl precursors, all of which branch from one biosynthetic pathway, to produce products required in vastly different amounts. Mechanistic issues surrounding the different oxidative isocylic ring cyclases will certainly be an interesting subject for future studies. Many of the answers likely lie in the continued biochemical characterization of enzymes *in vitro* combined with carefully planned genetic manipulations *in vivo*.

Acknowledgments

D.P.C. was funded by a European Commission Marie Skłodowska-Curie Global Fellowship (660652). Studies on green bacteria in the laboratory of D.A.B. have been generously supported by the Photosynthetic Systems Program, Division of Chemical Sciences, Geosciences, and Biosciences (CSGB), Office of Basic Energy Sciences of the United States Department of Energy (DE-FG02-94ER20137).

Conflict of interest

The authors declare that they have no conflicts of interest concerning the contents of this article.

References

Abicht, H. K., Martinez, J., Layer, G., Jahn, D., & Solioz, M. (2012). *Lactococcus lactis* HemW (HemN) is a haem-binding protein with a putative role in haem trafficking. *Biochemical Journal*, *442*, 335–343. https://doi.org/10.1042/BJ20111618.

Addlesee, H. A., Fiedor, L., & Hunter, C. N. (2000). Physical mapping of *bchG*, *orf427*, and *orf177* in the photosynthesis gene cluster of *Rhodobacter sphaeroides*: Functional assignment of the bacteriochlorophyll synthetase gene. *Journal of Bacteriology*, *182*, 3171–3182. https://doi.org/10.1128/JB.182.11.3175-3182.2000.

Addlesee, H. A., & Hunter, C. N. (2002). *Rhodospirillum rubrum* possesses a variant of the *bchP* gene, encoding geranylgeranyl-bacteriopheophytin reductase. *Journal of Bacteriology*, *184*, 1578–1586. https://doi.org/10.1128/JB.184.6.1578-1586.2002.

Airs, R. L., Borrego, C. M., Garcia-Gil, J., & Keely, B. J. (2001). Identification of the bacteriochlorophyll homologues of *Chlorobium phaeobacteroides* strain UdG6053 grown at low light intensity. *Photosynthesis Research*, *70*, 221–230. https://doi.org/10.1023/A:1015146304441.

Albus, C. A., Salinas, A., Czarnecki, O., Kahlau, S., Rothbart, M., Thiele, W., et al. (2012). LCAA, a novel factor required for magnesium protoporphyrin monomethylester cyclase accumulation and feedback control of aminolevulinic acid biosynthesis in tobacco. *Plant Physiology*, *160*, 1923–1936. https://doi.org/10.1104/pp.112.206045.

Blankenship, R. E. (1994). Protein-structure, electron-transfer and evolution of prokaryotic photosynthetic reaction centers. *Antonie van Leeuwenhoek International Journal of General and Molecular Microbiology*, *65*, 311–329. https://doi.org/10.1007/BF00872216.

Bobe, F. W., Pfennig, N., Swanson, K. L., & Smith, K. M. (1990). Red shift of absorption maxima in *Chlorobiineae* through enzymic methylation of their antenna bacteriochlorophylls. *Biochemistry*, *29*, 4340–4348. https://doi.org/10.1021/bi00470a012.

Bollivar, D. W., Jiang, Z. Y., Bauer, C. E., & Beale, S. I. (1994). Heterologous expression of the *bchM* gene product from *Rhodobacter capsulatus* and demonstration that it encodes S-adenosyl-L-methionine:Mg-protoporphyrin IX methyltransferase. *Journal of Bacteriology*, *176*, 5290–5296. https://doi.org/10.1128/jb.176.17.5290-5296.1994.

Bollivar, D. W., Suzuki, J. Y., Beatty, J. T., Dobrowolski, J. M., & Bauer, C. E. (1994). Directed mutational analysis of bacteriochlorophyll *a* biosynthesis in *Rhodobacter capsulatus*. *Journal of Molecular Biology*, *237*, 622–640. https://doi.org/10.1006/jmbi.1994.1260.

Bollivar, D. W., Wang, S., Allen, J. P., & Bauer, C. E. (1994). Molecular genetic analysis of terminal steps in bacteriochlorophyll *a* biosynthesis: Characterization of a *Rhodobacter capsulatus* strain that synthesizes geranylgeraniol-esterified bacteriochlorophyll *a*. *Biochemistry*, *33*, 12763–12768. https://doi.org/10.1021/bi00209a006.

Borrego, C. M., Garcia-Gil, J., Cristina, X. P., Vila, X., & Abella, C. A. (1998). Occurrence of new bacteriochlorophyll *d* forms in natural populations of green photosynthetic sulfur bacteria. *FEMS Microbiology Ecology*, *26*, 257–267. https://doi.org/10.1016/S0168-6496(98)00038-5.

Borrego, C. M., Gerola, P. D., Miller, M., & Cox, R. P. (1999). Light intensity effects on pigment composition and organization in the green sulfur bacterium *Chlorobium tepidum*. *Photosynthesis Research*, *59*, 159–166. https://doi.org/10.1023/A:1006161302838.

Boynton, T. O., Daugherty, L. E., Dailey, T. A., & Dailey, H. A. (2009). Identification of *Escherichia coli* HemG as a novel, menadione-dependent flavodoxin with protoporphyrinogen oxidase activity. *Biochemistry*, *48*, 6705–6711. https://doi.org/10.1021/bi900850y.

Boynton, T. O., Gerdes, S., Craven, S. H., Neidle, E. L., Phillips, J. D., & Dailey, H. A. (2011). Discovery of a gene involved in a third bacterial protoporphyrinogen oxidase activity through comparative genomic analysis and functional complementation. *Applied and Environmental Microbiology*, 77, 4795–4801. https://doi.org/10.1128/AEM.00171-11.

Broch-Due, M., & Ormerod, J. G. (1978). Isolation of a BChl *c* mutant from *Chlorobium* with BChl *d* by cultivation at low light intensity. *FEMS Microbiology Letters*, *3*, 305–308. https://doi.org/10.1111/j.1574-6968.1978.tb01953.x.

Bröcker, M. J., Virus, S., Ganskow, S., Heathcote, P., Heinz, D. W., Schubert, W.-D., et al. (2008). ATP-driven reduction by dark-operative protochlorophyllide oxidoreductase from *Chlorobium tepidum* mechanistically resembles nitrogenase catalysis. *Journal of Biological Chemistry*, *283*, 10559–10567. https://doi.org/10.1074/jbc.M708010200.

Bröcker, M. J., Wätzlich, D., Saggu, M., Lendzian, F., Moser, J., & Jahn, D. (2010). Biosynthesis of (Bacterio) chlorophylls ATP-dependent transient subunit interaction and electron transfer of dark operative protochlorophyllide oxidoreductase. *Journal of Biological Chemistry*, *285*, 8268–8277. https://doi.org/10.1074/jbc.M109.087874.

Bröcker, M. J., Wätzlich, D., Uliczka, F., Virus, S., Saggu, M., Lendzian, F., et al. (2008). Substrate recognition of nitrogenase-like dark operative protochlorophyllide oxidoreductase from *Prochlorococcus marinus*. *Journal of Biological Chemistry*, *283*, 29873–29881. https://doi.org/10.1074/jbc.M805206200.

Bruce, B. D., Fuller, R. C., & Blankenship, R. E. (1982). Primary photochemistry in the facultatively aerobic green photosynthetic bacterium *Chloroflexus-aurantiacus*. *Proceedings of the National Academy of Sciences of the United States of America*, *79*, 6532–6536. https://doi.org/10.1073/pnas.79.21.6532.

Bryant, D. A., & Canniffe, D. P. (2018). How nature designs light-harvesting antenna systems: Design principles and functional realization in chlorophototrophic prokaryotes. *Journal of Physics B: Atomic, Molecular and Optical Physics, 51*, 033001. https://doi.org/10.1088/1361-6455/aa9c3c.

Bryant, D. A., & Frigaard, N. U. (2006). Prokaryotic photosynthesis and phototrophy illuminated. *Trends in Microbiology, 14*, 488–496. https://doi.org/10.1016/j.tim.2006.09.001.

Bryant, D. A., Garcia Costas, A. M., Maresca, J. A., Gomez Maqueo Chew, A., Klatt, C. G., Bateson, M. M., et al. (2007). *Candidatus* Chloracidobacterium thermophilum: An aerobic phototrophic acidobacterium. *Science, 317*, 523–526. https://doi.org/10.1126/science.1143236.

Bryant, D. A., & Liu, Z. (2013). Green bacteria. Insights into green bacterial evolution through genomic analyses. *Advances in Botanical Research, 66*, 99–150. Elsevier. https://doi.org/10.1016/B978-0-12-397923-0.00004-7.

Bryant, D. A., Liu, Z., Li, T., Zhao, F., Garcia Costas, A. M., Klatt, C. G., et al. (2012). *Comparative and functional genomics of anoxygenic green bacteria from the taxa* Chlorobi, Chloroflexi, *and* Acidobacteria. *In R. Burnap & W. Vermaas (Eds.), Functional genomics and evolution of photosynthetic systems: Vol. 33. Advances in photosynthesis and respiration* (pp. 47–102). Dordrecht: Springer Netherlands. https://doi.org/10.1007/978-94-007-1533-2_3.

Bryant, D. A., Vassilieva, E. V., Frigaard, N. U., & Li, H. (2002). Selective protein extraction from *Chlorobium tepidum* chlorosomes using detergents. Evidence that CsmA forms multimers and binds bacteriochlorophyll *a*. *Biochemistry, 41*, 14403–14411. https://doi.org/10.1021/bi026599s.

Burke, D. H., Alberti, M., & Hearst, J. E. (1993). *bchFNBH* bacteriochlorophyll synthesis genes of *Rhodobacter capsulatus* and identification of the third subunit of light-independent protochlorophyllide reductase in bacteria and plants. *Journal of Bacteriology, 175*, 2414–2422. https://doi.org/10.1128/jb.175.8.2414-2422.1993.

Busch, M. S. A., Mueh, F., Madjet, M. E.-A., & Renger, T. (2011). The eighth bacteriochlorophyll completes the excitation energy funnel in the FMO protein. *Journal of Physical Chemistry Letters, 2*, 93–98. https://doi.org/10.1021/jz101541b.

Canniffe, D. P., Chidgey, J. W., & Hunter, C. N. (2014). Elucidation of the preferred routes of C8-vinyl reduction in chlorophyll and bacteriochlorophyll biosynthesis. *Biochemical Journal, 462*, 433–440. https://doi.org/10.1042/BJ20140163.

Canniffe, D. P., & Hunter, C. N. (2014). Engineered biosynthesis of bacteriochlorophyll *b* in *Rhodobacter sphaeroides*. *Biochimica et Biophysica Acta, 1837*, 1611–1616. https://doi.org/10.1016/j.bbabio.2014.07.011.

Canniffe, D. P., Jackson, P. J., Hollingshead, S., Dickman, M. J., & Hunter, C. N. (2013). Identification of an 8-vinyl reductase involved in bacteriochlorophyll biosynthesis in *Rhodobacter sphaeroides* and evidence for the existence of a third distinct class of the enzyme. *Biochemical Journal, 450*, 397–405. https://doi.org/10.1042/BJ20121723.

Canniffe, D. P., Thweatt, J. L., Gomez Maqueo Chew, A., Hunter, C. N., & Bryant, D. A. (2018). A paralog of a bacteriochlorophyll biosynthesis enzyme catalyzes the formation of 1,2-dihydrocarotenoids in green sulfur bacteria. *Journal of Biological Chemistry, 293*, 15233–15242. https://doi.org/10.1074/jbc.RA118.004672.

Cardona, T. (2015). A fresh look at the evolution and diversification of photochemical reaction centers. *Photosynthesis Research, 126*, 111–134. https://doi.org/10.1007/s11120-014-0065-x.

Charles, P., Kalendra, V., He, Z., Khurshov, V., van der Est, A., Golbeck, J. H., et al. (2019). Elucidating the role of zinc-bacteriochlorophyll *a'* in the primary photochemistry of *Chloroacidobacterium thermophilum* reaction centers, 2070-Pos, Biophysical Society 63rd Annual Meeting [Abstract Issue]. *Biophysical Journal, 116*, 419a. https://doi.org/10.1016/j.bpj.2018.11.2258.

Chen, G. E., Canniffe, D. P., Barnett, S. F. H., Hollingshead, S., Brindley, A. A., Vasilev, C., et al. (2018). Complete enzyme set for chlorophyll biosynthesis in *Escherichia coli*. *Science Advances*, *4*, eaaq1407. https://doi.org/10.1126/sciadv.aaq1407.

Chen, G. E., Canniffe, D. P., & Hunter, C. N. (2017). Three classes of oxygen-dependent cyclase involved in chlorophyll and bacteriochlorophyll biosynthesis. *Proceedings of the National Academy of Sciences of the United States of America*, *114*, 6280–6285. https://doi.org/10.1073/pnas.1701687114.

Cohen-Bazire, G., Pfennig, N., & Kunisawa, R. (1964). The fine structure of green bacteria. *The Journal of Cell Biology*, *22*, 207–225. https://doi.org/10.1083/jcb.22.1.207.

Cole, J. R., Wang, Q., Fish, J. A., Chai, B., McGarrell, D. M., Sun, Y., et al. (2014). Ribosomal database project: Data and tools for high throughput rRNA analysis. *Nucleic Acids Research*, *42*, D633–D642. https://doi.org/10.1093/nar/gkt1244.

Collins, A. M., Wen, J., & Blankenship, R. E. (2012). Chapter 4 photosynthetic light-harvesting complexes. In *Molecular solar fuels* (pp. 85–106): The Royal Society of Chemistry. https://doi.org/10.1039/9781849733038-00085.

Dailey, T. A., & Dailey, H. A. (1996). Human protoporphyrinogen oxidase: Expression, purification, and characterization of the cloned enzyme. *Protein Science*, *5*, 98–105.

Dailey, H. A., Dailey, T. A., Gerdes, S., Jahn, D., Jahn, M., O'Brian, M. R., et al. (2017). Prokaryotic heme biosynthesis: Multiple pathways to a common essential product. *Microbiology and Molecular Biology Reviews*, *81*, e00048-16. https://doi.org/10.1128/MMBR.00048-16.

Dubinina, G. A., & Gorlenko, V. M. (1975). New filamentous photosynthetic green bacteria containing gas vacuoles. *Microbiology*, *44*, 452–458.

Eisen, J. A., Nelson, K. E., Paulsen, I. T., Heidelberg, J. F., Wu, M., Dodson, R. J., et al. (2002). The complete genome sequence of *Chlorobium tepidum* TLS, a photosynthetic, anaerobic, green-sulfur bacterium. *Proceedings of the National Academy of Sciences of the United States of America*, *99*, 9509–9514. https://doi.org/10.1073/pnas.132181499.

Fenna, R. E., & Matthews, B. W. (1975). Chlorophyll arrangement in a bacteriochlorophyll protein from *Chlorobium-limicola*. *Nature*, *258*, 573–577. https://doi.org/10.1038/258573a0.

Fodje, M. N., Hansson, A., Hansson, M., Olsen, J. G., Gough, S., Willows, R. D., et al. (2001). Interplay between an AAA module and an integrin I domain may regulate the function of magnesium chelatase. *Journal of Molecular Biology*, *311*, 111–122. https://doi.org/10.1006/jmbi.2001.4834.

Frigaard, N. U., & Bryant, D. A. (2004). Seeing green bacteria in a new light: Genomics-enabled studies of the photosynthetic apparatus in green sulfur bacteria and filamentous anoxygenic phototrophic bacteria. *Archives of Microbiology*, *182*, 265–276. https://doi.org/10.1007/s00203-004-0718-9.

Frigaard, N. U., & Bryant, D. A. (2006). *Chlorosomes: Antenna organelles in photosynthetic green bacteria. In J. M. Shively (Ed.), Complex intracellular structures in prokaryotes* (pp. 79–114). Berlin, Heidelberg: Springer Berlin Heidelberg. https://doi.org/10.1007/7171_021.

Frigaard, N. U., & Dahl, C. (2009). Sulfur metabolism in phototrophic sulfur bacteria. In *Vol. 54. Advances in microbial physiology* (pp. 103–200): Elsevier Masson SAS. https://doi.org/10.1016/S0065-2911(08)00002-7.

Frigaard, N. U., Gomez Maqueo Chew, A., Li, H., Maresca, J. A., & Bryant, D. A. (2003). *Chlorobium tepidum*: Insights into the structure, physiology, and metabolism of a green sulfur bacterium derived from the complete genome sequence. *Photosynthesis Research*, *78*, 93–117. https://doi.org/10.1023/B:PRES.0000004310.96189.b4.

Frigaard, N. U., Li, H., Milks, K. J., & Bryant, D. A. (2004). Nine mutants of *Chlorobium tepidum* each unable to synthesize a different chlorosome protein still assemble functional chlorosomes. *Journal of Bacteriology*, *186*, 646–653. https://doi.org/10.1128/JB.186.3.646-653.2004.

Frigaard, N. U., Voigt, G. D., & Bryant, D. A. (2002). *Chlorobium tepidum* mutant lacking bacteriochlorophyll *c* made by inactivation of the *bchK* gene, encoding bacteriochlorophyll *c* synthase. *Journal of Bacteriology, 184*, 3368–3376. https://doi.org/10.1128/JB.184.12.3368-3376.2002.

Fuchs, G. (2011). Alternative pathways of carbon dioxide fixation: Insights into the early evolution of life? *Annual Review of Microbiology, 65*, 631–658. https://doi.org/10.1146/annurev-micro-090110-102801.

Garcia Costas, A. M., Liu, Z., Tomsho, L. P., Schuster, S. C., Ward, D. M., & Bryant, D. A. (2012). Complete genome of *Candidatus* Chloracidobacterium thermophilum, a chlorophyll-based photoheterotroph belonging to the phylum *Acidobacteria*. *Environmental Microbiology, 14*, 177–190. https://doi.org/10.1111/j.1462-2920.2011.02592.x.

Garcia Costas, A. M., Tsukatani, Y., Rijpstra, W. I. C., Schouten, S., Welander, P. V., Summons, R. E., et al. (2012). Identification of the bacteriochlorophylls, carotenoids, quinones, lipids, and hopanoids of "*Candidatus* Chloracidobacterium thermophilum" *Journal of Bacteriology, 194*, 1158–1168. https://doi.org/10.1128/JB.06421-11.

Garcia Costas, A. M., Tsukatani, Y., Romberger, S. P., Oostergetel, G. T., Boekema, E. J., Golbeck, J. H., et al. (2011). Ultrastructural analysis and identification of envelope proteins of "*Candidatus* Chloracidobacterium thermophilum" chlorosomes. *Journal of Bacteriology, 193*, 6701–6711. https://doi.org/10.1128/JB.06124-11.

Garrity, G. M., Holt, J. G., Castenholz, R. W., Pierson, B. K., Keppen, O. I., & Gorlenko, V. M. (2001). *Phylum BVI. Chloroflexi phy. Nov. In Bergey's manual of systematic bacteriology* (2nd ed., pp. 427–446). https://doi.org/10.1007/978-0-387-21609-6_23.

Gaubier, P., Wu, H. J., Laudié, M., Delseny, M., & Grellet, F. (1995). A chlorophyll synthetase gene from *Arabidopsis thaliana*. *Molecular & General Genetics, 249*, 58–64. https://doi.org/10.1007/BF00290236.

Gibson, L. C. D., & Hunter, C. N. (1994). The bacteriochlorophyll biosynthesis gene, *bchM*, of *Rhodobacter-sphaeroides* encodes *S*-adenosyl-L-methionine—Mg protoporphyrin-IX methyltransferase. *FEBS Letters, 352*, 127–130. https://doi.org/10.1016/0014-5793(94)00934-1.

Gisriel, C., Sarrou, I., Ferlez, B., Golbeck, J. H., Redding, K. E., & Fromme, R. (2017). Structure of a symmetric photosynthetic reaction center-photosystem. *Science, 357*, 1021–1025. https://doi.org/10.1126/science.aan5611.

Glaeser, J., Bañeras, L., Rütters, H., & Overmann, J. (2002). Novel bacteriochlorophyll *e* structures and species-specific variability of pigment composition in green sulfur bacteria. *Archives of Microbiology, 177*, 475–485. https://doi.org/10.1007/s00203-002-0416-4.

Gomez Maqueo Chew, A., & Bryant, D. A. (2007a). Chlorophyll biosynthesis in bacteria: The origins of structural and functional diversity. *Annual Review of Microbiology, 61*, 113–129. https://doi.org/10.1146/annurev.micro.61.080706.093242.

Gomez Maqueo Chew, A., & Bryant, D. A. (2007b). Characterization of a plant-like protochlorophyllide a divinyl reductase in green sulfur bacteria. *Journal of Biological Chemistry, 282*, 2967–2975. https://doi.org/10.1074/jbc.M609730200.

Gomez Maqueo Chew, A., Frigaard, N. U., & Bryant, D. A. (2004). Identification of BchV, a C-3^1 hydratase specific for hypermethylated bacteriochlorophyll *c* in *Chlorobium tepidum*. In A. van der Est & D. Bruce (Eds.), *Photosynthesis: Fundamental aspects to global perspectives research* (pp. 875–877). .

Gomez Maqueo Chew, A., Frigaard, N. U., & Bryant, D. A. (2007). Bacteriochlorophyllide *c* C-8^2 and C-12^1 methyltransferases are essential for adaptation to low light in *Chlorobaculum tepidum*. *Journal of Bacteriology, 189*, 6176–6184. https://doi.org/10.1128/JB.00519-07.

Gomez Maqueo Chew, A., Frigaard, N. U., & Bryant, D. A. (2008). Identification of the *bchP* gene, encoding geranylgeranyl reductase in *Chlorobaculum tepidum*. *Journal of Bacteriology, 190*, 747–749. https://doi.org/10.1128/JB.01430-07.

Gomez Maqueo Chew, A., Frigaard, N. U., & Bryant, D. A. (2009). Mutational analysis of three *bchH* paralogs in (bacterio-)chlorophyll biosynthesis in *Chlorobaculum tepidum*. *Photosynthesis Research*, *101*, 21–34. https://doi.org/10.1007/s11120-009-9460-0.

Gorlenko, V. M., & Korotkov, S. A. (1979). Morphological and physiological features of the new filamentous gliding green bacteria *Oscillochloris-trichoides* nov. comb. *Vol. 6* (pp. 848–857). Izvestiya Akademii Nauk Sssr Seriya Biologicheskaya.

Gough, S. P., Petersen, B. O., & Duus, J. Ø. (2000). Anaerobic chlorophyll isocyclic ring formation in *Rhodobacter capsulatus* requires a cobalamin cofactor. *Proceedings of the National Academy of Sciences of the United States of America*, *97*, 6908–6913. https://doi.org/10.1073/pnas.97.12.6908.

Grouzdev, D. S., Kuznetsov, B. B., Keppen, O. I., Krasil'nikova, E. N., Lebedeva, N. V., & Ivanovsky, R. N. (2015). Reconstruction of bacteriochlorophyll biosynthesis pathways in the filamentous anoxygenic phototrophic bacterium *Oscillochloris trichoides* DG-6 and evolution of anoxygenic phototrophs of the order Chloroflexales. *Microbiology*, *161*, 120–130. https://doi.org/10.1099/mic.0.082313-0.

Grouzdev, D. S., Rysina, M. S., Bryantseva, I. A., Gorlenko, V. M., & Gaisin, V. A. (2018). Draft genome sequences of '*Candidatus* Chloroploca asiatica' and '*Candidatus* Viridilinea mediisalina', candidate representatives of the *Chloroflexales* order: Phylogenetic and taxonomic implications. *Standards in Genomic Sciences*, *13*, 24. https://doi.org/10.1186/s40793-018-0329-8.

Harada, J., Miyago, S., Mizoguchi, T., Azai, C., Inoue, K., Tamiaki, H., et al. (2008). Accumulation of chlorophyllous pigments esterified with the geranylgeranyl group and photosynthetic competence in the CT2256-deleted mutant of the green sulfur bacterium *Chlorobium tepidum*. *Photochemical and Photobiological Sciences*, *7*, 1179–1187. https://doi.org/10.1039/b802435a.

Harada, J., Mizoguchi, T., Satoh, S., Tsukatani, Y., Yokono, M., Noguchi, M., et al. (2013). Specific gene *bciD* for C7-methyl oxidation in bacteriochlorophyll *e* biosynthesis of brown-colored green sulfur bacteria. *PLoS One*, *8*, e60026. https://doi.org/10.1371/journal.pone.0060026.

Harada, J., Mizoguchi, T., Tsukatani, Y., Noguchi, M., & Tamiaki, H. (2012). A seventh bacterial chlorophyll driving a large light-harvesting antenna. *Scientific Reports*, *2*(671). https://doi.org/10.1038/srep00671.

Harada, J., Mizoguchi, T., Tsukatani, Y., Yokono, M., Tanaka, A., & Tamiaki, H. (2014). Chlorophyllide *a* oxidoreductase works as one of the divinyl reductases specifically involved in bacteriochlorophyll *a* biosynthesis. *Journal of Biological Chemistry*, *289*, 12716–12726. https://doi.org/10.1074/jbc.M113.546739.

Harada, J., Saga, Y., Yaeda, Y., Oh-Oka, H., & Tamiaki, H. (2005). *In vitro* activity of C-20 methyltransferase, BchU, involved in bacteriochlorophyll *c* biosynthetic pathway in green sulfur bacteria. *FEBS Letters*, *579*, 1983–1987. https://doi.org/10.1016/j.febslet.2005.01.087.

Harada, J., Teramura, M., Mizoguchi, T., Tsukatani, Y., Yamamoto, K., & Tamiaki, H. (2015). Stereochemical conversion of C3-vinyl group to 1-hydroxyethyl group in bacteriochlorophyll *c* by the hydratases BchF and BchV: Adaptation of green sulfur bacteria to limited-light environments. *Molecular Microbiology*, *98*, 1184–1198. https://doi.org/10.1111/mmi.13208.

Hartzler, D. A., Niedzwiedzki, D. M., Bryant, D. A., Blankenship, R. E., Pushkar, Y., & Savikhin, S. (2014). Triplet excited state energies and phosphorescence spectra of (bacterio)chlorophylls. *Journal of Physical Chemistry B*, *118*, 7221–7232. https://doi.org/10.1021/jp500539w.

Haskamp, V., Karrie, S., Mingers, T., Barthels, S., Alberge, F., Magalon, A., et al. (2018). The radical SAM protein HemW is a heme chaperone. *Journal of Biological Chemistry*, *293*, 2558–2572. https://doi.org/10.1074/jbc.RA117.000229.

Heising, S., Richter, L., Ludwig, W., & Schink, B. (1999). *Chlorobium ferrooxidans* sp. nov., a phototrophic green sulfur bacterium that oxidizes ferrous iron in coculture with a "*Geospirillum*" sp. strain. *Archives of Microbiology, 172*, 116–124. https://doi.org/10.1007/s002030050748.

Hinchigeri, S. B., Hundle, B., & Richards, W. R. (1997). Demonstration that the *bchH* protein of *Rhodobacter capsulatus* activates S-adenosyl-L-methionine:Magnesium protoporphyrin methyltransferase. *FEBS Letters, 407*, 337–342. https://doi.org/10.1016/S0014-5793(97)00371-2.

Hollingshead, S., Kopecná, J., Jackson, P. J., Canniffe, D. P., Davison, P. A., Dickman, M. J., et al. (2012). Conserved chloroplast open-reading frame *ycf54* is required for activity of the magnesium protoporphyrin monomethylester oxidative cyclase in *Synechocystis* PCC 6803. *The Journal of Biological Chemistry, 287*, 27823–27833. https://doi.org/10.1074/jbc.M112.352526.

Hörtensteiner, S., & Kräutler, B. (2011). Chlorophyll breakdown in higher plants. *Biochimica et Biophysica Acta, 1807*, 977–988. https://doi.org/10.1016/j.bbabio.2010.12.007.

Huang, R. Y.-C., Wen, J., Blankenship, R. E., & Gross, M. L. (2012). Hydrogen-deuterium exchange mass spectrometry reveals the interaction of Fenna-Matthews-Olson protein and chlorosome CsmA protein. *Biochemistry, 51*, 187–193. https://doi.org/10.1021/bi201620y.

Huster, M. S., & Smith, K. M. (1990). Biosynthetic studies of substituent homologation in bacteriochlorophylls *c* and *d*. *Biochemistry, 29*, 4348–4355. https://doi.org/10.1021/bi00470a013.

Iino, T., Mori, K., Uchino, Y., Nakagawa, T., Harayama, S., & Suzuki, K. I. (2010). *Ignavibacterium album* gen. nov., sp. nov., a moderately thermophilic anaerobic bacterium isolated from microbial mats at a terrestrial hot spring and proposal of *Ignavibacteria* classis nov., for a novel lineage at the periphery of green sulfur bacteria. *International Journal of Systematic and Evolutionary Microbiology, 60*, 1376–1382. https://doi.org/10.1099/ijs.0.012484-0.

Imhoff, J. F. (2017). *Anoxygenic phototrophic bacteria from extreme environments. In Modern topics in the phototrophic prokaryotes: Environmental and applied aspects* (pp. 427–480). https://doi.org/10.1007/978-3-319-46261-5_13.

Imhoff, J. F., & Thiel, V. (2010). Phylogeny and taxonomy of *Chlorobiaceae*. *Photosynthesis Research, 104*, 123–136. https://doi.org/10.1007/s11120-009-9510-7.

Islam, M. R., Aikawa, S., Midorikawa, T., Kashino, Y., Satoh, K., & Koike, H. (2008). *slr1923* of *Synechocystis* sp. PCC6803 is essential for conversion of 3,8-divinyl(proto)-chlorophyll(ide) to 3-monovinyl(proto)chlorophyll(ide). *Plant Physiology, 148*, 1068–1081. https://doi.org/10.1104/pp.108.123117.

Ito, H., Yokono, M., Tanaka, R., & Tanaka, A. (2008). Identification of a novel vinyl reductase gene essential for the biosynthesis of monovinyl chlorophyll in *Synechocystis* sp. PCC6803. *Journal of Biological Chemistry, 283*, 9002–9011. https://doi.org/10.1074/jbc.M708369200.

Ivanovsky, R. N., Fal, Y. I., Berg, I. A., Ugolkova, N. V., Krasilnikova, E. N., Keppen, O. I., et al. (1999). Evidence for the presence of the reductive pentose phosphate cycle in a filamentous anoxygenic photosynthetic bacterium, *Oscillochloris trichoides* strain DG-6. *Microbiology, 147*, 1743–1748. https://doi.org/10.1099/13500872-145-7-1743.

Jaschke, P. R., Hardjasa, A., Digby, E. L., Hunter, C. N., & Beatty, J. T. (2011). A *bchD* (magnesium chelatase) mutant of *Rhodobacter sphaeroides* synthesizes zinc bacteriochlorophyll through novel zinc-containing intermediates. *Journal of Biological Chemistry, 286*, 20313–20322. https://doi.org/10.1074/jbc.M110.212605.

Jensen, P. E., Gibson, L. C. D., & Hunter, C. N. (1999). ATPase activity associated with the magnesium-protoporphyrin IX chelatase enzyme of *Synechocystis* PCC6803: Evidence for ATP hydrolysis during Mg^{2+} insertion. *Biochemical Journal, 339*, 127–134. https://doi.org/10.1042/0264-6021:3390127.

Johnson, E. T., & Schmidt-Dannert, C. (2008). Characterization of three homologs of the large subunit of the magnesium chelatase from *Chlorobaculum tepidum* and interaction with the magnesium protoporphyrin IX methyltransferase. *Journal of Biological Chemistry, 283*, 27776–27784. https://doi.org/10.1074/jbc.M804486200.

Jones, M. S., & Jones, O. T. (1970). Ferrochelatase of *Rhodopseudomonas spheroides*. *The Biochemical Journal, 119*, 453–462. https://doi.org/10.1042/bj1190453.

Kadnikov, V. V., Mardanov, A. V., Podosokorskaya, O. A., Gavrilov, S. N., Kublanov, I. V., Beletsky, A. V., et al. (2013). Genomic analysis of *Melioribacter roseus*, facultatively anaerobic organotrophic bacterium representing a novel deep lineage within *Bacteriodetes/Chlorobi* group. *PLoS One, 8*, e53047. https://doi.org/10.1371/journal.pone.0053047.

Kaneko, T., Tanaka, A., Sato, S., Kotani, H., Sazuka, T., Miyajima, N., et al. (1995). Sequence analysis of the genome of the unicellular cyanobacterium *Synechocystis* sp. strain PCC6803. I. Sequence features in the 1 Mb region from map positions 64% to 92% of the genome. *DNA Research, 2*, 153–166. https://doi.org/10.1093/dnares/2.4.153.

Kato, K., Tanaka, R., Sano, S., Tanaka, A., & Hosaka, H. (2010). Identification of a gene essential for protoporphyrinogen IX oxidase activity in the cyanobacterium *Synechocystis* sp. PCC6803. *Proceedings of the National Academy of Sciences of the United States of America, 107*, 16649–16654. https://doi.org/10.1073/pnas.1000771107.

Kenner, G. W., Rimmer, J., Smith, K. M., & Unsworth, J. F. (1978). Pyrroles and related compounds.39. Structural and biosynthetic-studies of *Chlorobium* chlorophylls-660 (bacteriochlorophylls-*c*)—Incorporations of methionine and porphobilinogen. *Journal of the Chemical Society-Perkin Transactions 1*, 845–852. https://doi.org/10.1039/p19780000845.

Keppen, O. I., Baulina, O. I., & Kondratieva, E. N. (1994). *Oscillochloris trichoides* neotype strain DG-6. *Photosynthesis Research, 41*, 29–33. https://doi.org/10.1007/BF02184143.

Kiesel, S., Wätzlich, D., Lange, C., Reijerse, E., Broecker, M. J., Ruediger, W., et al. (2015). Iron-sulfur cluster-dependent catalysis of chlorophyllide *a* oxidoreductase from *Roseobacter denitrificans*. *Journal of Biological Chemistry, 290*, 1141–1154. https://doi.org/10.1074/jbc.M114.617761.

Klatt, C. G., Bryant, D. A., & Ward, D. M. (2007). Comparative genomics provides evidence for the 3-hydroxypropionate autotrophic pathway in filamentous anoxygenic phototrophic bacteria and in hot spring microbial mats. *Environmental Microbiology, 9*, 2067–2078. https://doi.org/10.1111/j.1462-2920.2007.01323.x.

Klatt, C. G., William, P., Herrgard, M. J., Jay, Z. J., Rusch, D. B., & Miller, S. R. (2013). Community structure and function of high-temperature chlorophototrophic microbial mats inhabiting diverse geothermal environments. *Frontiers in Microbiology, 4*(106). https://doi.org/10.3389/fmicb.2013.00106.

Kobayashi, K., Masuda, T., Tajima, N., Wada, H., & Sato, N. (2014). Molecular phylogeny and intricate evolutionary history of the three isofunctional enzymes involved in the oxidation of protoporphyrinogen IX. *Genome Biology and Evolution, 6*, 2141–2155. https://doi.org/10.1093/gbe/evu170.

Lange, C., Kiesel, S., Peters, S., Virus, S., Scheer, H., Jahn, D., et al. (2015). Broadened substrate specificity of 3-hydroxyethyl bacteriochlorophyllide *a* dehydrogenase (BchC) indicates a new route for the biosynthesis of bacteriochlorophyll *a*. *Journal of Biological Chemistry, 290*, 19697–19709. https://doi.org/10.1074/jbc.M115.660555.

Larsen, K. L., Miller, M., & Cox, R. P. (1995). Incorporation of exogenous long-chain alcohols into bacteriochlorophyll *c* homologs by *Chloroflexus aurantiacus*. *Archives of Microbiology, 163*, 119–123. https://doi.org/10.1007/BF00381785.

Larson, C. R., Seng, C. O., Lauman, L., Matthies, H. J., Wen, J., Blankenship, R. E., et al. (2011). The three-dimensional structure of the FMO protein from *Pelodictyon phaeum* and the implications for energy transfer. *Photosynthesis Research, 107*, 139–150. https://doi.org/10.1007/s11120-010-9604-2.

Layer, G., Grage, K., Teschner, T., Schünemann, V., Breckau, D., Masoumi, A., et al. (2005). Radical S-adenosylmethionine enzyme coproporphyrinogen III oxidase

HemN: Functional features of the [4Fe-4S] cluster and the two bound S-adenosyl-L-methionines. *Journal of Biological Chemistry*, *280*, 29038–29046. https://doi.org/10.1074/jbc.M501275200.

Letunic, I., & Bork, P. (2016). Interactive tree of life (iTOL) v3: An online tool for the display and annotation of phylogenetic and other trees. *Nucleic Acids Research*, *44*, W242–W245.

Li, H., & Bryant, D. A. (2009). Envelope proteins of the CsmB/CsmF and CsmC/CsmD motif families influence the size, shape, and composition of chlorosomes in *Chlorobaculum tepidum*. *Journal of Bacteriology*, *191*, 7109–7120. https://doi.org/10.1128/JB.00707-09.

Li, H., Frigaard, N. U., & Bryant, D. A. (2006). Molecular contacts for chlorosome envelope proteins revealed by cross-linking studies with chlorosomes from *Chlorobium tepidum*. *Biochemistry*, *45*, 9095–9103. https://doi.org/10.1021/bi060776y.

Li, H., Frigaard, N. U., & Bryant, D. A. (2013). Proteins in chlorosomes: CsmI and CsmJ participate in light-dependent control of energy transfer in chlorosomes of *Chlorobaculum tepidum*. *Biochemistry*, *52*, 1321–1330. https://doi.org/10.1021/bi301454g.

Liu, Z., & Bryant, D. A. (2011a). Identification of a gene essential for the first committed step in the biosynthesis of bacteriochlorophyll *c*. *Journal of Biological Chemistry*, *286*, 22393–22402. https://doi.org/10.1074/jbc.M111.249433.

Liu, Z., & Bryant, D. A. (2011b). Multiple types of 8-vinyl reductases for (Bacterio)chlorophyll biosynthesis occur in many green sulfur bacteria. *Journal of Bacteriology*, *193*, 4996–4998. https://doi.org/10.1128/JB.05520-11.

Liu, Z., & Bryant, D. A. (2012). Bacteriochlorophyll biosynthesis and assembly in green chlorophototrophic bacteria: Theme and variations. In K. M. Kadish, K. M. Smith, & R. Guilard (Eds.), *Vol 20. Handbook of porphyrin science with applications to chemistry, physics, materials science, engineering, biology and medicine, Vol 20. Chlorophylls and related systems* (pp. 107–140). Singapore: World Scientific Publishing Company, Pt Ltd.

Liu, Z., Frigaard, N. U., Vogl, K., Iino, T., Ohkuma, M., Overmann, J., et al. (2012). Complete genome of *Ignavibacterium album*, a metabolically versatile, flagellated, facultative anaerobe from the phylum *Chlorobi*. *Frontiers in Microbiology*, *3*, 185. https://doi.org/10.3389/fmicb.2012.00185.

Liu, Z., Klatt, C. G., Ludwig, M., Rusch, D. B., Jensen, S. I., Kühl, M., et al. (2012). *Candidatus* Thermochlorobacter aerophilum: An aerobic chlorophotoheterotrophic member of the phylum *Chlorobi* defined by metagenomics and metatranscriptomics. *ISME Journal*, *6*, 1869–1882. https://doi.org/10.1038/ismej.2012.24.

Lundqvist, J., Elmlund, H., Wulff, R. P., Berglund, L., Elmlund, D., Emanuelsson, C., et al. (2010). ATP-induced conformational dynamics in the AAA+ motor unit of magnesium chelatase. *Structure*, *18*, 354–365. https://doi.org/10.1016/j.str.2010.01.001.

Majumder, E. L. W., Olsen, J. D., Qian, P., Collins, A. M., Hunter, C. N., & Blankenship, R. E. (2016). Supramolecular organization of photosynthetic complexes in membranes of *Roseiflexus castenholzii*. *Photosynthesis Research*, *127*, 117–130. https://doi.org/10.1007/s11120-015-0179-9.

Maresca, J. A., Gomez Maqueo Chew, A., Ponsatí, M. R., Frigaard, N. U., Ormerod, J. G., & Bryant, D. A. (2004). The *bchU* gene of *Chlorobium tepidum* encodes the C-20 methyltransferase in bacteriochlorophyll *c* biosynthesis. *Journal of Bacteriology*, *186*, 2558–2566. https://doi.org/10.1128/JB.186.9.2558-2566.2004.

Maresca, J. A., Romberger, S. P., & Bryant, D. A. (2008). Isorenieratene biosynthesis in green sulfur bacteria requires the cooperative actions of two carotenoid cyclases. *Journal of Bacteriology*, *190*, 6384–6391. https://doi.org/10.1128/JB.00758-08.

Matsuo, E., & Inagaki, Y. (2018). Patterns in evolutionary origins of heme, chlorophyll *a* and isopentenyl diphosphate biosynthetic pathways suggest non photosynthetic periods prior to plastid replacements in dinoflagellates. *PeerJ*, *6*, e5345 https://doi.org/10.7717/peerj.5345.

McGlynn, P., & Neil Hunter, C. (1993). Genetic analysis of the *bchC* and *bchA* genes of *Rhodobacter sphaeroides*. *Molecular & General Genetics: MGG*, *236*, 227–234. https://doi.org/10.1007/BF00277117.

McLean, S., & Hunter, C. N. (2009). An enzyme-coupled continuous spectrophotometric assay for magnesium protoporphyrin IX methyltransferases. *Analytical Biochemistry*, *394*, 223–228. https://doi.org/10.1016/j.ab.2009.07.036.

Milder, M. T. W., Brueggemann, B., van Grondelle, R., & Herek, J. L. (2010). Revisiting the optical properties of the FMO protein. *Photosynthesis Research*, *104*, 257–274. https://doi.org/10.1007/s11120-010-9540-1.

Mizoguchi, T., Harada, J., & Tamiaki, H. (2006). Structural determination of dihydro- and tetrahydrogeranylgeranyl groups at the 17-propionate of bacteriochlorophylls-*a*. *FEBS Letters*, *580*, 6644–6648. https://doi.org/10.1016/j.febslet.2006.11.020.

Mizoguchi, T., Harada, J., & Tamiaki, H. (2012). Characterization of chlorophyll pigments in the mutant lacking 8-vinyl reductase of green photosynthetic bacterium *Chlorobaculum tepidum*. *Bioorganic and Medicinal Chemistry*, *20*, 6803–6810. https://doi.org/10.1016/j.bmc.2012.09.054.

Mizoguchi, T., Harada, J., Yamamoto, K., & Tamiaki, H. (2015). Inactivation of *bciD* and *bchU* genes in the green sulfur bacterium *Chlorobaculum limnaeum* and alteration of photosynthetic pigments in the resultant mutants. *Journal of Photochemistry and Photobiology A: Chemistry*, *313*, 52–59. https://doi.org/10.1016/j.jphotochem.2015.06.012.

Möbius, K., Arias-Cartin, R., Breckau, D., Haennig, A.-L., Riedmann, K., Biedendieck, R., et al. (2010). Heme biosynthesis is coupled to electron transport chains for energy generation. *Proceedings of the National Academy of Sciences of the United States of America*, *107*, 10436–10441. https://doi.org/10.1073/pnas.1000956107.

Montaño, G. A., Bowen, B. P., LaBelle, J. T., Woodbury, N. W., Pizziconi, V. B., & Blankenship, R. E. (2003). Characterization of *Chlorobium tepidum* chlorosomes: A calculation of bacteriochlorophyll *c* per chlorosome and oligomer modeling. *Biophysical Journal*, *85*, 2560–2565. https://doi.org/10.1016/S0006-3495(03)74678-5.

Munoz, R., Yarza, P., Ludwig, W., Euzéby, J., Amann, R., Schleifer, K. H., et al. (2011). Release LTPs104 of the all-species living tree. *Systematic and Applied Microbiology*, *34*, 169–170. https://doi.org/10.1016/j.syapm.2011.03.001.

Muraki, N., Nomata, J., Ebata, K., Mizoguchi, T., Shiba, T., Tamiaki, H., et al. (2010). X-ray crystal structure of the light-independent protochlorophyllide reductase. *Nature*, *465*, 110–114. https://doi.org/10.1038/nature08950.

Nadson, G. A. (1906). The morphology of inferior algae. III. *Chlorobium limicola* Nads., the green chlorophyll bearing microbe. *Bulletin Du Jardin Impérial Botanique de St-Pétersbourg*, *6*, 190.

Nagaraj, V. A., Arumugam, R., Prasad, D., Rangarajan, P. N., & Padmanaban, G. (2010). Protoporphyrinogen IX oxidase from *Plasmodium falciparum* is anaerobic and is localized to the mitochondrion. *Molecular and Biochemical Parasitology*, *174*, 44–52. https://doi.org/10.1016/j.molbiopara.2010.06.012.

Nagata, N., Tanaka, R., Satoh, S., & Tanaka, A. (2005). Identification of a vinyl reductase gene for chlorophyll synthesis in *Arabidopsis thaliana* and implications for the evolution of *Prochlorococcus* species. *The Plant Cell*, *17*, 233–240. https://doi.org/10.1105/tpc.104.027276.

Nakanishi, H., Nozue, H., Suzuki, K., Kaneko, Y., Taguchi, G., & Hayashida, N. (2005). Characterization of the *Arabidopsis thaliana* mutant *pcb2* which accumulates divinyl chlorophylls. *Plant and Cell Physiology*, *46*, 467–473. https://doi.org/10.1093/pcp/pci053.

Niedzwiedzki, D. M., Orf, G. S., Tank, M., Vogl, K., Bryant, D. A., & Blankenship, R. E. (2014). Photophysical properties of the excited states of bacteriochlorophyll *f* in solvents and in chlorosomes. *Journal of Physical Chemistry B*, *118*, 2295–2305. https://doi.org/10.1021/jp409495m.

Nielsen, J. T., Kulminskaya, N. V., Bjerring, M., Linnanto, J. M., Ratsep, M., Pedersen, M. O., et al. (2016). In situ high-resolution structure of the baseplate antenna complex in *Chlorobaculum tepidum*. *Nature Communications*, 7, 12454 https://doi.org/10.1038/ncomms12454.

Nishimori, R., Mizoguchi, T., Tamiaki, H., Kashimura, S., & Saga, Y. (2011). Biosynthesis of unnatural bacteriochlorophyll *c* derivatives esterified with alpha,omega-diols in the green sulfur photosynthetic bacterium *Chlorobaculum tepidum*. *Biochemistry*, *50*, 7756–7764. https://doi.org/10.1021/bi200994h.

Nomata, J., Mizoguchi, T., Tamiaki, H., & Fujita, Y. (2006). A second nitrogenase-like enzyme for bacteriochlorophyll biosynthesis: Reconstitution of chlorophyllide *a* reductase with purified X-protein (BchX) and YZ-protein (BchY-BchZ) from *Rhodobacter capsulatus*. *Journal of Biological Chemistry*, *281*, 15021–15028. https://doi.org/10.1074/jbc.M601750200.

Nomata, J., Swem, L. R., Bauer, C. E., & Fujita, Y. (2005). Overexpression and characterization of dark-operative protochlorophyllide reductase from *Rhodobacter capsulatus*. *Biochimica et Biophysica Acta*, *1708*, 229–237. https://doi.org/10.1016/j.bbabio.2005.02.002.

Novoderezhkin, V. I., Taisova, A. S., Fetisova, Z., Blankenship, R. E., Savikhin, S., Buck, D. R., et al. (1998). Energy transfers in the B808-866 antenna from the green bacterium *Chloroflexus aurantiacus*. *Biophysical Journal*, *74*, 2069–2075. https://doi.org/10.1016/S0006-3495(98)77913-5.

Ohashi, S., Iemura, T., Okada, N., Itoh, S., Furukawa, H., Okuda, M., et al. (2010). An overview on chlorophylls and quinones in the photosystem I-type reaction centers. *Photosynthesis Research*, *104*, 305–319. https://doi.org/10.1007/s11120-010-9530-3.

Oostergetel, G. T., van Amerongen, H., & Boekema, E. J. (2010). The chlorosome: A prototype for efficient light harvesting in photosynthesis. *Photosynthesis Research*, *104*, 245–255. https://doi.org/10.1007/s11120-010-9533-0.

Orf, G. S., & Blankenship, R. E. (2013). Chlorosome antenna complexes from green photosynthetic bacteria. *Photosynthesis Research*, *116*, 315–331. https://doi.org/10.1007/s11120-013-9869-3.

Orf, G. S., Tank, M., Vogl, K., Niedzwiedzki, D. M., Bryant, D. A., & Blankenship, R. E. (2013). Spectroscopic insights into the decreased efficiency of chlorosomes containing bacteriochlorophyll *f*. *Biochimica et Biophysica Acta*, *1827*, 493–501. https://doi.org/10.1016/j.bbabio.2013.01.006.

Oster, U., Bauer, C. E., & Rüdiger, W. (1997). Characterization of chlorophyll *a* and bacteriochlorophyll *a* synthases by heterologous expression in *Escherichia coli*. *Journal of Biological Chemistry*, *272*, 9671–9676. https://doi.org/10.1074/jbc.272.15.9671.

Permentier, H. P., Schmidt, K. A., Kobayashi, M., Akiyama, M., Hager-Braun, C., Neerken, S., et al. (2000). Composition and optical properties of reaction Centre core complexes from the green sulfur bacteria *Prosthecochloris aestuarii* and *Chlorobium tepidum*. *Photosynthesis Research*, *64*, 27–39. https://doi.org/10.1023/A:1026515027824.

Pierson, B. K., & Castenholz, R. W. (1974a). A phototrophic gliding filamentous bacterium of hot springs, *Chloroflexus aurantiacus*,gen. and sp. nov. *Archives of Microbiology*, *100*, 5–24. https://doi.org/10.1007/BF00446302.

Pierson, B. K., & Castenholz, R. W. (1974b). Studies of pigments and growth in *Chloroflexus aurantiacus*, a phototrophic filamentous bacterium. *Archives of Microbiology*, *100*, 283–305. https://doi.org/10.1007/BF00446324.

Pierson, B. K., & Thornber, J. P. (1983). Isolation and spectral characterization of photochemical reaction centers from the thermophilic green bacterium *Chloroflexus aurantiacus* strain J-10-f1. *Proceedings of the National Academy of Sciences of the United States of America*, *80*, 80–84. https://doi.org/10.1073/pnas.80.1.80.

Piñta, V., Picaud, M., Reiss-Husson, F., & Astier, C. (2002). *Rubrivivax gelatinosus acsF* (previously *orf358*) codes for a conserved, putative binuclear-iron-cluster-containing protein involved in aerobic oxidative cyclization of Mg-protoporphyrin IX monomethylester. *Journal of Bacteriology*, *184*, 746–753. https://doi.org/10.1128/JB.184.3.746-753.2002.

Podosokorskaya, O. A., Kadnikov, V. V., Gavrilov, S. N., Mardanov, A. V., Merkel, A. Y., Karnachuk, O. V., et al. (2013). Characterization of *Melioribacter roseus* gen. nov., sp. nov., a novel facultatively anaerobic thermophilic cellulolytic bacterium from the class *Ignavibacteria*, and a proposal of a novel bacterial phylum *Ignavibacteriae*. *Environmental Microbiology*, *15*, 1759–1771. https://doi.org/10.1111/1462-2920.12067.

Porra, R. J., Schafer, W., Gadon, N., Katheder, I., Drews, G., & Scheer, H. (1996). Origin of the two carbonyl oxygens of bacteriochlorophyll alpha—Demonstration of two different pathways for the formation of ring E in *Rhodobacter sphaeroides* and *Roseobacter denitrificans*, and a common hydratase mechanism for 3-acetyl group formation. *European Journal of Biochemistry*, *239*, 85–92. https://doi.org/10.1111/j.1432-1033.1996.0085u.x.

Price, M. N., Dehal, P. S., & Arkin, A. P. (2009). FastTree: Computing large minimum-evolution trees with profiles instead of a distance matrix. *Molecular Biology and Evolution*, *26*, 1641–1650. https://doi.org/10.1093/molbev/msp077.

Pruesse, E., Peplies, J., & Glöckner, F. O. (2012). SINA: Accurate high-throughput multiple sequence alignment of ribosomal RNA genes. *Bioinformatics*, *28*, 1823–1829.

Pšenčík, J., Butcher, S. J., & Tuma, R. (2014). Chlorosomes: Structure, function and assembly. In M. F. HohmannMarriott, Govindjee, & T. Sharkey (Eds.), *Structural basis of biological energy generation: Vol. 39. Advances in photosynthesis and respiration* (pp. 77–109). Dordrecht, Netherlands: Springer. https://doi.org/10.1007/978-94-017-8742-0_5.

Pšenčík, J., Collins, A. M., Liljeroos, L., Torkkeli, M., Laurinmäki, P., Ansink, H. M., et al. (2009). Structure of chlorosomes from the green filamentous bacterium *Chloroflexus aurantiacus*. *Journal of Bacteriology*, *191*, 6701–6708. https://doi.org/10.1128/JB.00690-09.

Pšenčík, J., Torkkeli, M., Zupčanová, A., Vácha, F., Serimaa, R. E., & Tuma, R. (2010). The lamellar spacing in self-assembling bacteriochlorophyll aggregates is proportional to the length of the esterifying alcohol. *Photosynthesis Research*, *104*, 211–219. https://doi.org/10.1007/s11120-010-9541-0.

Reid, J. D., & Hunter, C. N. (2004). Magnesium-dependent ATPase activity and cooperativity of magnesium chelatase from *Synechocystis* sp PCC6803. *Journal of Biological Chemistry*, *279*, 26893–26899. https://doi.org/10.1074/jbc.M400958200.

Reid, J. D., Siebert, C. A., Bullough, P. A., & Hunter, C. N. (2003). The ATPase activity of the ChlI subunit of magnesium chelatase and formation of a heptameric AAA(+) ring. *Biochemistry*, *42*, 6912–6920. https://doi.org/10.1021/bi034082q.

Reinbothe, C., El Bakkouri, M., Buhr, F., Muraki, N., Nomata, J., Kurisu, G., et al. (2010). Chlorophyll biosynthesis: Spotlight on protochlorophyllide reduction. *Trends in Plant Science*, *15*, 614–624. https://doi.org/10.1016/j.tplants.2010.07.002.

Rüdiger, W., Benz, J., & Guthoff, C. (1980). Detection and partial characterization of activity of chlorophyll synthetase in etioplast membranes. *European Journal of Biochemistry*, *109*, 193–200. https://doi.org/10.1111/j.1432-1033.1980.tb04784.x.

Saer, R. G., & Blankenship, R. E. (2017). Light harvesting in phototrophic bacteria: Structure and function. *Biochemical Journal*, *474*, 2107–2131. https://doi.org/10.1042/BCJ20160753.

Saga, Y., Hayashi, K., Hirota, K., Harada, J., & Tamiaki, H. (2015). Modification of the esterifying farnesyl chain in light-harvesting bacteriochlorophylls in green sulfur photosynthetic bacteria by supplementation of 9-decyn-1-ol, 9-decen-1-ol, and decan-1-ol. *Journal of Photochemistry and Photobiology A: Chemistry*, *313*, 44–51. https://doi.org/10.1016/j.jphotochem.2015.05.002.

Saga, Y., Hayashi, K., Mizoguchi, T., & Tamiaki, H. (2014). Biosynthesis of bacteriochlorophyll *c* derivatives possessing chlorine and bromine atoms at the terminus of esterifying chains in the green sulfur bacterium *Chlorobaculum tepidum*. *Journal of Bioscience and Bioengineering*, *118*, 82–87. https://doi.org/10.1016/j.jbiosc.2013.12.023.

Saga, Y., Hirota, K., Harada, J., & Tamiaki, H. (2015). *In vitro* enzymatic activities of bacteriochlorophyll *a* synthase derived from the green sulfur photosynthetic bacterium *Chlorobaculum tepidum*. *Biochemistry*, *54*, 4998–5005. https://doi.org/10.1021/acs.biochem.5b00311.

Saga, Y., Oh-oka, H., Hayashi, T., & Tamiaki, H. (2003). Presence of exclusively bacteriochlorophyll-*c* containing substrain in the culture of green sulfur photosynthetic bacterium *Chlorobium vibrioforme* strain NCIB 8327 producing bacteriochlorophyll-*d*. *Analytical Sciences*, *19*, 1575–1579. https://doi.org/10.2116/analsci.19.1575.

Saga, Y., Yamashita, H., & Hirota, K. (2016). Introduction of perfluoroalkyl chain into the esterifying moiety of bacteriochlorophyll *c* in the green sulfur photosynthetic bacterium *Chlorobaculum tepidum* by pigment biosynthesis. *Bioorganic and Medicinal Chemistry*, *24*, 4165–4170. https://doi.org/10.1016/j.bmc.2016.07.004.

Sarma, R., Barney, B. M., Hamilton, T. L., Jones, A., Seefeldt, L. C., & Peters, J. W. (2008). Crystal structure of the L protein of *Rhodobacter sphaeroides* light-independent protochlorophyllide reductase with MgADP nound: A homologue of the nitrogenase Fe protein. *Biochemistry*, *47*, 13004–13015. https://doi.org/10.1021/bi801058r.

Sasarman, A., Letowski, J., Czaika, G., Ramirez, V., Nead, M. A., Jacobs, J. M., et al. (1993). Nucleotide-sequence of the *hemG* gene involved in the protoporphyrinogen oxidase activity of *Escherichia-coli* k12. *Canadian Journal of Microbiology*, *39*, 1155–1161. https://doi.org/10.1139/m93-174.

Saunders, A. H., Golbeck, J. H., & Bryant, D. A. (2013). Characterization of BciB: A ferredoxin-dependent 8-vinyl- protochlorophyllide reductase from the green sulfur bacterium *Chloroherpeton thalassium*. *Biochemistry*, *52*, 8442–8451. https://doi.org/10.1021/bi401172b.

Sawicki, A., & Willows, R. D. (2010). BchJ and BchM interact in a 1:1 ratio with the magnesium chelatase BchH subunit of *Rhodobacter capsulatus*. *FEBS Journal*, *277*, 4709–4721. https://doi.org/10.1111/j.1742-4658.2010.07877.x.

Schoch, S., Oster, U., Mayer, K., Feick, R., & Rudiger, W. (1999). Subtrate specificity of overexpressed bacteriochlorophyll synthase from *Chloroflexus aurantiacus*. In J. H. ArgyroudiAkoyunoglou & H. Senger (Eds.). *Vol. 64. Chloroplast: From molecular biology to biotechnology* (pp. 213–216). Dordrecht, Netherlands: Springer.

Shepherd, M., McLean, S., & Hunter, C. N. (2005). Kinetic basis for linking the first two enzymes of chlorophyll biosynthesis. *FEBS Journal*, *272*, 4532–4539. https://doi.org/10.1111/j.1742-4658.2005.04873.x.

Sirijovski, N., Lundqvist, J., Rosenbäck, M., Elmlund, H., Al-Karadaghi, S., Willows, R. D., et al. (2008). Substrate-binding model of the chlorophyll biosynthetic magnesium chelatase BchH subunit. *Journal of Biological Chemistry*, *283*, 11652–11660. https://doi.org/10.1074/jbc.M709172200.

Skotnicová, P., Sobotka, R., Shepherd, M., Hájek, J., Hrouzek, P., & Tichý, M. (2018). The cyanobacterial protoporphyrinogen oxidase HemJ is a new *b*-type heme protein functionally coupled with coproporphyrinogen III oxidase. *Journal of Biological Chemistry*, *293*, 12394–12404. https://doi.org/10.1074/jbc.RA118.003441.

Smith, K. M., & Goff, D. A. (1985). Bacteriochlorophylls-*d* from *Chlorobium-vibrioforme* — Chromatographic separations and structural assignments of the methyl bacteriopheophorbides. *Journal of the Chemical Society-Perkin Transactions 1*, 1099–1113. https://doi.org/10.1039/p19850001099.

Staehelin, L. A., Golecki, J. R., & Drews, G. (1980). Supramolecular organization of chlorosomes (chlorobium vesicles) and of their membrane attachment sites in *Chlorobium limicola*. *Biochimica et Biophysica Acta*, *589*, 30–45. https://doi.org/10.1016/0005-2728(80)90130-9.

Staehelin, L. A., Golecki, J. R., Fuller, R. C., & Drews, G. (1978). Visualization of the supramolecular architecture of chlorosomes (Chlorobium type vesicles) in freeze-fractured cells of *Chloroflexus aurantiacus*. *Archives of Microbiology*, *119*, 269–277. https://doi.org/10.1007/BF00405406.

Steensgaard, D. B., Cox, R. P., & Miller, M. (1996). Manipulation of the bacteriochlorophyll *c* homolog distribution in the green sulfur bacterium *Chlorobium tepidum*. *Photosynthesis Research*, *48*, 385–393. https://doi.org/10.1007/BF00029471.

Suzuki, J. Y., Bollivar, D. W., & Bauer, C. E. (1997). Genetic analysis of chlorophyll biosynthesis. *Annual Review of Genetics*, *31*, 16–89. https://doi.org/10.1146/annurev.genet.31.1.61.

Suzuki, Y., Doi, M., & Shioi, Y. (2002). Two enzymatic reaction pathways in the formation of pyropheophorbide *a*. *Photosynthesis Research*, *74*, 225–233. https://doi.org/10.1023/A:1020919929608.

Tamiaki, H., Komada, J., Kunieda, M., Fukai, K., Yoshitomi, T., Harada, J., et al. (2011). *In vitro* synthesis and characterization of bacteriochlorophyll-*f* and its absence in bacteriochlorophyll-*e* producing organisms. *Photosynthesis Research*, *107*, 133–138. https://doi.org/10.1007/s11120-010-9603-3.

Tanaka, A., Ito, H., Tanaka, R., Tanaka, N. K., Yoshida, K., & Okada, K. (1998). Chlorophyll *a* oxygenase (CAO) is involved in chlorophyll *b* formation from chlorophyll *a*. *Proceedings of the National Academy of Sciences of the United States of America*, *95*, 12719–12723. https://doi.org/10.1073/pnas.95.21.12719.

Tang, K. H., Barry, K., Chertkov, O., Dalin, E., Han, C. S., Hauser, L. J., et al. (2011). Complete genome sequence of the filamentous anoxygenic phototrophic bacterium *Chloroflexus aurantiacus*. *BMC Genomics*, *12*(334). https://doi.org/10.1186/1471-2164-12-334.

Tang, K. H., Wen, J., Li, X., & Blankenship, R. E. (2009). Role of the AcsF protein in *Chloroflexus aurantiacus*. *Journal of Bacteriology*, *191*, 3580–3587. https://doi.org/10.1128/JB.00110-09.

Tank, M., & Bryant, D. A. (2015a). *Chloracidobacterium thermophilum* gen. nov., sp. nov.: An anoxygenic microaerophilic chlorophotoheterotrophic acidobacterium. *International Journal of Systematic and Evolutionary Microbiology*, *65*, 1426–1430. https://doi.org/10.1099/ijs.0.000113.

Tank, M., & Bryant, D. A. (2015b). Nutrient requirements and growth physiology of the photoheterotrophic Acidobacterium, *Chloracidobacterium thermophilum*. *Frontiers in Microbiology*, *6*(226). https://doi.org/10.3389/fmicb.2015.00226.

Tank, M., Garcia Costas, A. M., & Bryant, D. A. (2019). Genus: Chloracidobacterium. In W. B. Whitman (Ed.), *Bergey's manual of systematics of bacteria and archaea*. New York: Wiley. supervising editor. In press.

Tank, M., Thiel, V., Ward, D. M., & Bryant, D. A. (2017). *A panoply of phototrophs: An overview of the thermophilic chlorophototrophs of the microbial mats of alkaline siliceous hot springs in Yellowstone National Park, WY, USA. In Modern topics in the phototrophic prokaryotes: Environmental and applied aspects* (pp. 87–137). https://doi.org/10.1007/978-3-319-46261-5_3.

Teramura, M., Harada, J., Mizoguchi, T., Yamamoto, K., & Tamiaki, H. (2016). *In vitro* assays of BciC showing C13^2-demethoxycarbonylase activity requisite for biosynthesis of chlorosomal chlorophyll pigments. *Plant and Cell Physiology*, *57*, 1048–1057. https://doi.org/10.1093/pcp/pcw045.

Teramura, M., Harada, J., & Tamiaki, H. (2016). *In vitro* stereospecific hydration activities of the 3-vinyl group of chlorophyll derivatives by BchF and BchV enzymes involved in bacteriochlorophyll *c* biosynthesis of green sulfur bacteria. *Photosynthesis Research*, *130*, 33–45. https://doi.org/10.1007/s11120-016-0220-7.

Teramura, M., Harada, J., & Tamiaki, H. (2018). *In vitro* enzymatic assays of photosynthetic bacterial 3-vinyl hydratases for bacteriochlorophyll biosyntheses. *Photosynthesis Research*, *135*, 319–328. https://doi.org/10.1007/s11120-017-0415-6.

Teramura, M., Harada, J., & Tamiaki, H. (2019). *In vitro* demethoxycarbonylation of various chlorophyll analogs by a BciC enzyme. *Photosynthesis Research, 139*(1–3), 163–171. https://doi.org/10.1007/s11120-018-0573-1.

Thiel, V., Tank, M., & Bryant, D. A. (2018). Diversity of chlorophototrophic bacteria revealed in the omics era. *Annual Review of Plant Biology, 69*, 21–49. https://doi.org/10.1146/annurev-arplant-042817-040500.

Thweatt, J. L., Ferlez, B. H., Golbeck, J. H., & Bryant, D. A. (2017). BciD is a radical S-adenosyl-L-methionine (SAM) enzyme that completes bacteriochlorophyllide *e* biosynthesis by oxidizing a methyl group into a formyl group at C-7. *Journal of Biological Chemistry, 292*, 1361–1373. https://doi.org/10.1074/jbc.M116.767665.

Tronrud, D. E., Wen, J., Gay, L., & Blankenship, R. E. (2009). The structural basis for the difference in absorbance spectra for the FMO antenna protein from various green sulfur bacteria. *Photosynthesis Research, 100*, 79–87. https://doi.org/10.1007/s11120-009-9430-6.

Tsukatani, Y., Harada, J., Mizoguchi, T., & Tamiaki, H. (2013). Bacteriochlorophyll homolog compositions in the *bchU* mutants of green sulfur bacteria. *Photochemical and Photobiological Sciences, 12*, 2195–2201. https://doi.org/10.1039/c3pp50253h.

Tsukatani, Y., Mizoguchi, T., Thweatt, J., Tank, M., Bryant, D. A., & Tamiaki, H. (2016). Glycolipid analyses of light-harvesting chlorosomes from envelope protein mutants of *Chlorobaculum tepidum*. *Photosynthesis Research, 128*, 235–241. https://doi.org/10.1007/s11120-016-0228-z.

Tsukatani, Y., Romberger, S. P., Golbeck, J. H., & Bryant, D. A. (2012). Isolation and characterization of homodimeric type-I reaction center complex from *Candidatus* chloracidobacterium thermophilum, an aerobic chlorophototroph. *Journal of Biological Chemistry, 287*, 5720–5732. https://doi.org/10.1074/jbc.M111.323329.

Tsukatani, Y., Wen, J., Blankenship, R. E., & Bryant, D. A. (2010). Characterization of the FMO protein from the aerobic chlorophototroph, *Candidatus* Chloracidobacterium thermophilum. *Photosynthesis Research, 104*, 201–209. https://doi.org/10.1007/s11120-009-9517-0.

Tsukatani, Y., Yamamoto, H., Harada, J., Yoshitomi, T., Nomata, J., Kasahara, M., et al. (2013). An unexpectedly branched biosynthetic pathway for bacteriochlorophyll *b* capable of absorbing near-infrared light. *Scientific Reports, 3*(1217). https://doi.org/10.1038/srep01217.

Tsukatani, Y., Yamamoto, H., Mizoguchi, T., Fujita, Y., & Tamiaki, H. (2013). Completion of biosynthetic pathways for bacteriochlorophyll *g* in *Heliobacterium modesticaldum*: The C8-ethylidene group formation. *Biochimica et Biophysica Acta, 1827*, 1200–1204. https://doi.org/10.1016/j.bbabio.2013.06.007.

van Gemerden, H., & Mas, J. (1995). Ecology of phototrophic sulfur bacteria. In R. E. Blankenship, M. T. Madigan, & C. E. Bauer (Eds.), *Anoxygenic photosynthetic bacteria: Vol. 2. Advances in photosynthesis and respiration* (pp. 49–85)Springer.

Vogl, K., Tank, M., Orf, G. S., Blankenship, R. E., & Bryant, D. A. (2012). Bacteriochlorophyll *f*: Properties of chlorosomes containing the "forbidden chlorophyll" *Frontiers in Microbiology, 3*(298) https://doi.org/10.3389/fmicb.2012.00298.

Wada, K., Yamaguchi, H., Harada, J., Niimi, K., Osumi, S., Saga, Y., et al. (2006). Crystal structures of BchU, a methyltransferase involved in bacteriochlorophyll *c* biosynthesis, and its complex with S-adenosylhomocysteine: Implications for reaction mechanism. *Journal of Molecular Biology, 360*, 839–849. https://doi.org/10.1016/j.jmb.2006.05.057.

Wahlund, T. M., & Madigan, M. T. (1993). Nitrogen fixation by the thermophilic green sulfur bacterium *Chlorobium tepidum*. *Journal of Bacteriology, 175*, 474–478. https://doi.org/10.1128/jb.175.2.474-478.1993.

Wahlund, T. M., & Tabita, F. R. (1997). The reductive tricarboxylic acid cycle of carbon dioxide assimilation: Initial studies and purification of ATP-citrate lyase from the green sulfur bacterium *Chlorobium tepidum*. *Journal of Bacteriology*, *179*, 4859–4867. https://doi.org/10.1128/jb.179.15.4859-4867.1997.

Walker, C. J., Mansfield, K. E., Smith, K. M., & Castelfranco, P. A. (1989). Incorporation of atmospheric oxygen into the carbonyl functionality of the protochlorophyllide isocyclic ring. *Biochemical Journal*, *257*, 599–602. https://doi.org/10.1042/bj2570599.

Wang, P., Wan, C., Xu, Z., Wang, P., Wang, W., Sun, C., et al. (2013). One divinyl reductase reduces the 8-vinyl groups in various intermediates of chlorophyll biosynthesis in a given higher plant species, but the isozyme differs between species. *Plant Physiology*, *161*, 521–534. https://doi.org/10.1104/pp.112.208421.

Wätzlich, D., Broecker, M. J., Uliczka, F., Ribbe, M., Virus, S., Jahn, D., et al. (2009). Chimeric nitrogenase-like enzymes of (bacterio)chlorophyll biosynthesis. *Journal of Biological Chemistry*, *284*, 15530–15540. https://doi.org/10.1074/jbc.M901331200.

Wen, J., Tsukatani, Y., Cui, W., Zhang, H., Gross, M. L., Bryant, D. A., et al. (2011). Structural model and spectroscopic characteristics of the FMO antenna protein from the aerobic chlorophototroph, *Candidatus* Chloracidobacterium thermophilum. *Biochimica et Biophysica Acta*, *1807*, 157–164. https://doi.org/10.1016/j.bbabio.2010.09.008.

Wen, J., Zhang, H., Gross, M. L., & Blankenship, R. E. (2009). Membrane orientation of the FMO antenna protein from *Chlorobaculum tepidum* as determined by mass spectrometry-based footprinting. *Proceedings of the National Academy of Sciences of the United States of America*, *106*, 6134–6139. https://doi.org/10.1073/pnas.0901691106.

Willows, R. D., Gibson, L. C. D., Kanangara, C. G., Hunter, C. N., & von Wettstein, D. (1996). Three separate proteins constitute the magnesium chelatase of *Rhodobacter sphaeroides*. *European Journal of Biochemistry*, *235*, 438–443. https://doi.org/10.1111/j.1432-1033.1996.00438.x.

Woodbury, N. W., Becker, M., Middendorf, D., & Parson, W. W. (1985). Picosecond kinetics of the initial photochemical electron-transfer reaction in bacterial photosynthetic reaction centers. *Biochemistry*, *24*, 7516–7521. https://doi.org/10.1021/bi00347a002.

Xin, Y., Lin, S., Montaño, G. A., & Blankenship, R. E. (2005). Purification and characterization of the B808-866 light-harvesting complex from green filamentous bacterium *Chloroflexus aurantiacus*. *Photosynthesis Research*, *86*, 155–163. https://doi.org/10.1007/s11120-005-5103-2.

Xin, Y., Shi, Y., Niu, T., Wang, Q., Niu, W., Huang, X., et al. (2018). Cryo-EM structure of the RC-LH core complex from an early branching photosynthetic prokaryote. *Nature Communications*, *9*(1568). https://doi.org/10.1038/s41467-018-03881-x.

Xiong, J., Fischer, W. M., Inoue, K., Nakahara, M., & Bauer, C. E. (2000). Molecular evidence for the early evolution of photosynthesis. *Science*, *289*, 1724–1730. https://doi.org/10.1126/science.289.5485.1724.

Zeng, Y., Feng, F., Medova, H., Dean, J., & Koblizek, M. (2014). Functional type 2 photosynthetic reaction centers found in the rare bacterial phylum *Gemmatimonadetes*. *Proceedings of the National Academy of Sciences of the United States of America*, *111*, 7795–7800. https://doi.org/10.1073/pnas.1400295111.

Zeng, Y., & Koblížek, M. (2017). *Phototrophic* Gemmatimonadetes: *A new "purple" branch on the bacterial tree of life*. *In Modern topics in the phototrophic prokaryotes: Environmental and applied aspects* (pp. 163–192). . https://doi.org/10.1007/978-3-319-46261-5_5.

Zill, J. C., He, Z., Tank, M., Ferlez, B. H., Canniffe, D. P., Lahav, Y., et al. (2018). ^{15}N photo-CIDNP MAS NMR analysis of reaction centers of *Chloracidobacterium thermophilum*. *Photosynthesis Research*, *137*, 295–305. https://doi.org/10.1007/s11120-018-0504-1.

Zsebo, K. M., & Hearst, J. E. (1984). Genetic-physical mapping of a photosynthetic gene cluster from *R. capsulata*. *Cell*, *37*, 937–947. https://doi.org/10.1016/0092-8674(84)90428-8.

CHAPTER THREE

Chlorophylls *c*—Occurrence, synthesis, properties, photosynthetic and evolutionary significance

Beata Myśliwa-Kurdziel[a], Dariusz Latowski[a], Kazimierz Strzałka[a,b,*]

[a]Department of Plant Physiology and Biochemistry, Faculty of Biochemistry, Biophysics and Biotechnology, Jagiellonian University in Krakow, Krakow, Poland
[b]Malopolska Centre of Biotechnology, Jagiellonian University in Krakow, Krakow, Poland
*Corresponding author: e-mail address: kazimierzstrzalka@gmail.com

Contents

Abstract

The term chlorophylls *c* (Chls *c*) cover a family of chlorophylls whose common structural feature is a fully unsaturated tetrapyrrole backbone. They are widely distributed within numerous photosynthesizing groups of organisms, mainly eukaryotic algae belonging to the Chromista kingdom. The most common Chls *c* are Chl *c1*, Chl *c2* and Chl *c3* but several other Chl *c* species, named Chl *c*-type compounds have been also described. Chls *c* are important constituents of light-harvesting complexes localized in the thylakoid membranes and associated with photosystems, where they serve, together with specific carotenoids, as energetic antennae. Chls *c*, which have an intense Soret band and blue-shifted Q bands in comparison to chlorophyll *a* (Chl *a*), are especially efficient in harvesting blue and green light which penetrates through deep waters better than light of other wavelengths. Thus, Chls *c* provide energy for photosynthesis to the organisms occupying underwater habitats.

Synthesis of Chls *c* follows many steps of the standard pathway that leads to Chl *a* formation. The branching point towards Chl *a* or Chls *c* is at protochlorophyllide (Pchlide), which can be converted either to Chlide *a*, which is the intermediate for

Advances in Botanical Research, Volume 90
ISSN 0065-2296
https://doi.org/10.1016/bs.abr.2019.04.002

the synthesis of chlorine-type Chls, or to Chl *c2*, which is an intermediate in the synthesis of other members of the Chl *c* family. Data on the occurrence of Chl *c* species as well as of other Chls is a useful tool in the taxonomy of algae, allowing us to assess the relationship among the chlorophylls and their phylogeny.

Abbreviations

acpPC chlorophyll *a*-chlorophyll *c2*-peridinin-protein complexes
ALA 5-aminolevulinic acid
Alx alloxanthin
Chl chlorophyll
Chlide chlorophyllide
Ddx diadinoxanthin
Dtx diatoxanthin
DV divinyl
EET excitation energy transfer
FCP fucoxanthin-chlorophyll proteins
FDSC first derivative spectrum chromatogram
Fx fucoxanthin
Hx heteroxanthin
ICT intramolecular charge-transfer
LHC light harvesting antenna complexes
MV monovinyl
Pchlide protochlorophyllide
PCP peridinin-chlorophyll protein
Per peridinin
PPP photoprotective pigments
Proto IX protoporphyrin IX
PS photosystem
PSP photosynthetic pigments
Px prasinoxanthin
UHPLC ultra-high performance liquid chromatography
Vx violaxanthin
XLH xanthophytes' light-harvesting complexes

1. Introduction

In the 19th century, in addition to chlorophylls (Chls) *a* and *b*, a new photosynthetic pigment, characteristic of algae was described and named chlorofucine—today known as one of Chls *c* (Sorby, 1873). The presence of Chls *c* in diatoms and brown algae was finally confirmed by its extraction and absorption characteristics in 1942 (Strain & Manning, 1942). A few years later, Granick pointed out the pheoporphyrin nature of this pigment (Granick, 1949). Further research revealed the heterogeneity of Chls *c* and led to the identification of three main Chls: Chl *c1*, Chl *c2* and Chl *c3*, and over a dozen of their derivatives that belong to the Chl *c* family (for review see Chen & Blankenship, 2018; Larkum, Ritchie, & Raven, 2018; Zapata, Garrido, & Jeffrey, 2006).

To date the presence of Chls *c* has been shown mainly in the photoautotrophic species of one of the largest kingdoms, called Chromista, which obtained its chloroplasts by secondary endosymbiosis (Cavalier-Smith, 2018; Falkowski & Knoll, 2007). The photoautotrophic species of Chromista inhabit a large range of contrasting adaptive zones and include the vast majority of marine, freshwater as well as soil algae. The size, shape and biology of photoautotrophic Chromista species are also very diverse. Generally, they include unicellular algae with two main photoautotrophic groups. One of them are the diatoms with about 100,000 species, whose silica frustules were once essential for polishing astronomical telescope mirrors or making dynamite. Dinoflagellates are the second group with about 2000 species, which achieve densities of 10^7–10^8 cells per liter (Taylor, Hoppenrath, & Saldarriaga, 2008) and power coral reefs or kill shellfish, the most abundant predators in soil. These two groups of algae are thought to be the most important groups of marine phytoplankton. However, Chromista, apart from single-cell algae, include also a large group of multicellular brown algae with about 2000 species, some of which may be even longer than a blue whale and reach 60 m in length. They form underwater kelp forests with a high level of biodiversity and some of the brown algae are also used by humans as food and an important source of human dietary supplements (Cavalier-Smith, 2018; Cock, Peters, & Coelho, 2011; D'Orazio et al., 2012). Because of their evolutionary success, photoautotrophic Chromista play a crucial role in creating a balanced aquatic ecosystem due to carbon fixation as well as oxygen and biomass production (Reinfelder, 2011; Yool & Tyrrell, 2003) and thus, they are immensely important for climate stability, agriculture and water or soil biology (Cavalier-Smith, 2018). Some Chl *c* family members have also been detected in several prokaryotic photoautotrophs (Larkum, Scaramuzzi, Cox, Hiller, & Turner, 1994; Zapata et al., 2006). In all these organisms, Chl *c* plays an important and unique role as accessory pigments while Chl *a* is always present in their photosynthetic reaction centers (reviewed by Björn, Papageorgiou, Blankenship, & Govindjee, 2009; Croce & Van Amerongen, 2014; Hallegraeff, Blackburn, Roy, Vernet, & Wright, 2012; Scheer, 2006). However, Chl *d* has been found in the photosynthetic reaction center of the photosystem (PS) I in *Acaryochloris marina*—an oxygenic photosynthetic prokaryotic organism (Hu et al., 1998). To date no Chl *c* has been found in plants and the remaining algae groups (Chen, 2014).

2. Chlorophylls *c* family

The Chl *c* family belongs to a group of closed tetrapyrroles, which are the oldest prosthetic group of vital importance in living organisms (reviewed by Mochizuki et al., 2010). The structure of their molecules consist of four pyrroles (A–D, Fig. 1) which form a closed ring, substituted with

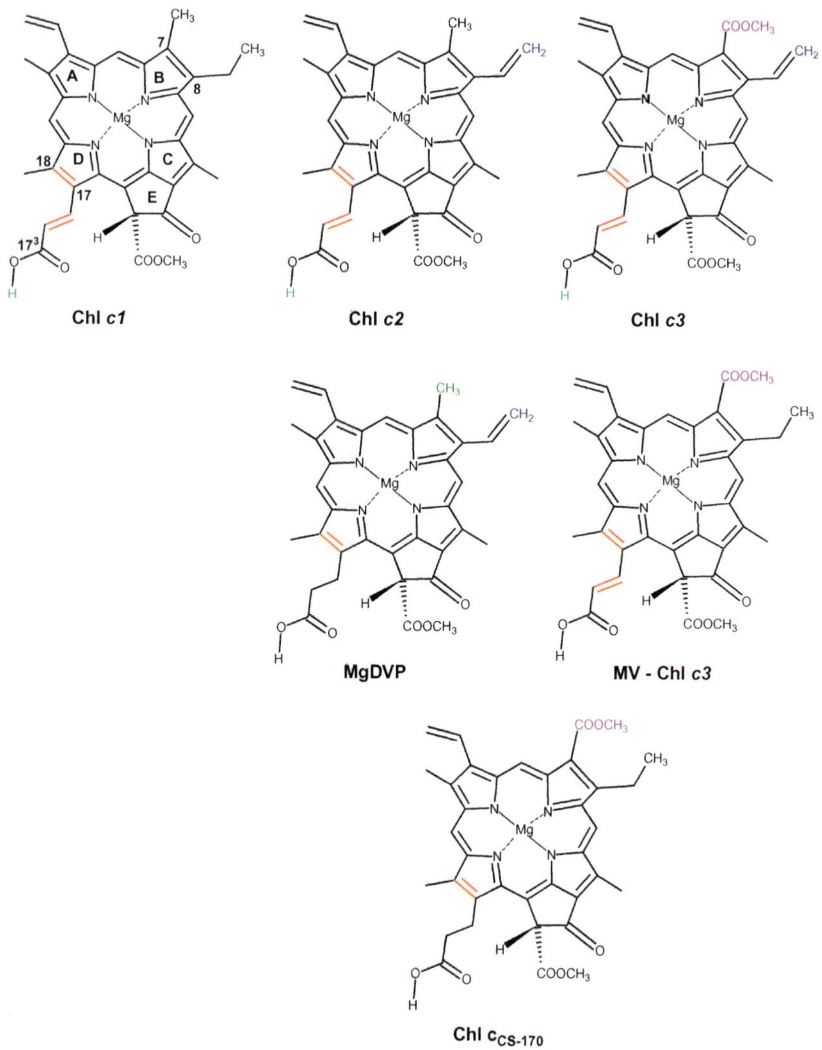

Fig. 1 Chemical structures of the main representatives of the Chl *c* family. The characteristic double bonds of the family are marked in red, i.e., C17=C18 and C17[1]=C17[2]. The latter is reduced in DV-Pchlide. The position where esterification is present in some members of the family is marked green. Other colors show different substituents at sites C7, C8.

a cyclopentanone ring (E, Fig. 1), and have the ion Mg^{2+} inside the tetrapyrrole backbone (reviewed in Chen, 2014; Scheer, 1991, 2006). Chls *c* are classified as Mg-phytoporphyrins due to the fact that their tetrapyrrole backbone is fully unsaturated. On the contrary, other Chls, namely Chl *a*, Chl *b*, Chl *d* and Chl *f*, are chlorins, having one double bond less in the tetrapyrrole ring D (i.e., C17—C18 bond, see Fig. 1). Moreover, they also possess a phytol substituent of the tetrapyrrole ring at $C17^3$, which is usually not the case for Chls *c*.

In general, the Chl *c* family can be divided into three categories. Two groups of polar compounds include Chls *c* that have an acrylic acid (unsaturated) or propionic acid (saturated) substituents at C17, in the pyrrole ring D (Fig. 1). Moreover, their structures show variations in their side chains at C7 and C8, in the ring B (Álvarez, Rodríguez, Riobó, Garrido, & Vaz, 2013; Álvarez, Zapata, Garrido, & Vaz, 2012).

The most common and representatives of Chls *c* family, which were identified first, are Chl *c1*, Chl *c2*, and Chl *c3* (Fig. 1; for a description of how they were extracted and identified see Garrido & Zapata, 2006; Hallegraeff et al., 2012; Jeffrey & Vesk, 1997; Zapata et al., 2006; and the references therein). They are all free carboxylic acids that have a free 17-acrylate residue (ring D), whereas they differ in substituents at C7 or C8, at the pyrrole ring B (Fig. 1). Chl *c1* is a 7-methyl-8-ethyl-porphyrin whereas Chl *c2* is the 8-vinyl analogue of Chl *c1* and Chl *c3* is 7-methoxycarbonyl analogue of Chl *c2*. Another known Chls *c* family member, MV-Chl *c3*, is a 7-methoxycarbonyl analogue of Chl *c1* and also an [8-ethyl]-Chl *c3*, the latter was used in the original description (Garrido & Zapata, 2006). Additional members of the Chls *c* family, sometimes referred to as Chls *c*-type compounds, were identified due to the development of chromatographic methods (Garrido & Zapata, 2006; Jeffrey & Vesk, 1997; Zapata et al., 2006). In particular, two other compounds are known that have a propionic substituent at C17. The first is divinyl protochlorophyllide (DV-Pchlide), termed also Mg-2,4-divinyl phaeoporphyrin *a*5 monomethyl ester (Zapata et al., 2006). Interestingly, this pigment is an intermediate in the biosynthesis of chlorin-type Chls (Chen, 2014). The next is, as originally described, Chl $c_{CS\text{-}170}$, which was identified later as [7-methoxycarbonyl-8-vinyl] Pchlide *a* (Álvarez et al., 2013).

The stereochemistry of these compounds has been confirmed by various methods including chiral HPLC, chemical modification and ^1H NMR, and circular dichroism spectroscopies (Álvarez et al., 2013, 2012; Helfrich et al., 2003; Mizoguchi, Kimura, & Tamiaki, 2010; Mizoguchi,

Kimura, Yoshitomi, & Hitoshi, 2011; Mizoguchi et al., 2009). It was found that the C17 acrylate residue forms a *trans*-isomer in the double bond and the 13^2-chiral centers are *R*-configured. The latter is a rule for other Chl molecules in the native organism.

The third group of Chls *c* consists of the compounds that have the acrylic chain at C17 esterified either with a lipid, monogalactosyl diacylglycerol (MGDG) or with phytol, or with an as yet unidentified substituent (Fawley, 1989; Zapata, Fraga, Rodríguez, & Garrido, 2012; Zapata et al., 2004). The former include Chl *c2*-MGDG (18:4/14:0) (Garrido, Otero, Maestro, & Zapata, 2000) and Chl *c2*-MGDG (14:0/14:0) (Zapata et al., 2001). The latter comprise several compounds described by indicating the most similar known compound and specifying the source organism, like Chl *c1*-like *Cryptoperidinium*-type—from dinoflagellates, as well as Chl *c2*-like *Pavlova gyrans*-type, Chl *c1*-like *Prymnesium parvum*-type (Zapata et al., 2006), and Chl *c1*-like *Exanthema christis*—type, which has recently been found (Zapata et al., 2012)—all from haptophytes.

The Chl *c*-containing groups of organisms have been examined systematically. Zapata showed the presence of various Chls *c* in 37 species (65 strains) of Haptophyta (Zapata et al., 2004), in 18 species (51 strains) of the pennate diatom *Pseudo-nitzschia* (Zapata, Rodríguez, Fraga, Barra, & Ruggiero, 2011) and in 64 dinoflagellate species (122 strains) (Zapata et al., 2012). Recently, a new method for analyzing the pigment content from marine phytoplankton using ultra-high performance liquid chromatography (UHPLC) with the first derivative spectrum chromatogram (FDSC) technique has been developed (Suzuki, Kamimura, & Hooker, 2015), which allowed a really good and fast pigment resolution.

3. Excitation properties of chlorophylls *c*

All Chls absorb electromagnetic radiation due to the fully conjugated double bonds of the tetrapyrrole ring, which in the case of some Chls *c* is extended to the acrylic substituent at C17 (Fig. 1). In general, the absorption spectra of Chls consist of two bands, one in the blue range (Soret band) and the other in the red range (Q_x and Q_y bands), corresponding to S_0-S_2 and S_0-S_1 transitions, respectively (for reviews see Björn et al., 2009; Chen, 2014). Like other Chls, *c*-type Chl is green, which is a simple consequence of their absorption properties due to a broad gap in the absorption spectrum; they reflect and scatter most of the green light. Similarly to other Chls, Chl *c* shows red fluorescence which is blue-shifted as compared to chlorine-types Chls.

Table 1 Absorption characteristic of the main chlorophylls *c* in diethyl ether.

	Absorption maximum (nm)			Band ratio	
Chl-type	**Soret (III)**	**Q (II)**	**Q (I)**	**II/I**	**III/I**
Chl *c1*	445.7	577.9	627.8	>1	~9
Chl *c2*	449.3	580.3	629.1	≅1	~12
Chl *c3*	451.8	583.9	625.4	<1	~30
DV-Pchlide	438.5	574.2	624.2	<1	~9

Adapted after Zapata, M., Garrido, J., & Jeffrey, S. W. (2006). Chlorophyll c pigments: Current status. In B. Grimm, R. Porra, W. Rüdiger, & H. Scheer (Eds.), *Chlorophylls and bacteriochlorophylls* (pp. 39–53, Vol. 255). Advances in photosynthesis and respiration, Springer, Dordrecht. https://doi.org/10.1007/1-4020-4516-6_3.

The characteristic feature of the absorption spectra of Chls *c* is the intense Soret band, which is >10 times as high compared to the absorption in Q bands as well as narrow Q bands (Larkum et al., 2018; Zapata et al., 2006). Structural alterations among Chl *c1*, Chl *c2* and Chl *c3* (Fig. 1) imply important differences in their absorption spectra concerning the position of the maxima and band ratios, summarized in Table 1 (see Zapata et al., 2006), where the spectra are shown. The saturation of C17 residue in DV-Pchlide changes mainly the position of the Soret band. Other members of the Chl *c* family are similar to that of the parent compound, i.e., they have the same substituents at C7 and C8. It should also be noted that both Q bands of Chls *c* are the most blue-shifted compared to other Chls (Björn et al., 2009; Zapata et al., 2006).

A similar influence of C7 and C8 substituents on the absorption properties have recently been observed for semi-synthetic Chls in dichloromethane. A red-shift of both the Soret and the Q_y absorption bands with an increase of π-conjugation in a molecule was noticed (Xu, Kinoshita, Matsubara, & Tamiaki, 2016). The shift was observed for DV as compared to MV compounds, i.e., DV-Pchlide vs MV-Pchlide as well as Chl *c2* vs Chl *c1*. The presence of acrylic acid at C17 in Chl *c1*, and Chl *c2* enhanced the shift and caused a broadening of the absorption bands. Ultimately, the extent of the red-shift followed the order Chl *c2* > Chl *c1* > DV-Pchlide > MV-Pchlide (Xu et al., 2016). The same order was observed for the emission maxima (Xu et al., 2016).

Analyses of the photophysical properties of Chls *c* in model systems are scarce; there is a lack of systematic research. A study of Chl *c2* in tetrahydrofuran showed the Soret and Q_y absorption bands to be at 456 and

634 nm; the fluorescence maximum at 640 nm; a fluorescence lifetime of 5.9 ns and a phosphorescence lifetime of 833 ± 60 μs (Niedzwiedzki & Blankenship, 2010).

4. Chlorophylls *c*—Occurrence and evolutionary aspects

All eukaryotic photoautotrophs can generally be divided into three large groups depending on the type of Chls present in their cells. Organisms of all these groups contain Chl *a*, but they vary in the type of their second, accessory chlorophyllic pigment. This can either be Chl *b* in one group or (a) member(s) of the Chls *c* family in the other. The third group includes Rhodophyta and Glaucophyta, which possess exclusively Chl *a*, because they have lost Chl *b* (Larkum et al., 2018).

The first taxon which grouped almost all eukaryotic organisms with Chls *c* was established as the kingdom called Chromista by Cavalier-Smith (1981) (Fig. 2). It was created by combining two independent kingdoms of Chls *c*, which contained organisms, i.e., Cryptophyta and Chromophyta.

Initially Chromists were defined as organisms that possess Chls *c*-containing plastid(s) with stroma surrounded by four membranes and/or tripartite or bipartite rigid tubular hairs on one or both cilia. They were divided into three groups, i.e., Heterokonts (called also Stramenopiles), Haptophytes and Cryptomonads (Fig. 2, in green) (Csűrös, Rogozin, & Koonin, 2008). Alveolata, as the fourth group, was incorporated into Chromista by the International Society of Protistologists in 2005, and it was suggested that Chromista should be considered one of the six highest ranked eukaryotic groups called Chromalveolata (Fig. 2, in red) (Adl et al., 2005), although the name "Chromista" is still in common usage. Five years later, Cavalier-Smith also suggested moving Alveolata together with Rhizaria and Heliozoa

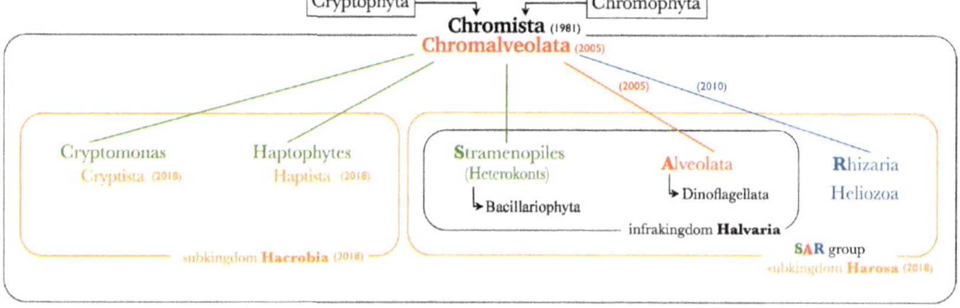

Fig. 2 The main groups of Chl *c*-containing algae with the most significant changes indicated by colors and years (for details see in the text).

into Chromista (Fig. 2, in blue) (Cavalier-Smith, 2010). Nowadays, it is generally, accepted that Alveolata together with Heterokonta form a reduced Chromalveolate clade called Halvaria (Burki et al., 2007; Burki, Shalchian-Tabrizi, & Pawlowski, 2008; Kim & Graham, 2008). Cavalier-Smith terms Halvaria as infrakingdom of Chromista with Alveolata and Heterokonta as two superphylas (Fig. 2) (Cavalier-Smith, 2018). In many analyses Heterokonta (Stramenopiles) and Alveolata are associated together with the Rhizaria, as one clade, which is called the "SAR" group derived from the initials of each group (Fig. 2) (Burki et al., 2007; Hampl et al., 2009). In Cavalier-Smith's most recent taxonomy (Cavalier-Smith, 2018), the SAR group is called Harosa, and is one of two Chromista sub-kingdoms. The other one is called Hacrobia and groups Haptophytes and Cryptomonads, in this taxonomy they are called Haptista and Cryptista, respectively (Fig. 2, in orange). Therefore, in all photosynthetic eukaryotes which belong to the Chromista kingdom, Chl *c* is the accessory pigment, with the exception of a small group of Rhizaria, called Chlorarachniophyta (Chlorarachniophyte or Chlorarachnea), which possesses Chl *b* instead of any of the Chls *c* species (Lee, 2008).

Chlorarachniophyta are the only group of Rhizaria capable of photosynthesis. It is postulated that Chlorarachniophyta, like all Chromista, evolved by symbiogenetic enslavement of a red alga by bikonts (Cavalier-Smith, 2000, 2018). However, all Rhizaria, including Chlorarachniophyta and several other groups of Chromista, lost entirely plastids or their ability to photosynthesize. It is suggested that solely in the Rhizaria group, three or four plastid losses have occurred. All Rhizaria retain heterotrophy, but all photosynthetic Chlorarachniophyta are thought to have enslaved algae again. With these green algae, Chlorarachniophyta have obtained chloroplasts with two extra surrounding membranes and Chl *a* and Chl *b* instead of any Chl *c* (Cavalier-Smith, 2006, 2018; Hirakawa, 2014; Hopkins et al., 2012; Suzuki, Hirakawa, Kofuji, Sugita, & Ishida, 2016). Similarly, some members of the genus *Lepidodinium*, which belongs to Dinoflagellates, have only Chl *a* and Chl *b* although all other photosynthetic Dinoflagellates contain Chl *c*. *Lepidodinium* has lost its photosynthetic ability rather more through the wastage of some enzymes required for Chl *a* synthesis than through the loss of plastids. Now, it is postulated that the genes for Chl *a* biosynthesis in *Lepidodinium* could have been reintroduced by transfer from Chlorarachniophytes as the prey of heterotrophic *Lepidodinium* (Matsuo & Inagaki, 2018). Interestingly, although >95% of photosynthetic dinoflagellates possess Chl *c2* as they have red algal-derived secondary chloroplasts, some groups of Dinoflagellates have replaced these typical chloroplasts with one from other microalgae, belonging to Haptophyta (Tengs et al., 2000; Zapata et al., 2012), or Bacillariophyta (Chesnick, Kooistra, Wellbrock, & Medlin, 1997; Riley & Wilson, 1967; Takano, Hansen, Fujita, & Horiguchi, 2008; Zhang, Liu, & Hu, 2011; Zhang et al., 2014). Thus, two additional Chl *c* types, i.e., Chl *c1* and Chl *c3*, are also identified in these groups (Yamada, Tanaka, & Horiguchi, 2015).

Among Heterokonta, Chls *c* were secondarily lost forever in a small group called Eustigmatophyte, with biotechnologically important genus *Nannochloropsis*, as well as in some species of Xanthophyte. Therefore these organisms possess only Chl *a* (Cavalier-Smith, 2018; Gardian, Tichý, & Vácha, 2011). Similarly, in the genus *Chromera* and *Vitrella* belonging to Chromeridae—one of the most recently discovered groups of Alveolata, Chls *c* were irretrievably lost (Pan, Šlapeta, Carter, & Chen, 2012).

Initially, Dinoflagellates were not included among Chromista, because they have three, not four, membranes surrounding their plastids. Nowadays, it is thought that all eukaryotic photoautotrophs which contain Chl *c* and possess plastids with stroma surrounded by more than two membranes belong to Chromista (Cavalier-Smith, 2018). These plastids are the effect of secondary or sometimes even tertiary and higher endosymbiotic events or/and horizontal eukaryote-to-eukaryote gene transfer (Qiu, Yoon, & Bhattacharya, 2013). Generally, Chl *c1* and sometimes Chl *c2* are present in diatoms, whereas Chl *c2* is the major type of Chl *c* in Dinoflagellates and Chl *c3* in haptophytes (Fawley, 1989; Jeffrey & Humphrey, 1975; Kraay, Zapata, & Veldhuis, 1992).

Although in the case of plastids that originally possessed Chl *c* the evolutionary scenario is still being discussed, it seems to be highly probable that the red alga was always the original ancestor of these plastids. Subsequently, in several groups of organisms, such as Chlorarachniophyta, Eustigmatophyte, some types of Xanthophytae or a genus, such as *Lepidodinium*, *Chromera* and *Vitrella*, chloroplasts or their key photosynthetic genes were first lost and subsequently complemented again by various evolutionary paths (Cavalier-Smith, 2018; Janouskovec, Horak, Obornik, Lukes, & Keeling, 2010). As far as we know, no special prokaryotic candidate has been identified as a primary endosymbiont to evolve into red algae chloroplasts. However, DV-Pchlide has been found in two cyanobacteria species, i.e., *A. marina* and *Prochlorococcus marinus* (Zapata et al., 2006). Besides DV-Pchlide *A. marina* contains Chl *d* as an additional Chl pigment (Miyashita et al., 1997;

Schliep, Chen, Larkum, & Quinnell, 2008). DV-Pchlide has also been found in some types of green eukaryotic algae which are examples of Archaeplastida, which also possessing Chl *a* and Chl *b*, such as *P. marinus*. These groups are called Prasinophyceae, with one of the most investigated species *Mantoniella squamata* (Fawley & Fawley, 2004; Kantz, Theriot, Zimmer, & Chapman, 1990; Latasa, Scharek, Le Gall, & Guillou, 2004; Rowan, 1998; Zingone et al., 2002), and Palmophyllophyceae (Leliaert et al., 2016).

Chls *c* are considered to be widespread, although not evenly distributed, photosynthetic pigments in photoautotrophs due to the large diversity and abundance of organisms containing Chl *c* (Larkum et al., 2018). Although the origin of Chl *c* is still unclear, today we may postulate that Chl *c2* was the first type of Chl *c* which arose on the evolutionary path. This seems to be confirmed by both the biosynthetic pathway in which Chl *c2* is a precursor to Chl *c1* and the occurrence of members of the Chl *c* family in photoautotrophs (Larkum, 2006; Larkum et al., 2018). Two groups of algae, such as cryptophytes and dinoflagellates, possess only Chl *c2*, whereas the other groups have Chl *c1*, Chl *c2* and in some cases also Chl *c3* (Larkum et al., 2018).

The spread of Chls *c* in the evolution of photosynthesis in a water environment can be explained taking into account the absorption properties of Chls *c* described above, namely the blue-shifted and intensive Soret band and relatively narrow and weak Q-band, compared to other chlorine-type Chls, in particular to Chl *a*. Therefore, Chls *c* supported by carotenoids could be favorable as the accessory pigments in blue light absorption, rather than red light, under special light conditions existing in a local algae habitat (Larkum et al., 2018). Efficient reduction of UV and red light, as the depth of the water increases, decreases the risk of damage to the photosynthetic structures and favors usage of blue light, which not only carries enough energy but also reaches the deepest levels under the water surface, because it is the least absorbed by water. This results in a low risk of harmful Chl excitation and consequently shifts the carotenoid's role from a photoprotective to a photosynthetic (Mendes et al., 2018; Stoń-Egiert, Majchrowski, Darecki, Kosakowska, & Ostrowska, 2012). Additionally, this conclusion seems to be in line with the conception of evolution of Chl *c*-containing organisms, leading to strong cooperation of Chls *c* with specific types of carotenoids (Larkum et al., 2018). On the other hand, Rhodophyta, which generally inhabit the deepest areas under water, totally lost the second type of Chl.

5. Light-harvesting antenna with chlorophylls *c*

In general, photosynthetic light-harvesting antenna systems (LHC) are composed of Chls and carotenoids with the absorption of light that covers a large part of the solar spectrum available in a local habitat. This helps the photosynthetic apparatus to adjust as well as possible to the spectral composition of available light (Björn et al., 2009; Larkum et al., 2018). The presence of Chl *c* in LHC is generally well correlated with carotenoids other than those associated with Chl *b* (Takaichi, 2011). Chls *c* participate in excitation energy transfer (EET) (Croce & Van Amerongen, 2014; Mirkovic et al., 2017), although the EET pathways are still not completely understood.

The optical properties of Chls *c*, as the LHC pigments, are to some extent similar to those of Chl *b* (Blankenship, 2014; Croce & Van Amerongen, 2014), which is a minor light-harvesting pigment in Chls *a*/*b* containing oxygenic phototrophs. The conjugation of the ring π system of Chls *c* is extended compared to Chl *a*, through the double bond of the *trans*-acrylate at C17 (Fig. 1), which serves the same role for the molecular electron density as the C7 formyl group in Chl *b*, and is important for the stability of the light-harvesting antenna (Eggink, LoBrutto, & Hoober, 2004). The Soret band of Chls *c* is red-shifted, whereas the Q bands are blue-shifted as compared to those of Chl *a*. The former extends the absorption band in the blue range (about 450 nm), which is advantageous for a water habitat (Larkum, 2006; Larkum et al., 2018). The latter enabled the transfer of singlet energy from the photoexcited states of Chl *c* to the ground state of Chl *a*. However, it is worth noting that the ratio of the intensity of the Soret band to that of the Q bands in Chl *c* is the highest among Chls due to the fully π-conjugated porphyrins (Xu et al., 2016), which is also important for survival under water.

LHC pigments can be divided into photosynthetic (PSP) and photoprotective pigments (PPP), according to their main role in a given system (Stoń-Egiert et al., 2012). PSP rapidly transfer their excitations to the lowest excited state of a nearby Chl *a* molecule, together forming a local funnel for EET to the reaction centers (Croce & Van Amerongen, 2014). In turn, PPP play a role in avoiding Chl *a* photooxidation under high light intensity. It has been shown that organisms living at increasing underwater depths decrease the amounts of PPP to keep the optimal photoprotective mechanisms at a local fluctuating light condition, and increase the PSP level to efficiently utilize the solar energy available (Stoń-Egiert et al., 2012).

The specific carotenoid composition of LHC containing Chls *c* imply the existence of different xanthophyll cycle from that based on violaxanthin (Vx), antheraxanthin and zeaxanthin, which is a classical photoprotective mechanism in Archeaplastida (Jahns, Latowski, & Strzalka, 2009). Many groups of Chl *c*-containing organisms, such as diatoms or haptophytes possess, as the major one, the xanthophyll cycle based on diadinoxanthin (Ddx) and diatoxanthin (Dtx), although the classical cycle is also observed under extreme light conditions (Lohr & Wilhelm, 2001). The molecular structure of all Chl *c*-containing LHCs is still unknown and little is known about their attribution to the PSs (Larkum et al., 2018).

As far as the major carotenoid bound to intrinsic LHC is concerned, organisms with Chls *c* can be divided into four main groups. The majority of Chls *c*-containing organisms, such as diatoms or brown algae, possess fucoxanthin (Fx), whereas in some groups of dinoflagellates peridinin (Per) is present. Two other carotenoids are alloxanthin (Alx) in Cryptophyta and heteroxanthin (Hx) in Xanthophyceae (Büchel, 2018). Of the carotenoids that belong to PSP, Fx and Per play an important role in extending the light-harvesting spectrum and thus ensuring the optimal absorption efficiency of the phytoplankton, making it predominant in highly-productive ecosystems (Kirk, 2010). Fx and Per are present in LHC in equal or even higher amounts than Chl (Mirkovic et al., 2017). These two carotenoids contain a conjugated carbonyl group that strongly affects the electronic properties of the molecules by a stabilizing the intramolecular charge-transfer (ICT) state in polar environments (Kosumi et al., 2014, 2011a, 2011b; for a review see Mirkovic et al., 2017; Polívka & Frank, 2010). The ICT state and the S1 state are electronically coupled, which results in a red-shift of the S2 absorption band to lower energies (to ~570 nm) as compared to the blue-range absorption band of LHC antenna in plants and green algae, and allows green light to be captured efficiently, which is important for underwater organisms (Polívka & Frank, 2010). Moreover, the coupled ITC/S1 state provides a good spectral overlap with the Qy band of Chl *a*, crucial for EET (Polívka & Frank, 2010).

Intrinsic LHC of thylakoid membranes, for example the Fx-Chl-protein (FCP) complexes of diatoms and brown algae, as well as Chl *a*-Chl *c2*-Per-protein complexes (acpPC) of some dinoflagellates, are related to the family of Chl *a*/*b* binding proteins (cab) from higher plants (Neilson & Durnford, 2010), characterized by three membrane-spanning α-helices. It is supposed that LHC of Chl *c*-containing algae is similarly arranged to the LHCII of higher plants (Durnford et al., 1999; Hiller, Wrench,

Gooley, Shoebridge, & Breton, 1993; Hiller, Wrench, & Sharples, 1995; Iglesias-Prieto, Govind, & Trench, 1991; Liu et al., 2004; Standfuss, Terwisscha van Scheltinga, Lamborghini, & Kühlbrandt, 2005). Additionally, as in higher plant LHCs, proteins of Chl *c*-containing algae are encoded in the nuclear genome, which suggested that these genes had been taken over by the host nucleus from the nucleus of the photosynthetic symbiont (Deane et al., 2000). Probably due to the complex evolution of Chl *c*-containing organisms, including sometimes the multiple loss and re-acquisition of their chloroplasts, LHCs are encoded by several multigene families (Büchel, 2018; Neilson & Durnford, 2010). In diatoms, FCPs serve as PS I antennae (Veith, Brauns, Weisheit, Mittag, & Büchel, 2009; Veith & Büchel, 2007). In Cryptophytes, their LHCs have also been shown to play the role of PS II antennae (Doust et al., 2004; Neilson & Durnford, 2010). It needs to be added that Cryptophytes, as the only Chl *c*-containing organisms, have lumenal phycobiliproteins, related to the phycobilisomes of cyanobacteria and red algae. Some dinoflagellates, in turn, possess unique water-soluble antenna systems that contain Per-Chl *a*-protein complexes (PCP) in addition to the acpPC (Larkum et al., 2018).

In LHCs of cryptophytes, Chl *c2* is present together with Chl *a* and Alx, whereas it is absent in their phycobiliproteins, which contain phycoerythrin and phycocyanin without any Chl molecules. LHC associated with PSII contains a higher Chl *a*:Chl *c2* ratio in respect to that associated with PS I, namely 22:5 and 17:5, respectively (Büchel, 2018). The role of Alx is still unclear—it is postulated to be rather PPP than PSP, because cryptophytes do not possess a xanthophyll cycle but under light stress they increase the level of Alx in their LHC (Mendes et al., 2018; Stoń-Egiert et al., 2012). It has been shown that ratios of Alx to Chl *a* rose by at least 100% as a results of transfer from low do high light conditions (Schlüter, Møhlenberg, Havskum, & Larsen, 2000). The level of Alx in LHC is also strongly dependent on depth of algae living (Mackey, Higgins, Mackey, & Holdsworth, 1998; Schlüter et al., 2000).

Among haptophytes, the LHCs of *Emiliania huxleyi* and *Isochrysis galbana* are the best studied. It is postulated that haptophytes like all heterokonts contain only intrinsic LHCs with Chl *a*, Chl *c3*, Chl *c2* esterified with MGDG and probably Fx together with its 19′-acyloxy derivatives and Ddx and Dtx, as the xanthophyll cycle pigments (Garrido & Zapata, 1998).

The most studied LHCs of heterokonts are examples of diatoms, brown algae and Xanthophyceae. Information about the LHCs of the remaining groups of heterokonts with Chls *c* is scarce; the only known LHCs concern

Chrysophyceae, which contain Chl *a*, Chl *c1* and Chl *c2*, although in some species Chl *c2* is not present. Among LHC carotenoids in Chrysophyceae, Fx and Vx engaged in the Vx cycle have also been detected (Lichtlé, Arsalane, Duval, & Passaquet, 1995).

In LHC of Xanthophyceae, no Fx is present and therefore their LHCs are also called xanthophytes' light-harvesting complexes (XLH). In XLH, the molar ratio of the pigment molecules Chl *a*:Ddx:Dtx:Hx:vaucheriaxanthin ester has been calculated to be 28:11:5:4:1. The level of Chl *c* is very low and, unlike algae with Fx or Per, Xanthophyceae have lower level of carotenoids than that of Chl; the Car:Chl ratio is about 1:2 (Büchel & Wilhelm, 1993; Gardian et al., 2011).

The LHCs of brown algae and diatoms contain Fx, as the main light-harvesting carotenoids, and thus their antenna are termed FCP. However, the FCPs of brown algae contain Vx and display Vx cycle, whereas in diatoms' FCP, Ddx and Dtx—engaged in Ddx cycle, are present (Ocampo-Alvarez, García-Mendoza, & Govindjee, 2013). Investigations of several brown algae revealed the pigment ratio 6:2:8 for Chl *a*:Chls *c*:Fx (De Martino et al., 2000; Pascal et al., 1998), or for Chl *a*:Chls *c*:Fx:Vx of 13:3:10:1 (Katoh, Mimuro, & Takaichi, 1989). Additionally, taking into account the interaction of Chls *c* with a local environment, two populations of Chls *c* have been found in brown algae, both participating in Chls *c* to Chl *a* EET (Mimuro, Katoh, & Kawai, 1990; Pascal et al., 1998).

FCPs of diatoms bind Chl *a*, Chl *c1* and Chl *c2*, Fx as well as Ddx and Dtx (Fawley, 1989; Jeffrey & Humphrey, 1975; Kraay et al., 1992). Two out of 10 suggested Chl-binding sites, a614 and a613, are ascribed to Chls *c* (Georgakopoulou et al., 2007). Generally, the pigment ratio depends on the isolation procedure, species and growth conditions, and is around 3–4 Chl *a*—1 of Chl *c*—3–4 Fx molecules. A higher amount of Fx has been observed in *Phaeodactylum tricornutum* (pennate diatoms) than in the *Cyclotella meneghiniana* (centric diatoms). The antenna complexes of these diatoms differ in terms of their molecular organization (Büchel, 2018).

The photophysical processes in FCP are currently being intensively investigated, mainly for *P. tricornutum* and *C. meneghiniana*. Fx can act both as PSP with extremely fast energy transfer to Chl *a* and as PPP being an efficient quencher of Chl *a* triplets (Di Valentin et al., 2010; Mirkovic et al., 2017). In the network of EET pathways, several components have been detected which have various time constants (Akimoto et al., 2014; Butkus et al., 2015; Gelzinis et al., 2015; Gildenhoff et al., 2010; Nagao, Yokono, Akimoto, & Tomo, 2013; Papagiannakis, Van Stokkum, Fey,

Büchel, & Van Grondelle, 2005; Songaila et al., 2013). Two ways of Fx-Chl *a* EET have been shown, the first one from the S2-excited state of Fx to the Qx state of Chl *a* (~100 fs, the fast one) and the other from the Fx S1/ICT state to the Qy state of Chl *a* (300–600 fs and 1–2 ps; slow components). Time constants for radiationless de-excitation assigned to the relaxation of Chl *a* Qy were of 5–10 ps and 1–3 ns, whereas it was 20 ps for carotenoid (Fx and Ddx) relaxation. The involvement of Chls *c* in EET has already been demonstrated, although the time constant is under discussion. Ultrafast (<100-fs time scale) and highly efficient (100%) EET has been observed from directly excited Chl *c2* to the Q_y band of Chl *a*, which proves the light-harvesting role of Chl *c2* in FCP in *C. meneghiniana* (Papagiannakis et al., 2005; Songaila et al., 2013). A much slower transfer (~600 fs) has also been determined (Akimoto et al., 2014). Apart from Fx-Chl *a*, Fx-Fx excitation energy transfer has also been detected (Gildenhoff et al., 2010). The EET between Fx and Chl *c* was not observed, although Chl *c* may be involved in the Fx-Chl *a* EET pathway (Gildenhoff et al., 2010; Papagiannakis et al., 2006).

In dinoflagellates, the two antenna systems acpPC and PCP exist. PCP probably form trimers located at the lumenal side of the thylakoid membrane (Hofmann et al., 1996) in a distance allowing Förster energy transfer from PCP to acpPC. Dinoflagellates can be divided into two major groups: the first one includes organisms which obtained their plastids directly from a secondary endosymbiosis and containing Per-containing complexes, whereas the other, after tertiary endosymbiosis with haptophytes, has Fx-containing complexes (Büchel, 2018).

A biochemical analysis of acpPC has mainly been performed for *Symbiodinium*, which is an endosymbiont in the endoderm of corals, and for *Amphidinium carterae*, which is used as a model organism for research of dinoflagellates. The antenna of *A. carterae* contain 10–12 molecules of Per, 7 Chls *a*, 4 Chls *c2*, and 2 Ddx (Hiller et al., 1993). *Symbiodinium* also has 2 molecules of Ddx, but only 6 molecules of Per, 6 Chls *c2*, and 4 Chls *a* (Niedzwiedzki, Jiang, Lo, & Blankenship, 2014).

Moreover, for *Symbiodinium* cells it has been shown that acpPC is more abundant than PCP (Iglesias-Prieto et al., 1991). In acpPC, Chl *c2* and Per molecules can both transfer the excitation energy to Chl *a* using the Per S2 state and S1/ICT (Bonetti et al., 2010; van Stokkum et al., 2009; Zigmantas, Hiller, Sundstrom, & Polivka, 2002). The transfer from Chl *c2* to Chl *a* is highly efficient (100%) and fast (~1.4–1.7 ps) (Di Valentin et al., 2010; Niedzwiedzki et al., 2014; Polívka et al., 2006; Šlouf et al., 2013).

The functional light-harvesting state of acpPC is a trimer similar to diatom FCP trimers, as suggested by bioinformatic analysis (Jiang et al., 2014). Ddx probably has a photoprotective role (Jovine, Johnsen, & Prézelin, 1995; Polívka et al., 2006), although it is inefficient in the Chl *a* triplet quenching in acpPC (Di Valentin et al., 2010; Niedzwiedzki et al., 2014). Interestingly, it has been shown that Per may also play a role in photoprotection (Di Valentin et al., 2010). Although Per-to-Chl-*c2* transfer has not been demonstrated so far, it is postulated that these pigments are located in very small distances in LHC of *A. carterae,* because Per is an effective quencher not only of the triplet state of Chl *a* but also of Chl *c2* (Di Valentin et al., 2010).

Another Chl *c* family member, DV-Pchlide, has been found in the LHCs of Prasinophyceae, such as *Ostreococcus tauri* and *M. squamata*, which are examples of green algae. The LHCs of prasinophytes have the non-typical xanthophyll prasinoxanthin (Px), which is postulated to occupy an analogous lutein position of higher plant LHCs (Goss, Wilhelm, & Garab, 2000). For every 2 molecules of Px, 1–2 molecules of DV-Pchlide as well as 1 dihydrolutein, 5 Chls *b*, 6 Chls *a* have been estimated (Swingley et al., 2010). The LHCs of prasinophytes are permanently associated with PSI and PSII, like a minor LHCII monomeric antenna, whereas in other green algae and plants they are mobile proteins that transfer excitation energy between the two PSs (Dekker & Boekema, 2005; Iwai, Takahashi, & Minagawa, 2008; Kargul et al., 2005; Tokutsu, Iwai, & Minagawa, 2009). The most abundant antenna proteins in *O. tauri*, Lhcp2, do not form trimers but multimeric complexes, consisting up to 6 Lhcp proteins (Büchel, 2003; Hobe, Förster, Klingler, & Paulsen, 1995; Koziol et al., 2007; Swingley et al., 2010).

6. Chlorophyll *c* biosynthesis

The biosynthesis of Chl *c* follows a multi-branched and conservative biosynthesis of tetrapyrroles (Fig. 3), which has been intensively studied for many years, resulting in the recognition of both metabolic intermediates and enzymes that catalyze particular reactions of the scheme (reviewed in Brzezowski, Richter, & Grimm, 2015; Masuda & Fujita, 2008; Rebeiz, 2014). Tetrapyrrole biosynthesis is precisely regulated at various levels to support the optimal level of different classes of tetrapyrroles, i.e., Chls, hemes, phytochromes, phycobilins (reviewed in Brzezowski et al., 2015; Czarnecki & Grimm, 2012; Fujita, Tsujimoto, & Aoki, 2015; Masuda & Fujita, 2008; Yin & Bauer, 2013).

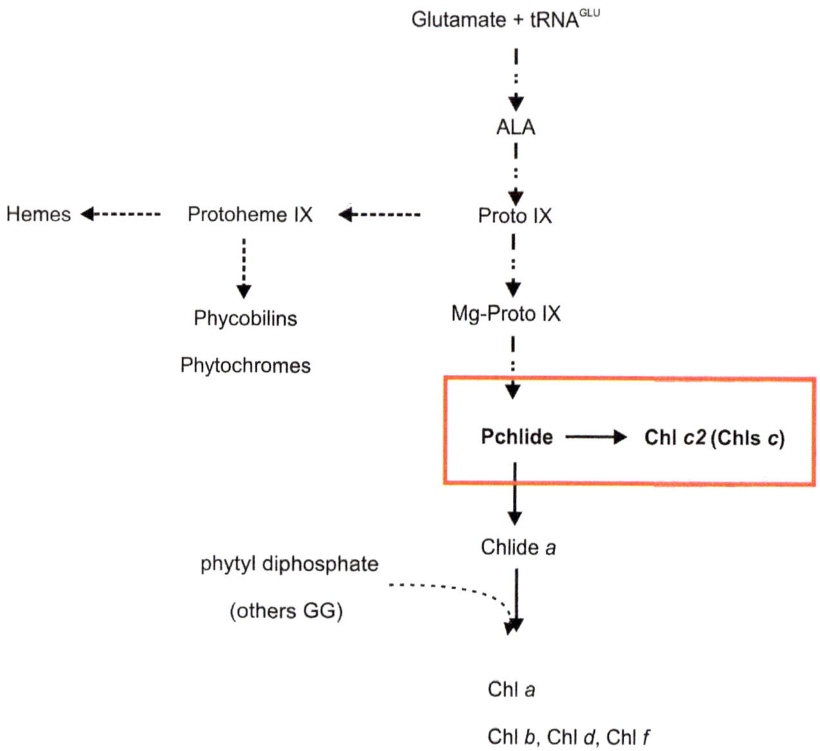

Fig. 3 The overall scheme of Chl *c* biosynthesis among other tetrapyrroles. Abbreviations: ALA, 5-aminolevulinic acid; Chlide, chlorophyllide; Pchlide, protochlorophyllide; proto IX, protoporphyrin IX; GG, geranylgeranyl.

The part of Chl *c* biosynthesis pathway shared with other tetrapyrroles (Fig. 3) starts from glutamate, which is activated by tRNA^{Glu} and then transformed into 5-aminolevulinic acid (ALA), which is the common precursor of all tetrapyrroles (Fig. 3). This three-step ALA biosynthesis (C_5 pathway) has been found in plants, algae, cyanobacteria and most photosynthetic bacteria (Beale, 1999; Huang, Wang, Gough, & Kannangara, 1984; Masuda & Fujita, 2008). The next part starting from ALA and resulting in protoporphyrin IX (Proto IX) formation consists of several reactions including the formation of porphobilinogen (a monopyrrole), followed by the head-to-tail condensation of four porphobilinogen, the ring closure and isomerization, decarboxylation of acetate side chains and two successive oxidation reactions. Proto IX is the substrate both for Chl and for heme synthesis via Mg-branch and Fe-branch, respectively. The Mg-branch starts with the incorporation of Mg^{2+} ion into the backbone of Proto IX, which is catalyzed by Mg chelatase. The incorporation of Mg is followed by the

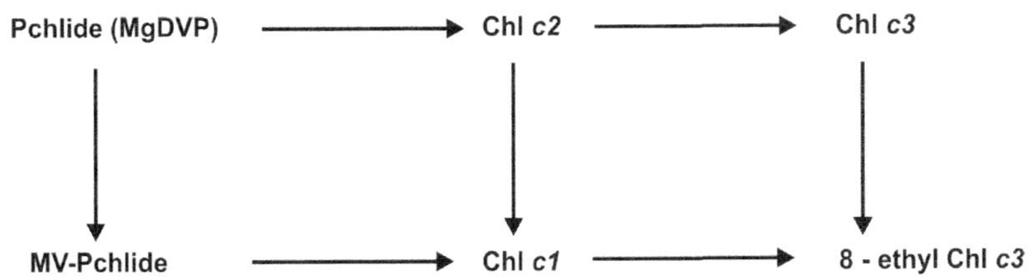

Fig. 4 A proposed scheme of the biosynthesis of the Chl *c* family.

methylation of Mg-protoporphyrin IX and the formation of the cyclopentanone ring, which is characteristic of phytoporphyrins. DV-Pchlide, which is formed in these reactions, is the substrate in the Chl *c* synthesis branch as well in the other branch, resulting in the synthesis of chlorine-type Chls.

Chl *c* biosynthesis, first proposed by Beale (1999), is one vaguely understood and it lacks experimental evidence (reviewed in Green, 2011; Zapata et al., 2006). It is assumed that DV-Pchlide is a key compound in Chl *c* biosynthesis (Álvarez et al., 2012; Ito & Tanaka, 2014); (Fig. 4).

DV-Pchlide is converted to MV-Pchlide by reducing the C8-vinyl group to an ethyl group, the reaction is catalyzed by 8-vinyl reductase. It has been suggested that MV- and DV-Pchlide are converted into Chl *c1* and Chl *c2*, respectively, by the dehydration of the propionyl side-chain to form a double bond of acrylate residue, although the respective oxidase (17^1 oxidase), which may be responsible for several reactions, is still hypothetical (Green, 2011). The mechanism of this dehydrogenation has until now only been observed in the synthesis of Chl *c1* analogs from Chl *a* (Xu & Tamiaki, 2014), while it has not been established in natural systems. However, it was also proposed that this reaction may occur at earlier stages of Chl biosynthesis (e.g., at the step of Proto IX synthesis) (Xu et al., 2016).

It was confirmed that Chl *c1* has an identical chemical structure to Pchlide *a* except the double bond in the side chain at C17, and that it is a competitive inhibitor of light-dependent Pchlide reductase (Helfrich et al., 2003), which catalyzes the light-triggered reduction of Pchlide to Chlide (for review see Gabruk & Mysliwa-Kurdziel, 2015).

7. Conclusions

Photosynthetic pigments which belong to the Chl *c* family are widely distributed among various groups of eukaryotic algae. Chls *c* are important constituents of pigment-protein light-harvesting complexes. Due to their specific structure and spectroscopic properties they allow light that penetrates underwater habitats to be harvested efficiently. The biosynthesis of

Chl *c* follows to a large extent that of Chl *a* and heme, although the reactions and enzymes involved in the final steps of their synthesis are not completely understood. Studies on the distribution of different species of the Chl *c* family within eukaryotic algae provide important information about their phylogenetic relationship and evolution.

Acknowledgment

This publication was financially supported by the Jagiellonian University (K/ZDS/008227).

References

Adl, S. M., Simpson, A. G. B., Farmer, M. A., Andersen, R. A., Anderson, R., Barta, J. R., et al. (2005). The new higher level classification of eukaryotes with emphasis on the taxonomy of protists. *Journal of Eukaryotic Microbiology*, *52*(5), 399–451. http://doi.org/https://doi.org/10.1111/j.1550-7408.2005.00053.

Akimoto, S., Teshigahara, A., Yokono, M., Mimuro, M., Nagao, R., & Tomo, T. (2014). Excitation relaxation dynamics and energy transfer in fucoxanthin-chlorophyll a/c-protein complexes, probed by time-resolved fluorescence. *Biochimica et Biophysica Acta*, *1837*, 1514–1521. https://doi.org/10.1016/j.bbabio.2014.02.002.

Álvarez, S., Rodríguez, F., Riobó, P., Garrido, J. L., & Vaz, B. (2013). Chlorophyll cCS-170 isolated from ostreococcus sp. is [7-methoxycarbonyl-8-vinyl]protochlorophyllide a. *Organic Letters*, *15*(17), 4430–4433. https://doi.org/10.1021/ol4019826.

Álvarez, S., Zapata, M., Garrido, J. L., & Vaz, B. (2012). Characterization of [8-ethyl]-chlorophyll c3 from *Emiliania huxleyi*. *Chemical Communications*, *48*(44), 5500–5502. https://doi.org/10.1039/C2CC31329D.

Beale, S. I. (1999). Enzymes of chlorophyll biosynthesis. *Photosynthesis Research*, *60*(1), 43–73. https://doi.org/10.1023/A:1006297731456.

Björn, L. O., Papageorgiou, G. C., Blankenship, R. E., & Govindjee (2009). A viewpoint: Why chlorophyll a? *Photosynthesis Research*, *99*(2), 85–98. https://doi.org/10.1007/s11120-008-9395-x.

Blankenship, R. E. (2014). *Molecular mechanisms of photosynthesis* (2nd ed.). Blackwell Science Ltd. https://doi.org/10.1002/9780470758472.

Bonetti, C., Alexandre, M. T. A., Van Stokkum, I. H. M., Hiller, R. G., Groot, M. L., Van Grondelle, R., et al. (2010). Identification of excited-state energy transfer and relaxation pathways in the peridinin-chlorophyll complex: An ultrafast mid-infrared study. *Physical Chemistry Chemical Physics*, *12*, 9256–9266. https://doi.org/10.1039/b923695c.

Brzezowski, P., Richter, A. S., & Grimm, B. (2015). Regulation and function of tetrapyrrole biosynthesis in plants and algae. *Biochimica et Biophysica Acta*, *1847*(9), 968–985. https://doi.org/10.1016/j.bbabio.2015.05.007.

Büchel, C. (2003). Fucoxanthin-chlorophyll proteins in diatoms: 18 and 19 kDa subunits assemble into different oligomeric states. *Biochemistry*, *42*(44), 13027–13034. https://doi.org/10.1021/bi0349468.

Büchel, C. (2018). Light-harvesting complexes in chlorophyll c-containing algae. In R. Croce, R. van Grondelle, H. van Amerongen, & I. van Stokkum (Eds.), *Light harvesting in photosynthesis* (1st ed., pp. 155–180). Taylor & Francis Group, LLC CRC Press is an imprint of Taylor & Francis Group, an Informa business.

Büchel, C., & Wilhelm, C. (1993). Isolation and characterization of a photosystem I-associated antenna (LHC I) and a photosystem I-core complex from the chlorophyll c-containing alga Pleurochloris meiringensis (Xanthophyceae). *Journal of Photochemistry and Photobiology, B: Biology*, *20*, 87–93. https://doi.org/10.1016/1011-1344(93)80136-W.

Burki, F., Shalchian-Tabrizi, K., Minge, M., Skjæveland, Å., Nikolaev, S. I., Jakobsen, K. S., et al. (2007). Phylogenomics reshuffles the eukaryotic supergroups. *PLoS One*, *2*(8), e790. https://doi.org/10.1371/journal.pone.0000790.

Burki, F., Shalchian-Tabrizi, K., & Pawlowski, J. (2008). Phylogenomics reveals a new "megagroup" including most photosynthetic eukaryotes. *Biology Letters*, *4*(4), 366–369. https://doi.org/10.1098/rsbl.2008.0224.

Butkus, V., Gelzinis, A., Augulis, R., Gall, A., Büchel, C., Robert, B., et al. (2015). Coherence and population dynamics of chlorophyll excitations in FCP complex: Two-dimensional spectroscopy study. *The Journal of Chemical Physics*, *142*(21) 212414. https://doi.org/10.1063/1.4914098.

Cavalier-Smith, T. (1981). Eukaryote kingdoms: Seven or nine? *Biosystems*, *14*(3), 461–481. https://doi.org/10.1016/0303-2647(81)90050-2.

Cavalier-Smith, T. (2000). Membrane heredity and early chloroplast evolution. *Trends in Plant Science*, *5*(4), 174–182. https://doi.org/10.1007/978-3-319-06947-0_19.

Cavalier-Smith, T. (2006). The tiny enslaved genome of a rhizarian alga. *Proceedings of the National Academy of Sciences of the United States of America*, *103*(25), 9379–9380. https://doi.org/10.1073/pnas.0603505103.

Cavalier-Smith, T. (2010). Kingdoms Protozoa and Chromista and the eozoan root of the eukaryotic tree. *Biology Letters*, *6*, 342–345. https://doi.org/10.1098/rsbl.2009.0948.

Cavalier-Smith, T. (2018). Kingdom Chromista and its eight phyla: A new synthesis emphasising periplastid protein targeting, cytoskeletal and periplastid evolution, and ancient divergences. *Protoplasma*, *255*(1), 297–357. https://doi.org/10.1007/s00709-017-1147-3.

Chen, M. (2014). Chlorophyll modifications and their spectral extension in oxygenic photosynthesis. *Annual Review of Biochemistry*, *83*, 317–340. https://doi.org/10.1146/annurev-biochem-072711-162943.

Chen, M., & Blankenship, R. E. (2018). Pigments: General properties and biosynthesis. In R. Croce, R. van Grondelle, H. van Amerongen, & I. van Stokkum (Eds.), *Light harvesting in photosynthesis* (1st ed., p. 7). Taylor & Francis Group, LLC CRC Press is an imprint of Taylor & Francis Group, an Informa business.

Chesnick, J. M., Kooistra, W. H. C. F., Wellbrock, U., & Medlin, L. K. (1997). Ribosomal RNA analysis indicates a benthic pennate diatom ancestry for the endosymbionts of the dinoflagellates Peridinium foliaceum and Peridinium balticum (Pyrrhophyta). *Journal of Eukaryotic Microbiology*, *44*, 314–320. https://doi.org/10.1111/j.1550-7408.1997.tb05672.x.

Cock, J., Peters, A., & Coelho, S. (2011). Brown algae. *Current Biology*, *21*(15), R573–R575.

Croce, R., & Van Amerongen, H. (2014). Natural strategies for photosynthetic light harvesting. *Nature Chemical Biology*, *10*, 492–501. https://doi.org/10.1038/nchembio.1555.

Csűrös, M., Rogozin, I. B., & Koonin, E. V. (2008). Extremely intron-rich genes in the alveolate ancestors inferred with a flexible maximum-likelihood approach. *Molecular Biology and Evolution*, *25*(5), 903–911.

Czarnecki, O., & Grimm, B. (2012). Post-translational control of tetrapyrrole biosynthesis in plants, algae, and cyanobacteria. *Journal of Experimental Botany*, *63*(4), 1675–1687. https://doi.org/10.1093/jxb/err437.

De Martino, A., Douady, D., Quinet-Szely, M., Rousseau, B., Crépineau, F., Apt, K., et al. (2000). The light-harvesting antenna of brown algae: Highly homologous proteins encoded by a multigene family. *European Journal of Biochemistry*, *267*, 5540–5549. https://doi.org/10.1046/j.1432-1327.2000.01616.x.

Deane, J. A., Fraunholz, M., Su, V., Maier, U. G., Martin, W., Durnford, D. G., et al. (2000). Evidence for nucleomorph to host nucleus gene transfer: Light-harvesting complex proteins from cryptomonads and chlorarachniophytes. *Protist*, *151*(3), 239–252. https://doi.org/10.1078/1434-4610-00022.

Dekker, J. P., & Boekema, E. J. (2005). Supramolecular organization of thylakoid membrane proteins in green plants. *Biochimica et Biophysica Acta*, *1706*, 12–39. https://doi.org/10.1016/j.bbabio.2004.09.009.

Di Valentin, M., Salvadori, E., Agostini, G., Biasibetti, F., Ceola, S., Hiller, R., et al. (2010). Triplet-triplet energy transfer in the major intrinsic light-harvesting complex of *Amphidinium carterae* as revealed by ODMR and EPR spectroscopies. *Biochimica et Biophysica Acta*, *1997*, 1759–1767. https://doi.org/10.1016/j.bbabio.2010.06.011.

D'Orazio, N., Gemello, E., Gammone, M. A., De Girolamo, M., Ficoneri, C., & Riccioni, G. (2012). Fucoxantin: A treasure from the sea. *Marine Drugs*, *10*, 604–616. https://doi.org/10.3390/md10030604.

Doust, A. B., Marai, C. N. J., Harrop, S. J., Wilk, K. E., Curmi, P. M. G., & Scholes, G. D. (2004). Developing a structure-function model for the cryptophyte phycoerythrin 545 using ultrahigh resolution crystallography and ultrafast laser spectroscopy. *Journal of Molecular Biology*, *344*, 135–153. https://doi.org/10.1016/j.jmb.2004.09.044.

Durnford, D. G., Deane, J. A., Tan, S., McFadden, G. I., Gantt, E., & Green, B. R. (1999). A phylogenetic assessment of the eukaryotic light-harvesting antenna proteins, with implications for plastid evolution. *Journal of Molecular Evolution*, *48*, 59–68. https://doi.org/10.1007/PL00006445.

Eggink, L., LoBrutto, R., & Hoober, K. (2004). Chlorophylls b and c: Why plants make them. series Ed. In W. M. Nelson (Ed.), *ACS symposium series*: *Vol. 887*. *Agricultural applications in green chemistry* (pp. 23–37). American Chemical Society. chapter 3. https://doi.org/10.1021/bk-2004-0887.ch003.

Falkowski, P., & Knoll, A. H. (2007). *Evolution of primary producers in the sea*. Elsevier. https://doi.org/10.1016/B978-0-12-370518-1.X5000-0.

Fawley, M. W. (1989). Detection of chlorophylls c1, c2 and c3 in pigment extracts of *Prymnesium parvum* (Prymnesiophyte). *Journal of Phycology*, *25*(3), 601–604. https://doi.org/10.1111/j.1529-8817.1989.tb00267.x.

Fawley, M. W., & Fawley, K. P. (2004). A simple and rapid technique for the isolation of DNA from microalgae. *Journal of Phycology*, *40*(1), 223–225. https://doi.org/10.1111/j.0022-3646.2004.03-081.x.

Fujita, Y., Tsujimoto, R., & Aoki, R. (2015). Evolutionary aspects and regulation of tetrapyrrole biosynthesis in cyanobacteria under aerobic and anaerobic environments. *Life (Basel, Switzerland)*, *5*(2), 1172–1203. https://doi.org/10.3390/life5021172.

Gabruk, M., & Mysliwa-Kurdziel, B. (2015). Light-dependent protochlorophyllide oxidoreductase: Phylogeny, regulation, and catalytic properties. *Biochemistry*, *54*(34), 5255–5262. https://doi.org/10.1021/acs.biochem.5b00704.

Gardian, Z., Tichý, J., & Vácha, F. (2011). Structure of PSI, PSII and antennae complexes from yellow-green alga Xanthonema debile. *Photosynthesis Research*, *108*, 25–32. https://doi.org/10.1007/s11120-011-9647-z.

Garrido, J. L., Otero, J., Maestro, M. A., & Zapata, M. (2000). The main nonpolar chlorophyll c from *Emiliania huxleyi* (Prymnesiophyceae) is a chlorophyll c2-monogalactosyldiacylglyceride ester: A mass spectrometry study. *Journal of Phycology*, *36*, 497–505. https://doi.org/10.1046/j.1529-8817.2000.99135.x.

Garrido, J. L., & Zapata, M. (1998). Detection of new pigments from *Emiliania huxleyi* (Prymnesiophyceae) by high-performance liquid chromatography, liquid chromatography-mass spectrometry, visible spectroscopy, and fast atom bombardment mass spectrometry. *Journal of Phycology*, *34*, 70–78. https://doi.org/10.1046/j.1529-8817.1998.340070.x.

Garrido, J., & Zapata, M. (2006). Chlorophyll analysis by new high performance liquid chromatography methods. In B. Grimm, R. J. Porra, W. Rüdiger, & H. Scheer (Eds.), *Advances in photosynthesis and respiration*: *Vol. 25*. *Chlorophylls and bacteriochlorophylls* (pp. 109–121). Dordrecht: Springer. https://doi.org/10.1007/1-4020-4516-6_8.

Gelzinis, A., Butkus, V., Songaila, E., Augulis, R., Gall, A., Büchel, C., et al. (2015). Mapping energy transfer channels in fucoxanthin-chlorophyll protein complex. *Biochimica et Biophysica Acta, 1847*, 241–247. https://doi.org/10.1016/j.bbabio.2014.11.004.

Georgakopoulou, S., Van Der Zwan, G., Bassi, R., Van Grondelle, R., Van Amerongen, H., & Croce, R. (2007). Understanding the changes in the circular dichroism of light harvesting complex II upon varying its pigment composition and organization. *Biochemistry, 46*(16), 4745–4754. https://doi.org/10.1021/bi062031y.

Gildenhoff, N., Amarie, S., Gundermann, K., Beer, A., Büchel, C., & Wachtveitl, J. (2010). Oligomerization and pigmentation dependent excitation energy transfer in fucoxanthin-chlorophyll proteins. *Biochimica et Biophysica Acta, 1797*, 543–549. https://doi.org/10.1016/j.bbabio.2010.01.024.

Goss, R., Wilhelm, C., & Garab, G. (2000). Organization of the pigment molecules in the chlorophyll a/b/c containing alga *Mantoniella squamata* (Prasinophyceae) studied by means of absorption, circular and linear dichroism spectroscopy. *Biochimica et Biophysica Acta, 1457*, 190–199. https://doi.org/10.1016/S0005-2728(00)00101-8.

Granick, S. (1949). The pheoporphyrin nature of chlorophyll c. *Journal of Biological Chemistry, 179*, 505.

Green, B. R. (2011). After the primary endosymbiosis: An update on the chromalveolate hypothesis and the origins of algae with Chl c. *Photosynthesis Research, 107*, 103–115. https://doi.org/10.1007/s11120-010-9584-2.

Hallegraeff, G. M., Blackburn, S. I., Roy, S., Vernet, M., & Wright, S. . (2012). Tribute to Shirley Jeffrey: 50 years of research on chlorophyll c. *Phycologia, 51*(2), 123–125. https://doi.org/10.2216/0031-8884-51.2.123.

Hampl, V., Hug, L., Leigh, J. W., Dacks, J. B., Lang, B. F., Simpson, A. G. B., et al. (2009). Phylogenomic analyses support the monophyly of excavata and resolve relationships among eukaryotic "supergroups" *Proceedings of the National Academy of Sciences of the United States of America, 106*(10), 3859–3864. https://doi.org/10.1073/pnas.0807880106.

Helfrich, M., Bommer, B., Oster, U., Klement, H., Mayer, K., Larkum, A. W. D., et al. (2003). Chlorophylls of the c family: Absolute configuration and inhibition of NADPH:protochlorophyllide oxidoreductase. *Biochimica et Biophysica Acta, 1605*(1–3), 97–103. https://doi.org/10.1016/S0005-2728(03)00081-1.

Hiller, R. G., Wrench, P. M., Gooley, A. P., Shoebridge, G., & Breton, J. (1993). The major intrinsic light-harvesting protein of Amphidinium: Characterization and relation to other light-harvesting proteins. *Photochemistry and Photobiology, 57*, 125–131. https://doi.org/10.1111/j.1751-1097.1993.tb02267.x.

Hiller, R. G., Wrench, P. M., & Sharples, F. P. (1995). The light-harvesting chlorophyll a-c-binding protein of dinoflagellates: A putative polyprotein. *FEBS Letters, 363*, 175–178. https://doi.org/10.1016/0014-5793(95)00297-M.

Hirakawa, Y. (2014). Complex plastids of chlorarachniophyte algae. *Perspectives in Phycology, 1*(2), 87–92.

Hobe, S., Förster, R., Klingler, J., & Paulsen, H. (1995). N-proximal sequence motif in light-harvesting chlorophyll a/b-binding protein is essential for the trimerization of light-harvesting chlorophyll a/b complex. *Biochemistry, 34*, 10224–10228. https://doi.org/10.1021/bi00032a016.

Hofmann, E., Wrench, P. M., Sharples, F. P., Hiller, R. G., Welte, W., & Diederichs, K. (1996). Structural basis of light harvesting by carotenoids: Peridinin-chlorophyll-protein from *Amphidinium carterae*. *Science, 272*, 1788–1791. https://doi.org/10.1126/science.272.5269.1788.

Hopkins, J. F., Spencer, D. F., Laboissiere, S., Neilson, J. A. D., Eveleigh, R. J. M., Durnford, D. G., et al. (2012). Proteomics reveals plastid- and periplastid-targeted proteins in the chlorarachniophyte alga *Bigelowiella natans*. *Genome Biology and Evolution, 4*, 1391–1406. https://doi.org/10.1093/gbe/evs115.

Hu, Q., Miyashita, H., Iwasaki, I., Kurano, N., Miyachi, S., Iwaki, M., et al. (1998). A photosystem I reaction center driven by chlorophyll d in oxygenic photosynthesis. *Proceedings of the National Academy of Sciences of the United States of America*, *95*(22), 13319–13323. https://doi.org/10.1073/pnas.95.22.13319.

Huang, D. D., Wang, W. Y., Gough, S. P., & Kannangara, C. G. (1984). Delta-aminolevulinic acid-synthesizing enzymes need an RNA moiety for activity. *Science*, *225*(4669), 1482–1484. https://doi.org/10.1126/science.6206568.

Iglesias-Prieto, R., Govind, N. S., & Trench, R. K. (1991). Apoprotein composition and spectroscopic characterization of the water-soluble peridinin-chlorophyll a-proteins from three symbiotic dinoflagellates. *Proceedings of the Royal Society B: Biological Sciences*, *246*, 275–283. https://doi.org/10.1098/rspb.1991.0155.

Ito, H., & Tanaka, A. (2014). Evolution of a new chlorophyll metabolic pathway driven by the dynamic changes in enzyme promiscuous activity. *Plant and Cell Physiology*, *55*(3), 593–603. https://doi.org/10.1093/pcp/pct203.

Iwai, M., Takahashi, Y., & Minagawa, J. (2008). Molecular remodeling of photosystem II during state transitions in *Chlamydomonas reinhardtii*. *Plant Cell*, *20*, 2177–2189. https://doi.org/10.1105/tpc.108.059352.

Jahns, P., Latowski, D., & Strzalka, K. (2009). Mechanism and regulation of the violaxanthin cycle: The role of antenna proteins and membrane lipids. *Biochimica et Biophysica Acta*, *1787*, 3–14. https://doi.org/10.1016/j.bbabio.2008.09.013.

Janouskovec, J., Horak, A., Obornik, M., Lukes, J., & Keeling, P. J. (2010). A common red algal origin of the apicomplexan, dinoflagellate, and heterokont plastids. *Proceedings of the National Academy of Sciences of the United States of America*, *107*(24), 10949–10954. https://doi.org/10.1073/pnas.1003335107.

Jeffrey, S. W. W., & Humphrey, G. F. F. (1975). New spectrophotometric equations for determining chlorophylls a, b, c1 and c2 in higher plants, algae and natural phytoplankton. *Biochemie und Physiologie der Pflanzen*, *167*(2), 191–194. https://doi.org/10.1016/S0015-3796(17)30778-3.

Jeffrey, S. W., & Vesk, M. (1997). Introduction to marine phytoplankton and their pigment signatures. In S. W. Jeffrey, R. F. C. Mantoura, & S. W. Wright (Eds.), *Phytoplankton pigments in oceanography: Guidelines to modern methods.* UNESCO Publishing. https://doi.org/10.1016/0891-5849(94)E0090-6.

Jiang, J., Zhang, H., Orf, G., Lu, Y., Xu, W., Harrington, L., et al. (2014). Evidence of functional trimeric chlorophyll a/c2-peridinin proteins in the dinoflagellate symbiodinium. *Biochimica et Biophysica Acta*, *1837*, 1904–1912.

Jovine, R. V. M., Johnsen, G., & Prézelin, B. B. (1995). Isolation of membrane bound light-harvesting-complexes from the dinoflagellates Heterocapsa pygmaea and Prorocentrum minimum. *Photosynthesis Research*, *44*, 127–138. https://doi.org/10.1007/BF00018303.

Kantz, T. S., Theriot, E. C., Zimmer, E. A., & Chapman, R. L. (1990). The pleurastrophyceae and micromonadophyceae: A cladistic analysis of nuclear rRNA sequence data. *Journal of Phycology*, *26*, 711–721. https://doi.org/10.1111/j.0022-3646.1990.00711.x.

Kargul, J., Turkina, M. V., Nield, J., Benson, S., Vener, A. V., & Barber, J. (2005). Light-harvesting complex II protein CP29 binds to photosystem I of *Chlamydomonas reinhardtii* under State 2 conditions. *FEBS Journal*, *272*, 4797–4806. https://doi.org/10.1111/j.1742-4658.2005.04894.x.

Katoh, T., Mimuro, M., & Takaichi, S. (1989). Light-harvesting particles isolated from a brown alga, *Dictyota dichotoma*. A supramolecular assembly of fucoxanthin-chlorophyll-protein complexes. *Biochimica et Biophysica Acta*, *976*, 233–240. https://doi.org/10.1016/S0005-2728(89)80235-X.

Kim, E., & Graham, L. E. (2008). EEF2 analysis challenges the monophyly of Archaeplastida and Chromalveolata. *PLoS One*, *3*(7), e2621. https://doi.org/10.1371/journal.pone.0002621.

Kirk, J. T. O. (2010). *Light and photosynthesis in aquatic ecosystems* (3rd ed.). Cambridge: Cambridge University Press. https://doi.org/10.1017/CBO9781139168212.

Kosumi, D., Fujii, R., Sugisaki, M., Oka, N., Iha, M., & Hashimoto, H. (2014). Characterization of the intramolecular transfer state of marine carotenoid fucoxanthin by femtosecond pump-probe spectroscopy. *Photosynthesis Research, 121*(1), 61–68. https://doi.org/10.1007/s11120-014-9995-6.

Kosumi, D., Kusumoto, T., Fujii, R., Sugisaki, M., Iinuma, Y., Oka, N., et al. (2011a). Ultrafast excited state dynamics of fucoxanthin: Excitation energy dependent intramolecular charge transfer dynamics. *Physical Chemistry Chemical Physics, 13*(22), 10762–10770. https://doi.org/10.1039/c0cp02568b.

Kosumi, D., Kusumoto, T., Fujii, R., Sugisaki, M., Iinuma, Y., Oka, N., et al. (2011b). Ultrafast S1and ICT state dynamics of a marine carotenoid probed by femtosecond one- and two-photon pump-probe spectroscopy. *Journal of Luminescence, 131*(3), 515–518. https://doi.org/10.1016/j.jlumin.2010.09.018.

Koziol, A. G., Borza, T., Ishida, K.-I., Keeling, P., Lee, R. W., & Durnford, D. G. (2007). Tracing the evolution of the light-harvesting antennae in chlorophyll a/b-containing organisms. *Plant Physiology, 143*, 1802–1816. https://doi.org/10.1104/pp.106.092536.

Kraay, G. W., Zapata, M., & Veldhuis, M. J. W. (1992). Separation of chlorophylls c1, c2 and c3 of marine phytoplankton by reversed-phase-C18-high-performance liquid chromatography. *Journal of Phycology, 28*, 708–712. https://doi.org/10.1111/j.0022-3646.1992.00708.x.

Larkum, A. W. D. (2006). The evolution of chlorophylls and photosynthesis. In B. Grimm, R. J. Porra, W. Rüdiger, & H. Scheer (Eds.), *Advances in photosynthesis and respiration: Vol. 25. Chlorophylls and bacteriochlorophylls* (pp. 261–282). Dordrecht: Springer. http://doi.org/13 978-1-4020-4515-8.

Larkum, A. W. D., Ritchie, R. J., & Raven, J. A. (2018). Living off the Sun: Chlorophylls, bacteriochlorophylls and rhodopsins. *Photosynthetica, 56*(1), 11–43.

Larkum, A. W. D., Scaramuzzi, C., Cox, G. C., Hiller, R. G., & Turner, A. G. (1994). Light-harvesting chlorophyll c-like pigment in Prochloron. *Proceedings of the National Academy of Sciences of the United States of America, 91*, 678–683. https://doi.org/10.1073/pnas.91.2.679.

Latasa, M., Scharek, R., Le Gall, F., & Guillou, L. (2004). Pigment suites and taxonomic groups in Prasinophyceae. *Journal of Phycology, 40*, 1149–1155. https://doi.org/10.1111/j.1529-8817.2004.03136.x.

Lee, R. E. (2008). *Phycology* (4th ed., p. 317). London: Cambridge University Press.

Leliaert, F., Tronholm, A., Lemieux, C., Turmel, M., Depriest, M. S., Bhattacharya, D., et al. (2016). Chloroplast phylogenomic analyses reveal the deepest-branching lineage of the Chlorophyta, Palmophyllophyceae class. nov. *Scientific Reports, 6*, 25367. https://doi.org/10.1038/srep25367.

Lichtlé, C., Arsalane, W., Duval, J. C., & Passaquet, C. (1995). Characterization of the light-harvesting complex of Giraudyopis stellifer (Chrysophyceae) and effects of light stress. *Journal of Phycology, 31*, 380–387. https://doi.org/10.1111/j.0022-3646.1995.00380.x.

Liu, Z., Yan, H., Wang, K., Kuang, T., Zhang, J., Gui, L., et al. (2004). Crystal structure of spinach major light-harvesting complex at 2.72 Å resolution. *Nature, 428*(6980), 287–292. http://doi.org/http://www.nature.com/nature/journal/v428/n6980/suppinfo/nature02373_S1.html.

Lohr, M., & Wilhelm, C. (2001). Xanthophyll synthesis in diatoms: Quantification of putative intermediates and comparison of pigment conversion kinetics with rate constants derived from a model. *Planta, 212*, 382–391. https://doi.org/10.1007/s004250000403.

Mackey, D. J., Higgins, H. W., Mackey, M. D., & Holdsworth, D. (1998). Algal class abundances in the western equatorial Pacific: Estimation from HPLC measurements of chloroplast pigments using CHEMTAX. *Deep-Sea Research Part I: Oceanographic Research Papers*, *45*, 1441–1468. https://doi.org/10.1016/S0967-0637(98)00025-9.

Masuda, T., & Fujita, Y. (2008). Regulation and evolution of chlorophyll metabolism. *Photochemical & Photobiological Sciences*, 7(10), 1131–1149. https://doi.org/10.1039/b807210h.

Matsuo, E., & Inagaki, Y. (2018). Patterns in evolutionary origins of heme, chlorophyll a and isopentenyl diphosphate biosynthetic pathways suggest non-photosynthetic periods prior to plastid replacements in dinoflagellates. *PeerJ*, *6*, e5345. https://doi.org/10.7717/peerj.5345.

Mendes, C. R. B., Tavano, V. M., Dotto, T. S., Kerr, R., de Souza, M. S., Garcia, C. A. E., et al. (2018). New insights on the dominance of cryptophytes in Antarctic coastal waters: A case study in Gerlache Strait. *Deep-Sea Research Part II*, *149*, 161–170. https://doi.org/10.1016/j.dsr2.2017.02.010.

Mimuro, M., Katoh, T., & Kawai, H. (1990). Spatial arrangement of pigments and their interaction in the fucoxanthin-chlorophyll ac protein assembly (FCPA) isolated from the brown alga *Dictyota dichotoma*. Analysis by means of polarized spectroscopy. *Biochimica et Biophysica Acta*, *1015*(3), 450–456. https://doi.org/10.1016/0005-2728(90)90078-I.

Mirkovic, T., Ostroumov, E. E., Anna, J. M., Van Grondelle, R., Govindjee, & Scholes, G. D. (2017). Light absorption and energy transfer in the antenna complexes of photosynthetic organisms. *Chemical Reviews*, *117*, 249–293. https://doi.org/10.1021/acs.chemrev.6b00002.

Miyashita, H., Adachi, K., Kurano, N., Ikemoto, H., Chihara, M., & Miyachi, S. (1997). Pigment composition of a novel oxygenic photosynthetic prokaryote containing chlorophyll d as the major chlorophyll. *Plant and Cell Physiology*, *38*(3), 274–281. https://doi.org/10.1093/oxfordjournals.pcp.a029163.

Mizoguchi, T., Kimura, Y., & Tamiaki, H. (2010). Exclusive observation of the (132R)-enantiomer of chlorophyll-c from a diatom Chaetoseros calcitrans. *Photochemistry and Photobiology*, *86*(2), 311–315. https://doi.org/10.1111/j.1751-1097.2009.00667.x.

Mizoguchi, T., Kimura, Y., Yoshitomi, T., & Hitoshi, T. (2011). The stereochemistry of chlorophyll-c3 from the haptophyte *Emiliania huxleyi*: The (132R)-enantiomers of chlorophylls-c are exclusively selected as the photosynthetically active pigments in chromophyte algae. *Biochimica et Biophysica Acta*, *1807*(11), 1463–1473.

Mizoguchi, T., Nagai, C., Kunieda, M., Kimura, Y., Okamura, A., & Tamiaki, H. (2009). Stereochemical determination of the unique acrylate moiety at the 17-position in chlorophylls-c from a diatom Chaetoseros calcitrans and its effect upon electronic absorption properties. *Organic and Biomolecular Chemistry*, 7(10), 2120–2126. https://doi.org/10.1039/b900802k.

Mochizuki, N., Tanaka, R., Grimm, B., Masuda, T., Moulin, M., Smith, A. G., et al. (2010). The cell biology of tetrapyrroles: A life and death struggle. *Trends in Plant Science*, *15*(9), 488–498. https://doi.org/10.1016/j.tplants.2010.05.012.

Nagao, R., Yokono, M., Akimoto, S., & Tomo, T. (2013). High excitation energy quenching in fucoxanthin chlorophyll a/c-binding protein complexes from the diatom Chaetoceros gracilis. *Journal of Physical Chemistry B*, *117*, 6888–6895. https://doi.org/10.1021/jp403923q.

Neilson, J. A. D., & Durnford, D. G. (2010). Structural and functional diversification of the light-harvesting complexes in photosynthetic eukaryotes. *Photosynthesis Research*, *106*, 57–71. https://doi.org/10.1007/s11120-010-9576-2.

Niedzwiedzki, D. M., & Blankenship, R. E. (2010). Singlet and triplet excited state properties of natural chlorophylls and bacteriochlorophylls. *Photosynthesis Research*, *106*(3), 227–238. https://doi.org/10.1007/s11120-010-9598-9.

Niedzwiedzki, D. M., Jiang, J., Lo, C. S., & Blankenship, R. E. (2014). Spectroscopic properties of the chlorophyll a-chlorophyll c2-peridinin-protein-complex (acpPC) from the coral symbiotic dinoflagellate Symbiodinium. *Photosynthesis Research*, *120*, 125–139. https://doi.org/10.1007/s11120-013-9794-5.

Ocampo-Alvarez, H., García-Mendoza, E., & Govindjee (2013). Antagonist effect between violaxanthin and de-epoxidated pigments in nonphotochemical quenching induction in the qE deficient brown alga *Macrocystis pyrifera*. *Biochimica et Biophysica Acta*, *1827*, 427–437. https://doi.org/10.1016/j.bbabio.2012.12.006.

Pan, H., Šlapeta, J., Carter, D., & Chen, M. (2012). Phylogenetic analysis of the light-harvesting system in *Chromera velia*. *Photosynthesis Research*, *111*, 19–28. https://doi.org/10.1007/s11120-011-9710-9.

Papagiannakis, E., Van Stokkum, I. H. M., Fey, H., Büchel, C., & Van Grondelle, R. (2005). Spectroscopic characterization of the excitation energy transfer in the fucoxanthin-chlorophyll protein of diatoms. *Photosynthesis Research*, *86*(1–2), 241–250. https://doi.org/10.1007/s11120-005-1003-8.

Papagiannakis, E., Vengris, M., Larsen, D. S., Van Stokkum, I. H. M., Hiller, R. G., & Van Grondelle, R. (2006). Use of ultrafast dispersed pump-dump-probe and pump-repump-probe spectroscopies to explore the light-induced dynamics of peridinin in solution. *Journal of Physical Chemistry B*, *110*(1), 512–521. https://doi.org/10.1021/jp053094d.

Pascal, A. A., Caron, L., Rousseau, B., Lapouge, K., Duval, J. C., & Robert, B. (1998). Resonance Raman spectroscopy of a light-harvesting protein from the brown alga *Laminaria saccharina*. *Biochemistry*, *37*, 2450–2457. https://doi.org/10.1021/bi9719657.

Polívka, T., & Frank, H. A. (2010). Molecular factors controlling photosynthetic light harvesting by carotenoids. *Accounts of Chemical Research*, *43*(8), 1125–1134. https://doi.org/10.1021/ar100030m.

Polívka, T., Van Stokkum, I. H. M., Zigmantas, D., Van Grondelle, R., Sundström, V., & Hiller, R. G. (2006). Energy transfer in the major intrinsic light-harvesting complex from *Amphidinium carterae*. *Biochemistry*, *45*(28), 8516–8526. https://doi.org/10.1021/bi060265b.

Qiu, H., Yoon, H. S., & Bhattacharya, D. (2013). Algal endosymbionts as vectors of horizontal gene transfer in photosynthetic eukaryotes. *Frontiers in Plant Science*, *4*(366), 1–8. https://doi.org/10.3389/fpls.2013.00366.

Rebeiz, C. A. (2014). *Chlorophyll biosynthesis and technological applications*. Springer.

Reinfelder, J. R. (2011). Carbon concentrating mechanisms in eukaryotic marine phytoplankton. *Annual Review of Marine Science*, *3*, 291–315. https://doi.org/10.1146/annurev-marine-120709-142720.

Riley, J., & Wilson, T. (1967). The pigments of some marine phytoplankton species. *Journal of the Marine Biological Association of the UK*, *47*, 351–361.

Rowan, R. (1998). Diversity and ecology of zooxanthellae on coral reefs. *Journal of Phycology*, *36*, 407–417. https://doi.org/10.1046/j.1529-8817.1998.340407.x.

Scheer, H. (1991). Chlorophylls. In H. Scheer (Ed.), *Chlorophylls* (pp. 3–33). Boca Raton: CRC Press.

Scheer. (2006). An overview of chlorophylls and bacteriochlorophylls: Biochemistry, biophysics, functions and applications. In B. Grimm, R. J. Porra, W. Rüdiger, & H. Scheer (Eds.), *Advances in photosynthesis and respiration: Vol. 25. Chlorophylls and bacteriochlorophylls* (pp. 1–26)Dordrecht: Springer. https://doi.org/10.1007/1-4020-4516-6_1.

Schliep, M., Chen, M., Larkum, A., & Quinnell, R. (2008). The function of MgDVP in a chlorophyll d-containing organism. In J. F. Allen, E. Gantt, J. H. Golbeck, & B. Osmond (Eds.), *Photosynthesis. Energy from the sun* (pp. 1125–1128). Dordrecht: Springer Netherlands.

Schlüter, L., Møhlenberg, F., Havskum, H., & Larsen, S. (2000). The use of phytoplankton pigments for identifying and quantifying phytoplankton groups in coastal areas: Testing the influence of light and nutrients on pigment/chlorophyll a ratios. *Marine Ecology Progress Series*, *192*, 49–63. https://doi.org/10.3354/meps192049.

Šlouf, V., Fuciman, M., Johanning, S., Hofmann, E., Frank, H. A., & Polívka, T. (2013). Low-temperature time-resolved spectroscopic study of the major light-harvesting complex of *Amphidinium carterae*. *Photosynthesis Research*, *117*(2013), 257–265. https://doi.org/10.1007/s11120-013-9900-8.

Songaila, E., Augulis, R., Gelzinis, A., Butkus, V., Gall, A., Büchel, C., et al. (2013). Ultrafast energy transfer from chlorophyll c2 to chlorophyll a in fucoxanthin-chlorophyll protein complex. *Journal of Physical Chemistry Letters*, *4*, 3590–3595. https://doi.org/10.1021/jz401919k.

Sorby, H. (1873). On comparative vegetable chromatology. *Proceedings of the Royal Society of London*, *21*, 442–483.

Standfuss, J., Terwisscha van Scheltinga, A. C., Lamborghini, M., & Kühlbrandt, W. (2005). Mechanisms of photoprotection and nonphotochemical quenching in pea light-harvesting complex at 2.5 Å resolution. *The EMBO Journal*, *24*(5), 919–928. https://doi.org/10.1038/sj.emboj.7600585.

Stoń-Egiert, J., Majchrowski, R., Darecki, M., Kosakowska, A., & Ostrowska, M. (2012). Influence of underwater light fields on pigment characteristics in the Baltic Sea—Results of statistical analysis. *Oceanologia*, *54*(1), 7–27. https://doi.org/10.5697/oc.54-1.007.

Strain, H. H., & Manning, M. (1942). Chlorofucine (chlorophyll γ), a green pigment of diatoms and brown algae. *Journal of Biological Chemistry*, *144*, 625–636.

Suzuki, S., Hirakawa, Y., Kofuji, R., Sugita, M., & Ishida, K. I. (2016). Plastid genome sequences of Gymnochlora stellata, Lotharella vacuolata, and Partenskyella glossopodia reveal remarkable structural conservation among chlorarachniophyte species. *Journal of Plant Research*, *129*, 581–590. https://doi.org/10.1007/s10265-016-0804-5.

Suzuki, K., Kamimura, A., & Hooker, B. (2015). Rapid and highly sensitive analysis of chlorophylls and carotenoids from marine phytoplankton using ultra-high performance liquid chromatography (UHPLC) with the first derivative spectrum chromatogram (FDSC) technique. *Marine Chemistry*, *176*, 96–109. https://doi.org/10.1016/j.marchem.2015.07.010.

Swingley, W. D., Iwai, M., Chen, Y., Ozawa, S. i., Takizawa, K., Takahashi, Y., et al. (2010). Characterization of photosystem I antenna proteins in the prasinophyte *Ostreococcus tauri*. *Biochimica et Biophysica Acta*, *1797*(8), 1458–1464. https://doi.org/10.1016/j.bbabio.2010.04.017.

Takaichi, S. (2011). Carotenoids in algae: Distributions, biosyntheses and functions. *Marine Drugs*, *9*(6), 1101–1118. https://doi.org/10.3390/md9061101.

Takano, Y., Hansen, G., Fujita, D., & Horiguchi, T. (2008). Serial replacement of diatom endosymbionts in two freshwater dinoflagellates, Peridiniopsis spp. (Peridiniales, Dinophyceae). *Phycologia*, *47*, 41–53. https://doi.org/10.2216/07-36.1.

Taylor, F. J. R., Hoppenrath, M., & Saldarriaga, J. F. (2008). Dinoflagellate diversity and distribution. *Biodiversity and Conservation*, *17*, 407–418. https://doi.org/10.1007/s10531-007-9258-3.

Tengs, T., Dahlberg, O. J., Shalchian-Tabrizi, K., Klaveness, D., Rudi, K., Delwiche, C. F., et al. (2000). Phylogenetic analyses indicate flint the 19′hexanoyloxy-fucoxanthin-containing dinoflagellates have tertiary plastids of haptophyte origin. *Molecular Biology and Evolution*, *17*, 718–729. https://doi.org/10.1093/oxfordjournals.molbev.a026350.

Tokutsu, R., Iwai, M., & Minagawa, J. (2009). CP29, a monomeric light-harvesting complex II protein, is essential for state transitions in *Chlamydomonas reinhardtii*. *Journal of Biological Chemistry*, *284*, 7777–7782. https://doi.org/10.1074/jbc.M809360200.

van Stokkum, I. H. M., Papagiannakis, E., Vengris, M., Salverda, J. M., Polívka, T., Zigmantas, D., et al. (2009). Inter-pigment interactions in the peridinin chlorophyll protein studied by global and target analysis of time resolved absorption spectra. *Chemical Physics*, *357*, 70–78. https://doi.org/10.1016/j.chemphys.2008.10.005.

Veith, T., Brauns, J., Weisheit, W., Mittag, M., & Büchel, C. (2009). Identification of a specific fucoxanthin-chlorophyll protein in the light harvesting complex of photosystem I in the diatom Cyclotella meneghiniana. *Biochimica et Biophysica Acta, 1787*, 905–912. https://doi.org/10.1016/j.bbabio.2009.04.006.

Veith, T., & Büchel, C. (2007). The monomeric photosystem I-complex of the diatom *Phaeodactylum tricornutum* binds specific fucoxanthin chlorophyll proteins (FCPs) as light-harvesting complexes. *Biochimica et Biophysica Acta, 1767*, 1428–1435. https://doi.org/10.1016/j.bbabio.2007.09.004.

Xu, M., Kinoshita, Y., Matsubara, S., & Tamiaki, H. (2016). Synthesis of chlorophyll-c derivatives by modifying natural chlorophyll-a. *Photosynthesis Research, 123*(3), 335–345. https://doi.org/10.1007/s11120-015-0190-1.

Xu, M., & Tamiaki, H. (2014). Transformation of natural chlorophyll-a into chlorophyll-c analogs possessing the 17-acrylate residue. *Chemistry Letters, 43*(12), 1864–1866.

Yamada, N., Tanaka, A., & Horiguchi, T. (2015). Pigment compositions are linked to the habitat types in dinoflagellates. *Journal of Plant Research, 128*(6), 923–932. https://doi.org/10.1007/s10265-015-0745-4.

Yin, L., & Bauer, C. E. (2013). Controlling the delicate balance of tetrapyrrole biosynthesis. *Philosophical Transactions of the Royal Society of London. Series B, Biological Sciences, 368*(1622), 20120262. https://doi.org/10.1098/rstb.2012.0262.

Yool, A., & Tyrrell, T. (2003). Role of diatoms in regulating the ocean's silicon cycle. *Global Biogeochemical Cycles, 17*(4), 1103. https://doi.org/10.1029/2002GB002018.

Zapata, M., Edvardsen, B., Rodrıguez, F., Maestro, M. Á. M. A., Garrido, J. L. J., Rodríguez, F., et al. (2001). Chlorophyll c2 monogalactosyldiacylglyceride ester (chl c2-MGDG). A novel marker pigment for Chrysochromulina species (Haptophyta). *Marine Ecology Progress Series, 219*, 85–98. https://doi.org/10.3354/meps219085.

Zapata, M., Fraga, S., Rodríguez, F., & Garrido, J. L. (2012). Pigment-based chloroplast types in dinoflagellates. *Marine Ecology Progress Series, 465*, 33–52. https://doi.org/10.3354/meps09879.

Zapata, M., Garrido, J., & Jeffrey, S. W. (2006). Chlorophyll c pigments: Current status. In B. Grimm, R. Porra, W. Rüdiger, & H. Scheer (Eds.), *Advances in photosynthesis and respiration: Vol. 25. Chlorophylls and bacteriochlorophylls* (pp. 39–53). Dordrecht: Springer. https://doi.org/10.1007/1-4020-4516-6_3.

Zapata, M., Jeffrey, S. W., Wright, S. W., Rodríguez, F., Garrido, J. L., & Clementson, L. (2004). Photosynthetic pigments in 37 species (65 strains) of haptophyta: Implications for oceanography and chemotaxonomy. *Marine Ecology Progress Series, 270*, 83–102. https://doi.org/10.3354/meps270083.

Zapata, M., Rodríguez, F., Fraga, S., Barra, L., & Ruggiero, M. V. (2011). Chlorophyll c pigment patterns in 18 species (51 strains) of the genus Pseudo-nitzschia (bacillariophyceae). *Journal of Phycology, 47*(6), 1274–1280. https://doi.org/10.1111/j.1529-8817.2011.01055.x.

Zhang, Q., Liu, G., & Hu, Z. (2011). Morphological differences and molecular phylogeny of freshwater blooming species, Peridiniopsis spp. (Dinophyceae) from China. *European Journal of Protistology, 47*, 149–160. https://doi.org/10.1016/j.ejop.2011.03.001.

Zhang, Q., Liu, G., Hu, Z., Zhang, Q., Liu, G., & Z, H. (2014). Description of a new freshwater bloom-forming dinoflagellate with a diatom endosymbiont, Peridiniopsis minima sp. nov. (Peridiniales, Dinophyceae) from China. *Algological Studies, 146*, 119–133.

Zigmantas, D., Hiller, R. G., Sundstrom, V., & Polivka, T. (2002). Carotenoid to chlorophyll energy transfer in the peridinin-chlorophyll-a-protein complex involves an intramolecular charge transfer state. *Proceedings of the National Academy of Sciences of the United States of America, 99*, 16760–16765. https://doi.org/10.1073/pnas.262537599.

Zingone, A., Borra, M., Brunet, C., Forlani, G., Kooistra, W. H. C. F., & Procaccini, G. (2002). Phylogenetic position of Crustomastix stigmatica sp. nov. and Dolichomastix tenuilepis in relation to the Mamiellales (Prasinophyceae, Chlorophyta). *Journal of Phycology, 38*, 1024–1039. https://doi.org/10.1046/j.1529-8817.2002.t01-1-02035.x.

CHAPTER FOUR

Chlorophylls *d* and *f*: Synthesis, occurrence, light-harvesting, and pigment organization in chlorophyll-binding protein complexes

Min Chen*
School of Life and Environmental Sciences, University of Sydney, Sydney, NSW, Australia
*Corresponding author: e-mail address: min.chen@sydney.edu.au

Contents

Abstract

Now is a very exciting time for those people who are interested in understanding oxygenic photosynthesis driven by longer wavelength radiation. The past two decades have seen the determination of molecular mechanisms of photosynthesis driven by long wavelengths. Apart from the characterization of additional potential of longer wavelength absorption from chlorophyll *a*, the two chlorophyll variants chlorophyll *d* and chlorophyll *f* were introduced with red-shifted absorption features compared to chlorophyll *a* and chlorophyll *b*, which are attributable to the position of the formyl substitutions. The maximal absorption of chlorophyll *d* is 697 nm, 32 nm red-shifted

Advances in Botanical Research, Volume 90
ISSN 0065-2296
https://doi.org/10.1016/bs.abr.2019.03.006

compared with that of chlorophyll *a* at 665 nm in 100% methanol; while the maximal absorption of chlorophyll *f* is 707 nm in methanol, red-shifted 42 nm compared with that of chlorophyll *a*. Chlorophyll *d* was reported to be widely distributed in the ecological niches with lower light intensities, but enriched by far-red wavelength light, but so far has been found only in one species of cyanobacteria, *Acaryochloris marina*. Chlorophyll *f* is co-occurring with chlorophyll *a* and its biosynthesis is regulated by far-red light conditions. It has been reported from various genera of cyanobacteria.

Abbreviations

CAO	chlorophyll *a* oxygenase
Chl	chlorophyll
ChlF	chlorophyll f synthase
ChlF^{7521}	chlorophyll *f* synthase (ChlF) from *Fischerella thermalis* PCC 7521
CHLF^{9212}	chlorophyll *f* synthase (ChlF) from *Chlorogloeopsis fritschii* PCC 9212
Chlide	chlorophyllide
DAS	decay-associated spectra
FaRLiP	far-red light photoacclimation
FRL	far-red light
ICP-MS	inductively coupled plasma mass spectrometry
IsiA	iron-stress induced protein A
LPOR	light-dependent NADPH:protochlorophyllide oxidoreductase
PAR	photosynthetic active radiation
PBS	phycobilisome
pheaide	phaeophorbide
RC	reaction center
PSI/PSII	photosystem I/photosystem II

1. Introduction

In photosynthetic organisms, sunlight is absorbed by various pigment-binding protein complexes. Chlorophylls (Chls) are essential molecules for charge separation and electron transfer processes in the reaction centers (RCs) of organisms performing oxygenic photosynthesis. Light absorption limitation of photosynthetic organisms is defined by their own chlorophylls. There are five types of chlorophylls and four of them (Chl *a*, Chl *b*, Chl *d* and Chl *f*) are Mg-chlorin-type chlorophylls with a phytol chain esterified at C17^{3}on ring D. Chl *b*, Chl *d*, and Chl *f* differ from Chl *a* in the formyl substitutions on ring A and ring B in the macrocycle, resulting in the increased hydrophilicity of chlorophylls (Fig. 1).

A

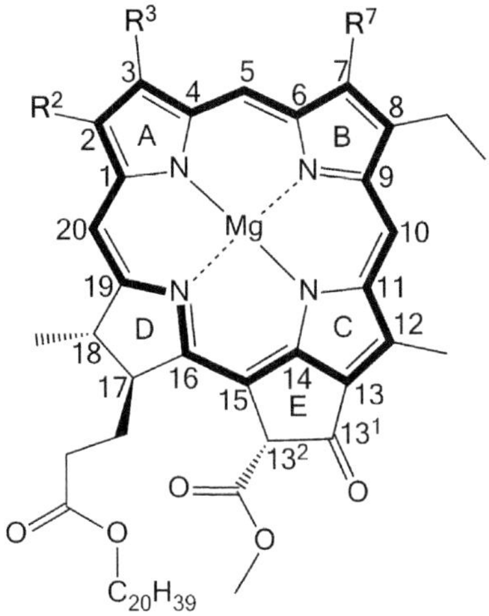

B

	R^2	R^3	R^7
Chlorophyll *a*	CH_3	$CHCH_2$	CH_3
Chlorophyll *b*	CH_3	$CHCH_2$	CHO
Chlorophyll *d*	CH_3	CHO	CH_3
Chlorophyll *f*	CHO	$CHCH_2$	CH_3

C

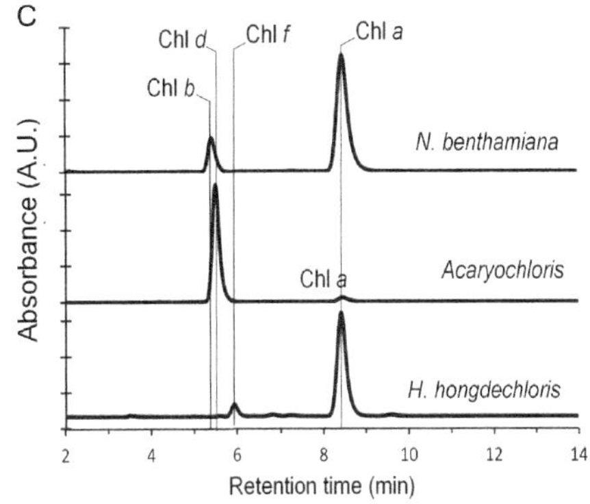

Fig. 1 Chemical structure of chlorophylls. (A) Numbering and subgroup schemes for chlorophylls; (B) the subgroups of Chl *a*, *b*, *d* and *f*; (C) Different HPLC retention times of purified chlorophylls. *Nicotiana banthamiana (N. benthamiana)* contains Chl *a* and Chl *b*; *Acaryochloris marina* (*Acaryochloris*) contains Chl *a* and Chl *d*; *Halomicronema hongdechloris* (*H. hongdechloris*) contains Chl *a* and Chl *f*. HPLC, reverse phase system used here and mobile phase is 100% methanol. *Adapted from Loughlin, P.C., Willows, R. D., & Chen, M. (2014) In vitro conversion of vinyl to formyl groups in naturally occurring chlorophylls.* Scientific Reports, 4, *6069.*

Chl *a*, *b* and *c* have been initially named in the order of their discovery. Subsequently, Chl *d* was discovered in 1943 (Manning & Strain, 1943), and more than six decades afterwards Chl *f* was discovered and reported in 2010 (Chen et al., 2010). The name Chl *e* was given to an extracted Chl pigment from the algae *Tribonema bombycinum* and *Vaucheria bamata* with absorption bands at 415 and 654 nm in methanol, but neither its chemical nature nor the organism producing it has been explored further (Holt, 1966).

This chapter will be a review of what is currently known about the structure and function of the photosynthesis driven by Chl *d* and Chl *f* as well as their distribution among photosynthetic organisms. Thus, the need for research on the biosynthesis and regulation mechanisms of Chl *d* and *f* will be emphasized.

2. Chemistry of chlorophyll *d* and chlorophyll *f*

2.1 Formyl substitution

2.1.1 Chlorophyll d

Chl *d* was firstly isolated from the pigment extraction of red algae and reported by Manning and Strain in 1943. Then, for more than half a century, people believed that Chl *d* was an artificial product created during experimental procedures (Schere, 1991). The discovery of *Acaryochloris marina*, a cyanobacterium that uses Chl *d* as its major photopigment, unambiguously confirmed the presence of Chl *d* and opened new episodes in oxygenic photosynthesis research (Miyashita et al., 1996). The chemical structure of Chl *d* was established by infra red absorption spectra from the various chemical derivatives (Fig. 1) (Holt & Morley, 1959). In 1959, Holt and Morley successfully produced Chl *d* by using Chl *a* as substrate. Combined with the difficulties of repeatable results of Manning and Strains' -algal pigment extractions, Chl *d* was considered as an artificial product during experimental procedures (Schere, 1991).

2.1.2 Chlorophyll f

Chl *f* is the newest member in the chlorophyll family and has the most red-shifted maximum absorption (Chen et al., 2010). The chemical structure of Chl *f* differs from that of Chl *a* in carrying a formyl group at the C2 position (Fig. 1) and is thus determined as [2-formyl]-Chl *a* (Chen et al., 2010; Willows, Li, Scheer, & Chen, 2013).

2.2 Spectral properties of Chl *d* and Chl *f*

The absorption spectra of Chls *a*, *d* and *f* are shown in Fig. 2A. Compared with Chl *a*, both Soret bands and the Q_Y band of Chl *d* are red-shifted as a result of the replaced vinyl group (-CH$=$$CH_2$) at C3 in Chl *a* by a formyl group (-CHO) in Chl *d*. Three absorption peaks at 400, 456 and 697 nm of purified Chl *d* in 100% methanol are well resolved (Fig. 2). Small shifts in the absorption peaks are observed using different solvents (Table 1).

By contrast, Chl *f* has two resolved absorption peaks at 406 and 707 nm in 100% methanol (Fig. 2). The spectral feature of Chl *f* is due to the replacement of an electron-releasing methyl group (-CH_3) at C2 in Chl *a* by a formyl group (-CHO) in Chl *f* (Fig. 1). The co-presence of the vinyl group at C3 and the formyl group at C2 on the ring A in the macrocycle causes the bigger gaps between the Soret bands and Q_Y band in Chl *f*

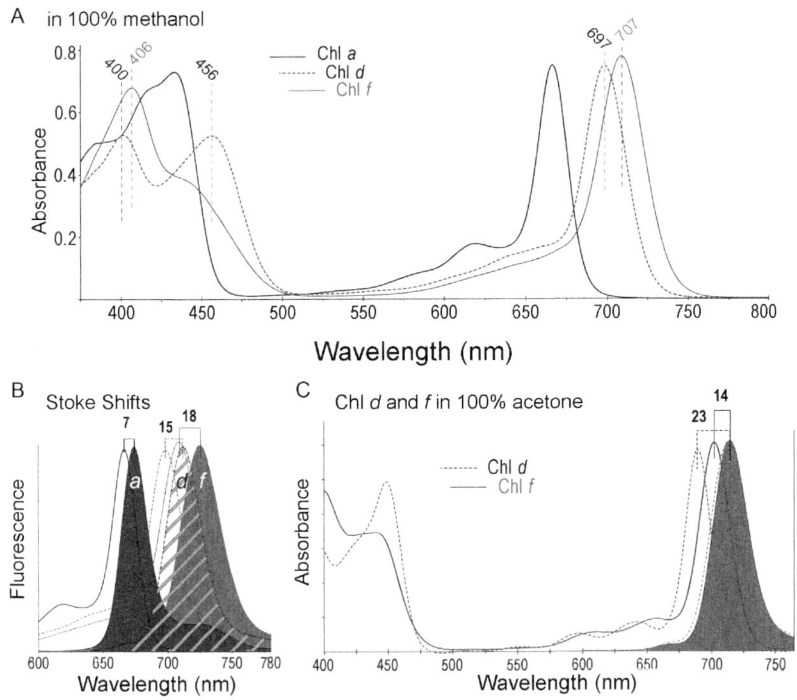

Fig. 2 The absorption and fluorescence spectra of Chl *d* and Chl *f* in methanol and acetone. (A), Absorption spectra in 100% methanol; (B), The Stock shifts in 100% methanol; (C), The spectra of Chl *d* and *f* in 100% acetone and the stock shifts. Chl *a*, *Solid black lines*; Chl *d*, *Dotted gray lines*; Chl *f*, *gray lines*. The shaded areas represent fluorescence emission spectra.

Table 1 Spectroscopic properties of Chl *d*, and *f* in various solvents.

Pigment	Solvent	Wavelength maxima (nm)[ε]		
Chl *d*	Methanol (100%)	440.0 [45.74]	455.5 [44.41]	697.0 [63.68]
	Acetone (100%)	393.0 [43.89]	477.0 [68.16]	688.0 [78.77]
	Acetone (90%)	393.0 [46.13]	449.5 [64.26]	691.0 [75.43]
	Diethylether	390.5 [50.06]	445.5 [88.38]	686.0 [104.50]
Chl *f*	Methanol (100%)	406.5 [66.92]		707.0 [71.11]
	Acetone (100%)	397.0 [59.31]	439.5 [48.04]	698.0 [77.97]
	Acetone (90%)	397.5 [65.08]	439.5 [46.97]	701.0 [77.89]
	Diethylether	395.5 [66.96]	439.5 [64.09]	694.5 [96.65]

Note: The wavelengths of the peaks are given in nm and extinction coefficients (ε in cm^{-1} mM^{-1}) appear in square brackets. Reference: (Li, et al., 2012)

(Chen & Scheer, 2013). The spectral profile of Chl *f* in 100% diethyl ether showed three well-resolved absorption peaks at 395.5, 439.5, and 695 nm and the Qy peak showed the most red-shifted spectrum when dissolved in methanol (Table 1).

The fluorescence maximum of Chl *a* in 100% methanol is approximately 8 nm red-shifted compared to its absorption maximum of 665.5 nm, which is called the Stokes shift. The fluorescence spectrum of Chl *d* in 100% methanol at room temperature shows its maximum at 712 nm (Nieuwenburg et al., 2003), a 15 nm Stokes shift (Fig. 2B). The Stokes shift of Chl *d* in 100% acetone shows 23 nm (Fig. 2C). The isolated Chl *f* in 100% methanol showed the maximum fluorescence emission at 725 nm (Chen et al., 2010), a Stokes shift of 18 nm in 100% methanol, and has a decreased Stokes shift of 14 nm in 100% acetone (Fig. 2). Chl *f* in the hexa-coordinating solvent, pyridine, has a Stokes shift of 17 nm at room temperature (Niedzwiedzki, Liu, Chen, & Blankenship, 2014).

The fluorescence quantum yield (Φ_F) represents the fraction of excited state to decay via fluorescence emission (Blankenship, 2014). The Φ_F of Chl *f* in pyridine is 0.16, the same as that of Chl *b*, but much lower than that of 0.35 in Chl *a* (Table 2, Niedzwiedzki et al., 2014). The resonant energy transfer from Chl *f* to Chl *a* either within the same antenna complexes, or in the reaction centers will require uphill energy transfer.

It was previously proposed that Chl *f* is accumulated in light-harvesting antenna proteins and simply complements Chl *a* in the light-harvesting

Table 2 Summary of fluorescence properties of chlorophylls.

Pigments	Solvents	λ_{max} (nm)	ΦF	References
Chl *a*	diethylether	668		White, Jones, Gibbs, and Butler (1972)
			0.32	Weber and Teale (1957)
	pyridine	677	0.35	Niedzwiedzki et al. (2014)
Chl *b*	diethylether	649		White et al. (1972)
			0.17	Weber and Teale (1957)
	pyridine	662	0.16	Niedzwiedzki et al. (2014)
Chl *d*	pyridine	705 nm		Niedzwiedzki et al. (2014)
Chl *f*	pyridine	722	0.16	Niedzwiedzki et al. (2014)

A missing fluorescence yield indicates it has not been determined, not that the fluorescence yield is missing.

process (Chen & Blankenship, 2011). However, taking under consideration its spectroscopic properties, Chl *f* does not seem to be really suitable for the role of accessory pigment. Resonant energy transfer to Chl *a* either within the same antenna complex or in the reaction center will require a significant uphill energy transfer. Mathematical simulations of decay-associated spectra (DAS) based on rate equation models for energy transfer between PBSs, Chl *a* and/or Chl *f* suggest that the long ns lifetime of Chl *f*, which would otherwise qualify the pigment as an energy trap, can indeed be an advantage for energy-harvesting. However, this is only effective if the Chl *f* to Chl *a* ratio does not exceed about 1:10, as seen *in vivo* under far-red light (FRL) illumination (Schmitt et al., 2018). Taken together with a relatively low fluorescence quantum yield, the overall performance of the Chl *f*-to-Chl *a* energy transfer process becomes very questionable. In order to understand the actual role of Chl *f* in photosynthesis of the cyanobacterium *Halomicronema hongdechloris*, further spectroscopic studies on Chl *f*-containing complexes are necessary.

2.3 Extinction coefficients

The extinction coefficient of chlorophyll is an important parameter for quantitative determination of the chlorophylls including isolated chlorophylls and total chlorophylls in pigment-protein complexes or photosynthetic organisms by the absorbance intensity at the given wavelength per mass density (g) or molar concentration (mol), respectively. The extinction coefficients for Chl *a* and Chl *b* was reported firstly in 1955 (Smith & Benitez, 1955) and updated by Porra et al. in 1989 by using atomic absorbance spectrophotometry (Porra, Thompson, & Kriedemann, 1989). First reported extinction coefficient of Chl *d* was diethyl ether (Smith & Benitez, 1955) and updated by Li et al. using Inductively Coupled Plasma Mass Spectrometry (ICP-MS) (Table 1) (Li, Scales, Blankenship, Willows, & Chen, 2012).

2.4 Other chlorophyll derivatives

All chlorophylls have similar chemical structure with one or a few side substitutions in the macrocycle. The modified side subunits change the spectral properties.

2.4.1 8-vinyl derivatives

The marine cyanobacteria, *Prochlorococcus* spp. use 8-vinyl Chl *a* and 8-vinyl Chl *b* due to an unreduced vinyl side group at the C8 position

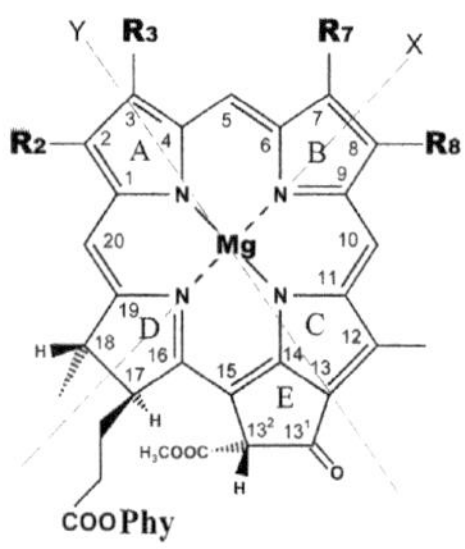

	R_2	R_3	R_7	R_8	λ nm (blue / red)
[8-vinyl]-Chl *a*	CH_3	C_2H_3	CH_3	C_2H_3	**442 / 666** [1]
[8-vinyl] Chl *b*	CH_3	C_2H_3	CHO	C_2H_3	**478 / 658** [1]
[8-vinyl] Chl *d*	CH_3	CHO	CH_3	C_2H_3	**467 / 695** [2]
[8-formyl] Chl *a*	CH_3	C_2H_3	CH_3	CHO	**465 / 675** [2]
[3,8-diformyl] Chl *a*	CH_3	CHO	CH_3	CHO	**487 / 687** [2]
[3,7-diformyl] Chl *a*	CH_3	CHO	CHO	C_2H_5	**483 / 674** [2]
[7-hydroxymethyl] Chl *a*	CH_3	C_2H_3	CHOH	C_2H_5	**436 / 660** [3]
[3-hydroxymethyl] Chl *a*	CH_3	CHOH	CH_3	C_2H_5	**429 / 658** [3]
[2-hydroxymethyl] Chl *a*	CHOH	C_2H_3	CH_3	C_2H_5	**433 / 671** [3]
[3,7-dihydroxymethyl] Chl *a*	CH_3	CHOH	CHOH	C_2H_5	**434 / 653** [3]
[2,3-dihydroxymethyl] Chl *a*	CHOH	CHOH	CH_3	C_2H_5	**429 / 665** [3]
[3-formyl-7-hydroxymethyl] Chl *a*	CH_3	CHO	CHOH	C_2H_5	**460 / 690** [3]
[2-hydroxymethyl-3-formyl] Chl *a*	CHOH	CHO	CH_3	C_2H_5	**403 / 705** [3]

Fig. 3 Chlorophyll derivatives and their absorption maxima (l nm). References for lnm: (1) Goericke & Repeta, 1992; (2) Loughlin et al., 2014; (3) Sawicki, Willows, & Chen, 2019.

(Fig. 2, modified position structure with table insert for position changes). The spectral profiles of 8-vinyl derivatives showed the similar Qy transition, but red-shifted Soret bands compared to their counterparts (Fig. 3, Loughlin et al., 2014).

2.4.2 Hydroxymethyl derivates

Reducing the formyl side group using reductive reagents, such as sodium borohydride ($NaBH_4$), a series of hydroxymethyl replacements were obtained (Fig. 3, Sawicki et al., 2019). The 7-hydroxymethyl Chl *a* is an intermediate product in the cycle of Chl *a* to/from Chl *b* (Ito, Ohtsuka, & Tanaka, 1996). Hydroxymethyl replacements at side group of Chl by using $NaBH_4$ resulted in 2-hydroxymethyl Chl *a*, 2-hydroxymethyl Chl *d* (2-hydroxymethyl-3-formyl Chl *a*, 2,3-dihydroxymethyl Chl *a*, 7-hydroxymethyl Chl *d* (7-hydroxymethyl-3-formyl Chl *a*), and 3,7-dihydroxymethyl Chl *a* (Fig. 3, Sawicki et al., 2019). The presence of the hydroxymethyl side group results in hydrophilic properties and shifted spectral absorption according to its position on the macrocycle (Fig. 3). The exchange of a formyl group into a hydroxymethyl group from C7 of Chl *b* resulted into a red-shifted of the Qy peak. In contrast, the replacement at C2 of Chl *f* or C3 of Chl *d* resulted in blue-shifted Qy peaks. Additionally, 3,7-dihydroxy-methyl phaeophorbide *a* (phaeide *a*), 3-hydroxymethyl-7-formyl phaeide *a* and 3-formyl-7-hydroxymethyl phaeide *a* were synthesized using $NaBH_4$ (Kusunoki & Tamiaki, 2018).

2.4.3 Formyl derivatives

In addition to the reported natural occurrence of formyl-substituted chlorophylls, Chl *b*, Chl *d* and Chl *f* (Fig. 1), 3,7 diformyl Chl *a* (Fig. 3) was artificially synthesized either via the *CAO* (chlorophyllide *a* oxygenase) gene transformant (Tsuchiya et al., 2012), or chemical oxidative cleavage of 3-vinyl side chain in Chl *b* (Loughlin et al., 2014). The 3,8 diformyl Chl *a* (Fig. 3) was synthesized using the same chemical oxidative cleavage of the vinyl side groups in 8-vinyl Chl *a*. There is a correlation of molecular hydrophilic properties with the presence of formyl substitution (Loughlin et al., 2014). The derivatives containing two formyl subgroups showed more hydrophilic properties compared with one formyl substitution derivatives/chlorophylls (Loughlin et al., 2014; Sawicki et al., 2019).

The synthesis of a formyl from a vinyl side group of synthesized methyl pyropheophorbide *a*, a Chl *a* structural analog, through C3-(1-bromoethyl)- and formyl-derivatives, led to the latter oxidation of the C3-vinyl of C3-methyl pyropheophorbide *a* to the formyl group of methyl pyropheophorbide d by $KMnO_4$, OsO_4 and O_3^- (Oba et al., 2011). The 3-vinyl group could be converted to the formyl group by the actions of homofunctionalized components in the presence of oxygen molecules (Oba et al., 2011).

3. Distribution

The recent advances in knowledge of Chl *d* and Chl *f*, and cyanobacteria containing them, challenges the physical limitation of oxygenic photosynthesis and their ecological significance (Chen, 2014; Chen & Blankenship, 2011; Chen et al., 2010; Gan & Bryant, 2015; Miyashita et al., 1996). The red-shifted absorption properties of Chl *d* and Chl *f* extended active photosynthetic spectral region into the FRL region, allowing the organisms containing them to thrive in low light, but increased FRL niches (Blankenship & Chen, 2013; Kühl, Chen, Ralph, Schreiber, & Larkum, 2005). *Acaryochloris marina* MBIC 11017 was first isolated and maintained as a free-living cyanobacterium in the laboratory in 1996. There are six strains of *Acaryochloris* that have been isolated from various ecological environments (Loughlin, Lin, & Chen, 2013). All reported isolated *Acaryochloris* spp. or *Acaryochloris*-like cyanobacteria use Chl *d* as their major photopigment, except *Acaryochloris* RCC1774. *Acaryochloris* RC1774 is related to the *Acaryochloris* genus based on phylogenetic 16s rDNA analysis, but it uses Chl *a* as its major photopigment instead of Chl *d* (Partensky et al., 2018). In additional, some previous studies have reported a trace amount of Chl *d* (approximately 1% of total detected chlorophylls) in various cyanobacterial species, when they were grown under solo FRL conditions.

Chl *f*-producing cyanobacteria have two striking differences from Chl *d*-containing cyanobacteria (*Acaryochloris.* spp): (1) Chl *f*-producing cyanobacteria belong to various cyanobacterial subgroups based on 16s rDNA phylogenetic analysis, different from the mono-genus Chl *d*-containing *Acaryochloris* (Gan & Bryant, 2015). (2) Biosynthesis of Chl *f* is tightly regulated by available FRL, which is only produced under FRL conditions in all newly reported Chl *f*-producing cyanobacteria. In contrast, the production of Chl *d* in *Acaryochloris* is independent of the light quality. When a large part of the incident solar light spectrum is filtered out, FRL enriched niches are created. Those unique light niches likely constitute a strong selective pressure favoring phototrophs containing Chl *d* and/or Chl *f* (Li & Chen, 2015).

The first reported species producing Chl *f* has been named *Halomicronema hongdechloris*, and belongs to a cyanobacterial phylum characterized by filamentous growth in the sub-branch of *Leptolyngbya* (Chen, Li, Birch, & Willows, 2012). Chl *f* was detected in microbial mats collected from the Nakabusa hot spring (Ohkubo & Miyashita, 2017) and stromatolite mats (Fisher et al., 2019).

4. Photosynthetic functions

4.1 Photosynthesis driven by Chl *d*

Since all known oxygenic photosynthetic organisms contain fractions of Chl *a*, this pigment seems to be necessary to serve as the primary donor in photosystem II (PSII) to provide a sufficient redox potential for water oxidation. *Acaryochloris* contains Chl *d* up to 99% of its total Chl content (Loughlin et al., 2013; Miyashita et al., 1996). It can replace all functions of Chl *a* in the photosystems for driving oxygenic photosynthesis to the FRL conditions (Mielke, Kiang, Blankenship, Gunner, & Mauzerall, 2011, Tomo, Allakhverdiev, & Mimuro, 2011). The function of Chl *d* as the major light-harvesting pigment in *Acaryochloris* was established in 2002 (Chen, Quinnell, & Larkum, 2002) and then confirmed that the Chl *d*-binding and 6-helix transmembrane protein complexes serve as main light-harvesting protein complexes to the PSI and PSII by forming supracomplexes (Chen, Bibby, Nield, Larkum, & Barber, 2005a, 2005b). Recently isolated Chl *d*-binding IsiA complexes showed of main absorption of 703 nm and fluorescence emission maximal of 712 nm (Li, Vella, & Chen, 2018). The Chl *d* molecule functions also as the special pair (P740) in PSI instead of Chl *a* (P700) (Akiyama et al., 2001; Hu et al., 1998). However, the controversy remains whether Chl *d* functions as the special pair in PSII of *A. marina*. There are still contrary opinions that either water oxidation occurs in a dimer of Chl *d* (Chen et al., 2005), or heterodimer formed from Chl *a* and Chl *d* (Schlodder et al., 2007), while other groups still propose a RC formed by a Chl *a* dimer (Tomo et al., 2007).

4.2 Photosynthesis driven by Chl *f*

The question as to what extent Chl *f* plays a role in photosynthetic energy transfer is still unresolved. The long lifetimes of Chl *f* fluorescence (Niedzwiedzki et al., 2014) and potential uphill energy transfer from Chl *f* to Chl *a* support its role in energy harvesting and also an energy carrier role for dissipating the energy to and from Chl *a* in the reaction centers (Nürnberg et al., 2018; Schmitt et al., 2018). Tomo et al. examined fast time-resolved fluorescence spectra of *H. hongdechloris* cells grown under white and FRL at 77 K (Tomo et al., 2014). White light-grown cells showed typical fluorescence components for Chl *a* from PSII (685 nm) and PSI (730 nm) immediately upon excitation, with PSII fluorescence decaying mainly with 150 ps and 520 ps. A further red-shifted fluorescence at 742 nm was observed in these cells, which appeared delayed after about 1 ns and was attributed to sensitized emission of a red Chl *a* in one of the photosystems, since Chl *f* is not produced under these conditions. Long-lived ns-fluorescence components at 695 nm, 730 nm and 745 nm remained indicating energetic equilibration among the Chl *a* pools (Tomo et al., 2014). In cells grown under FRL, a broad 748 nm fluorescence band with shoulders at 730 nm (attributed to a red Chl *a* of PSI, as mentioned above) and 760 nm (attributed to Chl *f*) were observed. The rise of the main 748 nm component with a time constant of more than 10 ns rather, Chl *f* was suggested to represent an energy trap, instead of an uphill energy transfer from Chl *f* to Chl *a* at 77 K (Tomo et al., 2014). Further investigation on isolated thylakoid membranes from *H. hongdechloris* cells grown under FRL growth conditions revealed that the equilibration of the Chl *a* and Chl *f* pools at physiological temperature could provide sufficient energy for the charge separation by Chl *a* in RCs. However, the tightly regulated Chl *f* biosynthesis plays an important role in maintaining the ratio of Chl *a* to Chl *f* because excitation energy could be trapped or dissipated by Chl *f* at high Chl *f* contents (Akimoto, Shinoda, Chen, Allakhverdiev, & Tomo, 2015). The energy transfer model based on fluorescence lifetime and fluorescence decay ratio suggested that the optimal Chl *f* to Chl *a* ratio of 1:10 is essential for energy transfer between phycobilisomes (PBSs), Chl *a* and/or Chl *f* (Schmitt et al., 2018).

Therefore, the long lifetime of Chl *f* allows the pigment to serve as intermediate energy store for uphill energy transfer to Chl *a* driven by an entropy effect, thereby increasing the efficiency of the photosynthesis reactions under FRL illumination.

5. Biochemistry and regulatory mechanism

5.1 Oxygen is precursor for the formation of the formyl group in Chl *d* and Chl *f*

The pathway of Chl *a* biosynthesis is a metabolically expensive process and contains 17 steps of enzymatic reactions (Chen, 2014). The latter stages lead to the synthesis of other chlorophylls, Chl *b*, Chl *d* and Chl *f*. Chl *b* biosynthesis is derived from Chl *a* or Chlide *a* by the oxidation of the methyl group into a formyl group via a 7-hydroxymethyl intermediate catalyzed by CAO (Oster, Tanaka, Tanaka, & Rudiger, 2000; Tanaka et al., 1998). Both Chl *d* and Chl *f* are made from Chl *a* although the mechanism of formation of Chl *d* and Chl *f* remains to be dissected.

The oxidation of the vinyl group at C3 position gives the formyl group in Chl *d* that requires O_2 molecule as a precursor (Schliep, Crossett, Willows, & Chen, 2010). Using potassium permanganate, Holt and Morley produced Chl *d* from Chl *a* with a yield of 10% (Holt & Morley, 1959). Recently, thiol-type compounds such as thiophenol, benzyl mercaptan, β-mercaptoethanol or dithiothreitol have been reported to be an efficient reductant at specifically converting the Chl vinyl group to a formyl group in the presence of molecular oxygen, which functions as the electron acceptor (Fukusumi et al., 2012; Loughlin et al., 2014; Oba et al., 2011). The mechanism of oxidation of Chl *a* with thiophenol/O_2 is proposed to involve a thiyl radical reacting with the vinyl group followed by oxygen to form a hydroperoxide intermediate, and oxidative cleavage to yield Chl *d* (Fukusumi et al., 2012; Oba et al., 2011). The phylogenetically unrelated enzymes papain (Koizumi et al., 2005) and horseradish peroxidase catalyzed with hydrogen peroxide as substrate (Hynninen, Kaartinen, & Kolehmainen, 2010) were able to produce a small amount of Chl *d* from Chl *a*. Hence, it is unclear which type of enzymatic reaction mechanism occurs in vivo. The two candidates suggested to be involved with Chl *d* biosynthesis: oxygenase (Schliep et al., 2010) or radical SAM proteins (Partensky et al., 2018) have no supporting biochemical evidence.

The mechanism of Chl *f* biosynthesis was predicted to have a similar mechanism as Chl *b* synthesis since both mechanism involve chemical conversion of a methyl (-CH_3) to a formyl (-CHO) group (Kräutler, 2011). However, some doubt remains because no apparent homolog of CAO exists in Chl *f*-producing organisms (Chen, Hernandez-Prieto, Loughlin, Li, & Willows, 2019). Using ^{18}O-labeling of O_2 and H_2O, the oxygen atom in the C2 formyl group of Chl *f* originates from molecular oxygen, supporting the oxygenase mechanism (Garg, Loughlin, Willows, & Chen, 2017). Recently, a paralog of the D1 protein of photosystem II (PsbA4, renamed ChlF) was identified as a FRL-induced photooxidoreductase with the ability to oxidize Chl(ide) *a* into Chl(ide) *f* suggesting a radical mechanism and molecular oxygen may not be required (Ho, Shen, Canniffe, Zhao, & Bryant, 2016). When the proposed radical reaction site of $ChlF^{9212}$ (chlorophyll *f* synthase from *Chlorogloeopsis fritschii* PCC 9212), Tyr at position 183, was replaced by Phe (removing the redox active Tyr), the ChlF activity was unexpectedly increased by 1.9 folds (Shen et al., 2019). The Tyr183 in $ChlF^{9212}$ plays no role in the photochemical activity of ChlF in Chl *f* biosynthesis of the transformed strain of *Synechococcus* PCC 7002. In addition, expression of heterogenous $ChlF^{7521}$ in *Synechococcus* PCC7002 lacking PsbD and PSII activity, confirmed that Chl *f* synthesis is independent to the active intact PSII in the cells and argues against the previous proposed radical mechanisms (Ho et al., 2016).

5.2 Biosynthetic regulation

Up to date, all Chl *f*-producing cyanobacteria possess a photo-regulatory process called "far-red light photoacclimation (FaRLiP)" (Ho et al., 2016). The genome comparative analysis revealed that 20-genes in the FaRLiP cluster are highly conserved in Chl *f*-producing cyanobacteria including the gene encoding Chl *f* synthase (*chlf*) (Ho et al., 2016). Hence, the biosynthesis of Chl *f* in Chl *f*-producing cyanobacteria is tightly regulated by a knot-less phytochrome (RfpA) and its associated transcriptional activators (RfpB and RfpC), which activate the expression of the FaRLiP gene cluster. In *H. hongdechloris*, the first reported cyanobacterium that can produce Chl *f* under FRL, Chl *f* is a red-light induced chlorophyll. Approximately 10–12.5% Chl *f is* detected in *H. hongdechloris* grown under FRL (Chen et al., 2012; Li, Lin, Loughlin, & Chen, 2014). The Chl *f* to Chl *a* ratio was decreased when switching culture light condition from FR to white light conditions and the content of Chl *f* reached almost zero upon continuous illumination in white light. The FRL induction of Chl *f* biosynthesis indicated that ChlF is a light-induced enzyme (Ho et al., 2016), which is the second light-induced enzyme found in the context of Chl biosynthesis in oxygenic photosynthetic organisms. Light-dependent NADPH:protochlorophyllide oxidoreductase (LPOR) catalyzes the trans-reduction of ring D of protochlorophyllide and is the reason for light-dependent greening of angiosperms (Björn, 2018).

6. Applications and future directions

The strategy of *H. hongdechloris* employing newly synthesized unique pigments and re-assembling photosynthetic apparatus to adapt to changing light conditions represents one natural model to improve photosynthetic efficiency. Photosynthesis used wavelengths from 400 to 700nm (visible light region), less than 50% of the solar spectrum (Fig. 4). Absorbing light in the far-red light region could potentially be accomplished by inserting cyanobacterial genes encoding the enzymes responsible for Chl *d* and Chl *f* synthesis into plants. Then in addition, alternative light-harvesting complexes are suggested to be introduced that bind the new chlorophylls, which could cover the gap where Chl *a* and *b* have less absorption efficiency (Blankenship & Chen, 2013).

It is noticeable that the predicted functions of Chl *d* and Chl *f* are distinct from the established "red chlorophylls" present in cyanobacteria as well as algae and plants, which are in fact Chl *a* species. The absorption maxima of these red Chl *a* pigments, which are mostly found in light-harvesting complexes (Chen & Blankenship, 2011), are shifted out to 740nm by their particular protein environment. Thus, the energetic profile is modulated (Schlodder et al., 2007).

In summary, the novel chlorophylls Chl *d* and Chl *f* offer a potential for extending the active photosynthetic spectral wavelength into far-red light, up to 760nm (Chen & Blankenship, 2011). The extended photosynthetic active radiation (PAR) will enhance the photosynthetic efficiency by accessing 19% additional photon flux in the region of 700–760nm. In addition of red-shifted chlorophylls, "red-chlorophyll *a*," which is available due to changes of the protein environments of the pigment-protein complexes, can be used to perform oxygenic photosynthesis instead of Chl substitutions (Durnford et al., 1999). Additionally, recent research on

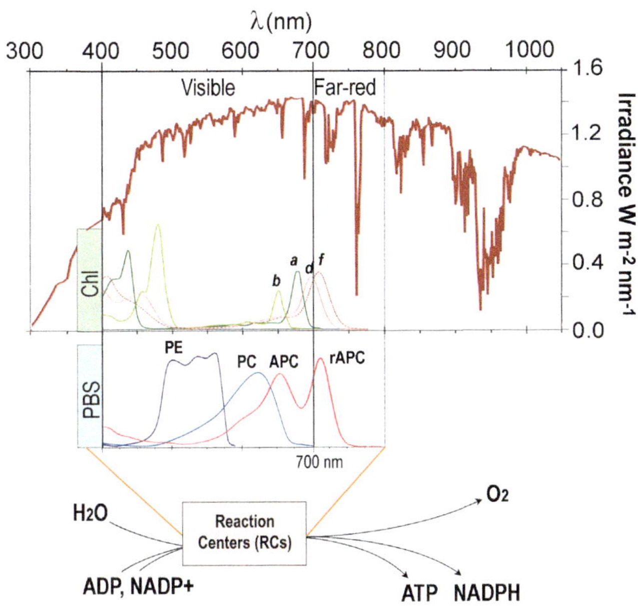

Fig. 4 Solar output spectra on the earth surface and spectra of several isolated photopigments in oxygenic photosynthetic organisms. The absorption spectra of chlorophylls and phycobiliproteins are indicated on the bottom. Solar spectra, which can be downloaded at http://rredc.nrel.gov/solar/spectra/am1.5/. The absorption spectra of chlorophylls *a*, *b*, *d*, *f* in methanol. The visible, far-red lights and oxygenic photosynthetic light reactions are indicated. *The absorption spectra of phycoerythrin (PE) and phycocyanin (PC) are adapted from Kiang, Siefert, Govindjee, and Blankenship (2007). The allophycocyanin (APC) and red-shifted allophycocyanin (rAPC) are adapted from Li, Y., Lin, Y., Garvey, C.J., Birch, D., Corkery, R.W., Loughlin, P.C., et al. (2016) Characterization of red-shifted phycobilisomes isolated from the chlorophyll* f*-containing cyanobacterium* Halomicronema hongdechloris*. Biochimica et Biophysica Acta, 1857, 107–114.*

re-assembled phycobiliproteins revealed that the protein environment is important for pigment absorption features (Li et al., 2016). *Re*-arranging the subunits of allophycocyanin enables absorption of far-red light, which is more far-red-shifted than that absorbed from Chl *a*-/Chl *b*-binding protein complexes (Ho, Gan, Shen, Zhao, & Bryant, 2017; Li et al., 2016).

References

Akimoto, S., Shinoda, T., Chen, M., Allakhverdiev, S. I., & Tomo, T. (2015). Energy transfer in the chlorophyll *f*-containing cyanobacterium, *Halomicronema hongdechloris*, analyzed by time-resolved fluorescence spectroscopies. *Photosynthesis Research*, *125*, 115–122.

Akiyama, M., Miyashita, H., Kise, H., Watanabe, T., Miyachi, S., & Kobayashi, M. (2001). Detection of chlorophyll *d* and pheophytin *a* in a chlorophyll *d*-dominating oxygenic photosynthetic prokaryote *Acaryochloris marina*. *Analytical Sciences*, *17*, 205–208.

Björn, L. O. (2018). Photoenzymes and related topics: An update. *Photochemistry and Photobiology, 94*, 459–465.

Blankenship, R. E. (2014). *Molecular mechanism pf photosynthesis* (2nd ed., pp. 249–286): Wiley Blackwell.

Blankenship, R. E., & Chen, M. (2013). Spectral expansion and antenna reduction can enhance photosynthesis for energy production. *Current Opinion in Chemical Biology, 17*(3), 457–461.

Chen, M. (2014). Chlorophyll modifications and their spectral extension in oxygenic photosynthesis. *Annual Review of Biochemistry, 83*, 317–340.

Chen, M., & Blankenship, R. E. (2011). Expanding the solar spectrum used by photosynthesis. *Trends in Plant Science, 16*, 427–431.

Chen, M., Bibby, T. S., Nield, J., Larkum, A. W. D., & Barber, J. (2005a). Iron deficiency induces a chlorophyll *d*-binding Pcb antenna system around photosystem I in *Acaryochloris marina. Biochim Biophys Acta, 1708*, 367–374.

Chen, M., Bibby, T. S., Nield, J., Larkum, A. W. D., & Barber, J. (2005b). Structure of a large photosystem II supercomplex from *Acaryochloris marina. FEBS Letters, 579*, 1306–1310.

Chen, M., Hernandez-Prieto, M. A., Loughlin, P. C., Li, Y., & Willows, R. D. (2019). Genome and proteome of the chlorophyll f-producing cyanobacterium *Halomicronema hongdechloris*: Adaptative proteomics shifts under different light conditions. *BMC Genomics, 20*(1), 207.

Chen, M., Li, Y., Birch, D., & Willows, R. D. (2012). A cyanobacterium that contains chlorophyll *f*—A red absorbing photopigment. *FEBS Letters, 586*, 3249–3254.

Chen, M., Quinnell, R. G., & Larkum, A. W. D. (2002). The major light harvesting pigment protein of *Acaryochloris marina. FEBS Letters, 514*, 149–152.

Chen, M., & Scheer, H. (2013). Extending the limits of natural photosynthesis and implications for technical light harvesting. *Journal of Porphyrins and Phthalocyanines, 17*, 1–15.

Chen, M., Schliep, M., Willows, R. D., Cai, Z., Neilan, B. A., & Scheer, H. (2010). A red-shifted chlorophyll. *Science, 329*, 1318–1319.

Chen, M., Telfer, A., Lin, S., Pascal, A., Larkum, A. W. D., Barber, J., et al. (2005). The nature of the photosystem II reaction centre in the chlorophyll *d*-containing prokaryote, *Acaryochloris marina. Photochemical & Photobiological Sciences, 4*(12), 1060–1064.

Durnford, D. G., Deane, J. A., Tan, S., McFadden, G. I., Gantt, E., & Green, B. R. (1999). Aphylogenetic assessment of the eukaryotic light-harvesting antenna proteins, with implications for plastid evolution. *Journal of Molecular Evolution, 58*, 59–68.

Fisher, A., Wangpraseurt, D., Larkum, A. W. D., Johnson, M., Kühl, M., Chen, M., et al. (2019). Correlation of bio-optical properties with photosynthetic pigment and microorganism distribution in microbial mats from Hamelin Pool, Australia. *FEMS Microbiology Ecology, 95*(1), 1–13. https://doi.org/10.1093/femsec/fiy219.

Fukusumi, T., Matsuda, K., Mizoguchi, T., Miyatake, T., Ito, S., Ikeda, T., et al. (2012). Non-enzymatic conversion of chloro- phyll-*a* into chlorophyll-*d* in vitro: A model oxidation pathway for chlorophyll-*d* biosynthesis. *FEBS Letters, 586*, 2338–2341.

Gan, F., & Bryant, D. A. (2015). Adaptive and acclimative responses of cyanobacteria to far-red light. *Environmental Microbiology, 17*, 3450–3465.

Garg, H., Loughlin, R. C., Willows, R. D., & Chen, M. (2017). The $C2^1$-formyl group in chlorophyll *f* originates from molecular oxygen. *The Journal of Biological Chemistry, 292*, 19279–19289.

Goericke, R., & Repeta, D. J. (1992). The pigment of *Prochlorococcus* marinus: The presence of divinyl chlorophyll a and b in a marina prokaryote. *Limnology and Oceanography, 37*, 425–433.

Ho, M.-Y., Gan, F., Shen, G., Zhao, C., & Bryant, D. A. (2017). Far-red light photoacclimation (FaRLiP) in *Synechococcus* sp. PCC 7335: I. Regulation of FaRLiP gene expression. *Photosynthesis Research, 131*, 173–186.

Ho, M.-Y., Shen, G., Canniffe, D. P., Zhao, C., & Bryant, D. A. (2016). Light- dependent chlorophyll *f* synthase is a highly divergent paralog of PsbA of photosystem II. *Science*, *353*, 213–227.
Holt, A. S. (1966). Recently characterized chlorophylls. In L. P. Vernon & G. R. Seely (Eds.), *The chlorophylls* (pp. 111–118). Academic Press INC.
Holt, A. S., & Morley, H. V. (1959). A proposed structure for chlorophyll *d*. *Canadian Journal of Chemistry*, *37*(3), 507–514.
Hu, Q., Miyashita, H., Iwasaki, I., Kurano, N., Miyachi, S., Iwaki, M., et al. (1998). A photosystem I reaction center driven by chloro- phyll *d* in oxygenic photosynthesis. *Proceedings of the National Academy of Sciences of the United States of America*, *95*(22), 13319–13323.
Hynninen, P. H., Kaartinen, V., & Kolehmainen, E. (2010). Horseradish peroxidase-catalyzed oxidation of chlorophyll *a* with hydrogen peroxide: Characterization of the products and mechanism of the reaction. *Biochimica et Biophysica Acta*, *1797*(5), 531–542.
Ito, H., Ohtsuka, T., & Tanaka, A. (1996). Conversion of chlorophyll *b* to chlorophyll *a* via 7-hydroxymethyl chlorophyll. *The Journal of Biological Chemistry*, *271*(3), 1475–1479.
Kiang, N. Y., Siefert, J., Govindjee, & Blankenship, R. E. (2007). Spectral signatures of photosynthesis. I. Review of Earth organisms. *Astrobiology*, *7*, 222–251.
Koizumi, H., Itoh, Y., Hosoda, S., Akiyama, M., Hoshino, T., Shiraiwa, Y., et al. (2005). Serendipitous discovery of Chl *d* formation from Chl *a* with papain. *Science and Technology of Advanced Materials*, *6*(6), 551–557.
Kräutler, B. (2011). A new factor in life's quest for energy. *Angewandte Chemie (International Ed. in English)*, *50*(11), 2439–2441.
Kühl, M., Chen, M., Ralph, P., Schreiber, U., & Larkum, A. W. D. (2005). A niche for cyanobacteria containing chlorophyll *d*. *Nature*, *433*, 820.
Kusunoki, M., & Tamiaki, H. (2018). Synthesis of 7-substituted chlorophyll-*a* derivatives as chlorophyll-*b* analogs with specific visible absorption bands. *Tetrahedron*, *74*(4), 453–464.
Li, Y., & Chen, M. (2015). Novel chlorophylls and new directions in photosynthesis research. *Functional Plant Biology*, *42*, 493–501.
Li, Y., Lin, Y., Garvey, C. J., Birch, D., Corkery, R. W., Loughlin, P. C., et al. (2016). Characterization of red-shifted phycobilisomes isolated from the chlorophyll *f*-containing cyanobacterium *Halomicronema hongdechloris*. *Biochimica et Biophysica Acta*, *1857*, 107–114.
Li, Y., Lin, Y., Loughlin, P. C., & Chen, M. (2014). Optimization and effects of different culture conditions on growth of *Halomicronema hongdechloris*—A filamentous cyanobacterium containing Chl *f*. *Frontiers in Plant Science*, *5*, 67.
Li, Y., Scales, N., Blankenship, R. E., Willows, R. D., & Chen, M. (2012). Extinction coefficient for red-shifted chlorophylls: Chlorophyll *d* and chlorophyll *f*. *Biochimica et Biophysica Acta*, *1817*, 1292–1298.
Li, Y., Vella, N., & Chen, M. (2018). Characterization of isolated pho- tosystem I from *Halomicronema hongdechloris*, a chlorophyll *f*-producing cyanobacterium. *Photosynthetica*, *56*, 306–315.
Loughlin, C. P., Lin, Y., & Chen, M. (2013). Chlorophyll *d* and *Acaryochloris marina*: Current status. *Photosynthesis Research*, *116*, 277–293.
Loughlin, P. C., Willows, R. D., & Chen, M. (2014). In vitro conversion of vinyl to formyl groups in naturally occurring chlorophylls. *Scientific Reports*, *4*, 6069.
Manning, W. M., & Strain, H. H. (1943). Chlorophyll *d*, a green pigment in red algae. *The Journal of Biological Chemistry*, *151*, 1–19.
Mielke, S., Kiang, N., Blankenship, R. E., Gunner, M., & Mauzerall, D. (2011). Efficiency of photosynthesis in a Chl *d*-utilizing cyanobacterium is comparable to or higher than that in Chl *a*-utilizing oxygenic species. *Biochimica et Biophysica Acta*, *1807*, 1231–1236.
Miyashita, H., Ikemoto, H., Kurano, N., Adachi, K., Chihara, M., & Miyachi, S. (1996). Chlorophyll *d* as a major pigment. *Nature*, *383*, 402–403.

Niedzwiedzki, D. M., Liu, H., Chen, M., & Blankenship, R. E. (2014). Excited state properties of chlorophyll *f* in organic solvents at ambient and cryogenic temperatures. *Photosynthesis Research, 121*, 25–34.

Nieuwenburg, P., Clarke, R. J., Cai, Z. L., Chen, M., Larkum, A. W. D., Cabral, N. M., et al. (2003). Examination of the photophysical processes of chlorophyll *d* leading to a clarification of proposed uphill energy transfer processes in cells of *Acaryochloris marina*. *Photochemistry and Photobiology, 77*, 628–637.

Nürnberg, D. J., Morton, J., Santabarbara, S., Telfer, A., Joliot, P., Antonaru, L. A., et al. (2018). Photochemistry beyond the red limit in chlorophyll *f*-containing photosystems. *Science, 360*, 1210–1213.

Oba, T., Uda, Y., Matsuda, K., Fukusumi, T., Ito, S., Hiratani, K., et al. (2011). A mild conversion from 3-vinyl- to 3-formyl-chlorophyll derivatives. *Bioorganic & Medicinal Chemistry Letters, 21*, 2489–2491.

Ohkubo, S., & Miyashita, H. (2017). A niche for cyanobacteria producing chlorophyll *f* within a microbial mat. *ISME Journal, 11*(10), 2368–2378.

Oster, U., Tanaka, R., Tanaka, A., & Rudiger, W. (2000). Cloning and functional expression of the gene encoding the key enzyme for chlorophyll *b* biosynthesis (CAO) from Arabidopsis thaliana. *The Plant Journal, 21*, 305–310.

Partensky, F., Six, C., Ratin, M., Garczarek, L., Vaulot, D., Probert, I., et al. (2018). A novel species of the marine cyanobacterium *Acaryochloris* with a unique pigment content and lifestyle. *Scientific Reports, 8*(1), 9142.

Porra, R., Thompson, W., & Kriedemann, P. (1989). Determination of accurate extinction coefficients and simultaneous equations for assaying chlorophylls a and b extracted with four different solvents: Verification of the concentration of chlorophyll standards by atomic absorption spectroscopy. *Biochimica et Biophysica Acta—Bioenergetics, 975*, 384–394.

Sawicki, A., Willows, R. D., & Chen, M. (2019). Spectral signature of five hydroxymethyl chlorophyll *a* derivates chemically derived from chlorophyll *d* or chlorophyll *f*. *Photosynthesis Research, 140*(1), 115–127. https://doi.org/10.1007/s11120-018-00611-8.

Schere, H. (1991). Structure and occurrence of chlorophylls. In H. Scheer (Ed.), *Chlorophylls* (pp. 3–30). CRC Press:

Schliep, M., Crossett, B., Willows, R. D., & Chen, M. (2010). ^{18}O labeling of chlorophyll *d* in *Acaryochloris marina* reveals that chlorophyll *a* and molecular oxygen are precursors. *The Journal of Biological Chemistry, 285*(37), 28450–28456.

Schlodder, E., Çetin, M., Eckert, H. J., Schmitt, F. J., Barber, J., & Telfer, A. (2007). Both chlorophylls *a* and *d* are essential for the photochemistry in photosystem II of the cyanobacteria, *Acaryochloris marina*. *Biochimica et Biophysica Acta, 1767*, 589–595.

Schmitt, F.-J., Campbell, Z. Y., Bui, M. V., Hüls, A., Tomo, T., Chen, M., et al. (2018). Photosynthesis supported by a chlorophyll *f*-dependent, entropy-driven uphill energy transfer in *Halomicronema hongdechloris* cells adapted to far-red light. *Photosynthesis Research, 139*, 185–201.

Shen, G., Canniffe, D. P., Ho, M.-Y., Kurashov, V., van der Est, A., Golbeck, J. H., et al. (2019). Characterization of chlorophyll *f* synthase heterologously produced in *Synechococcus* sp. PCC 7002. *Photosynthesis Research, 140*(1), 77–92. https://doi.org/10.1007/s11120-018-00610-9.

Smith, J. H. C., & Benitez, A. (1955). Chlorophylls: Analysis in plant materials. In K. Paech & M. V. Tracey (Eds.), *Vol. 4. Modern methods of plant analysis* (pp. 142–196). Berlin: Springer-Verlag.

Tanaka, A., Ito, H., Tanaka, R., Tanaka, N. K., Yoshida, K., & Okada, K. (1998). Chlorophyll *a* oxygenase (*CAO*) is involved in chlo- rophyll *b* formation from chlorophyll *a*. *Proceedings of the National Academy of Sciences of the United States of America, 95*(21), 12719–12723.

Tomo, T., Allakhverdiev, S. I., & Mimuro, M. (2011). Constitution and energetics of photosystem I and photosystem II in the chlorophyll *d*-dominated cyanobacterium *Acaryochloris marina*. *Journal of Photochemistry and Photobiology B, 104*, 333–340.

Tomo, T., Okubo, T., Akimoto, S., Yokono, M., Miyashita, H., Tsuchiya, T., et al. (2007). Identification of the special pair of photosystem II in a chlorophyll *d*-dominated cyanobacterium. *Proceedings of the National Academy of Sciences of the United States of America, 104*, 7283–7288.

Tsuchiya, T., Mizoguchi, T., Akimoto, S., Tomo, T., Tamiaki, H., & Mimuro, M. (2012). Metabolic engineering of the Chl *d*-dominated cyanobac- terium *Acaryochloris marina*: Production of a novel Chl species by the introduction of the chlorophyllide *a* oxygenase gene. *Plant & Cell Physiology, 53*(3), 518–527.

Weber, G., & Teale, F. W. J. (1957). Determination of the absolute quantum yield of fluorescent solutions. *Transactions of the Faraday Society, 53*, 646–655.

White, R. C., Jones, I. D., Gibbs, E., & Butler, L. S. (1972). Fluorometric estimation of chlorophyllides, pheophytins and pheophorbides in mixtures. *Journal of Agricultural and Food Chemistry, 20*, 773–778.

Willows, R. D., Li, Y., Scheer, H., & Chen, M. (2013). Structure of chlorophyll *f*. *Organic Letters, 15*, 1588–1590.

Further reading

Airs, R., Temperton, B., Sambles, C., Farnham, G., Skill, S. C., & Llewellyn, C. A. (2014). Chlorophyll *f* and chlorophyll *d* are produced in the cyanobacterium *Chlorogloeopsis fritschii* when cultured under natural light and near-infrared radiation. *FEBS Letters, 588*, 3770–3777.

Allakhverdiev, S. I., Kreslavski, V., Zharmukhamedov, S. K., Voloshin, R. A., Korol'kova, D. V., Tom, T., et al. (2016). Chlorophylls *d* and *f* and their role in primary photosynthetic processes of cyanobacteria. *Biochemistry, 81*, 201–212.

Gan, F., Shen, G., & Bryant, D. A. (2015). Occurrence of far-red light photoacclimation (FaRLiP) in diverse cyanobacteria. *Life, 5*, 4–24.

Gan, F., Zhang, S., Rockwell, N. C., Martin, S. S., Lagarias, J. C., & Bryant, D. A. (2014). Extensive remodeling of a cyanobacterial photosynthetic apparatus in far-red light. *Science, 345*, 1312–1317.

Ho, M.-Y., Gan, F., Shen, G., & Bryant, D. A. (2017). Far-red light photoac-climation (FaRLiP) in *Synechococcus* sp. PCC 7335. II. Characterization of phycobiliproteins produced during acclimation to far-red light. *Photosynthesis Research, 131*, 187–202.

Ho, M.-Y., Soulier, T., Canniffe, D. P., Shen, G., & Bryant, D. A. (2017). Light regulation of pigment and photosystem biosynthesis in cyanobacteria. *Current Opinion in Plant Biology, 37*, 24–33.

Holt, A. S. (1959). Reduction of chlorophyllides, chlorophylls and chlorophyll derivatives by sodium borohydride. *Plant Physiology, 34*(3), 310–314.

Lichtenthaler, H. K. (1987). Chlorophylls and carotenoids: Pigments of photosynthetic biomembranes. *Methods in Enzymology, 148*, 350.

Miyashita, H., Ohkubo, S., Komatsu, H., Sorimachi, Y., Fukayama, D., Fujinuma, D., et al. (2014). Discovery of chloro- phyll *d* in *Acaryochloris marina* and chlorophyll *f* in a unicellular cyanobacterium, strain KC1, isolated from Lake Biwa. *Journal of Physical Chemistry and Biophysics, 4*, 149.

Watanbe, T., Hongu, A., Honda, K., Nakazzato, M., Konno, M., & Saitoh, S. (1984). Preparation of chlorophylls and pheophytins by isocratic liquid chromatography. *Analytical Chemistry, 56*(2), 251–256.

Yoneda, A., Wittmann, B. J., King, J. D., Blankenship, R. E., & Dantas, G. (2016). Transcriptomic analysis illuminates genes involved in chlorophyll synthesis after nitrogen starvation in *Acaryochloris* sp. CCMEE 5410. *Photosynthesis Research, 129*, 171–182.

Zhang, Z.-C., Li, Z.-K., Yin, Y.-C., Li, Y., Jia, Y., Chen, M., & Qiu, B.-S. (2019). Widespread occurrence and unexpected diversity of red-shifted chlorophyll producing cyanobacteria in humid subtropical forest ecosystems. *Environmental microbiology*. https://doi.org/10.1111/1462-2920.14582.

CHAPTER FIVE

The Mg branch of chlorophyll synthesis: Biosynthesis of chlorophyll *a* from protoporphyrin IX

Robert D. Willows*
Department of Molecular Sciences, Macquarie University, North Ryde, NSW, Australia
*Corresponding author: e-mail address: robert.willows@mq.edu.au

Contents

Abstract

Chlorophyll biosynthesis and specific enzymes with the chlorophyll biosynthetic pathway have been reviewed extensively over the last 20 years. This chapter is intended to both summarize and update on previous reviews but it will primarily concentrate on the reaction mechanisms, enzymology and structure function relationships of the enzymes involved in synthesizing chlorophyll *a* from protoporphyrin IX.

Advances in Botanical Research, Volume 90
ISSN 0065-2296
https://doi.org/10.1016/bs.abr.2019.03.003

Abbreviations

AAA+ ATPases-associated with diverse cellular activities
DPOR light independent/dark operative protochlorophyllide oxidoreductase
EM electron microscopy
FNR ferredoxin-NADP$^+$ reductase
GG-PP geranylgeranyl-pyrophosphate
LPOR light dependent protochlorophyllide oxidoreductase
MgPME magnesium protoporphyrin IX monomethyl ester
MgPMTase *S*-adenosyl-L-methionine magnesium PPIX *O*-methyl transferase
MgPPIX magnesium protoporphyrin IX
NTRC NADPH-dependent thioredoxin reductase C
Pchlide 3,8-divinyl protochlorophyllide
Pfam protein family
PLB prolamellar body
PPIX protoporphyrin IX
SAH *S*-adenosyl-L-homocysteine
SAM *S*-adenosyl-L-methionine
TRX thioredoxin

1. Overview of chlorophyll synthesis

Numerous reviews of chlorophyll biosynthesis and specific enzymes within this branch of the pathway have been previously published (Bollivar, 2003; Chew & Bryant, 2007b; Fujita & Bauer, 2003; Gabruk & Mysliwa-Kurdziel, 2015; Rüdiger, 2003; Tanaka & Tanaka, 2007; Tripathy & Pattanayak, 2012; Willows, 2003; Willows & Hansson, 2003). Transcriptional regulation and gene structure have been dealt with in preceding chapters and the biosynthesis of chlorophylls *b*, *c*, *d* and *f* are covered elsewhere in this volume. This chapter is intended to both summarize and update on previous reviews and it will primarily describe the reaction mechanisms, enzymology and structure function relationships of the enzymes involved in synthesizing chlorophyll *a* from protoporphyrin IX (PPIX) shown in Fig. 1.

2. Magnesium chelatase

The first step in the synthesis of chlorophyll from PPIX is the insertion of magnesium into the tetrapyrrole macrocycle to make magnesium protoporphyrin IX (MgPPIX) (reviewed in Hansson, Lundqvist, Sirijovski, & Al-Karadaghi, 2014; Kannangara & von Wettstein, 2010; Walker &

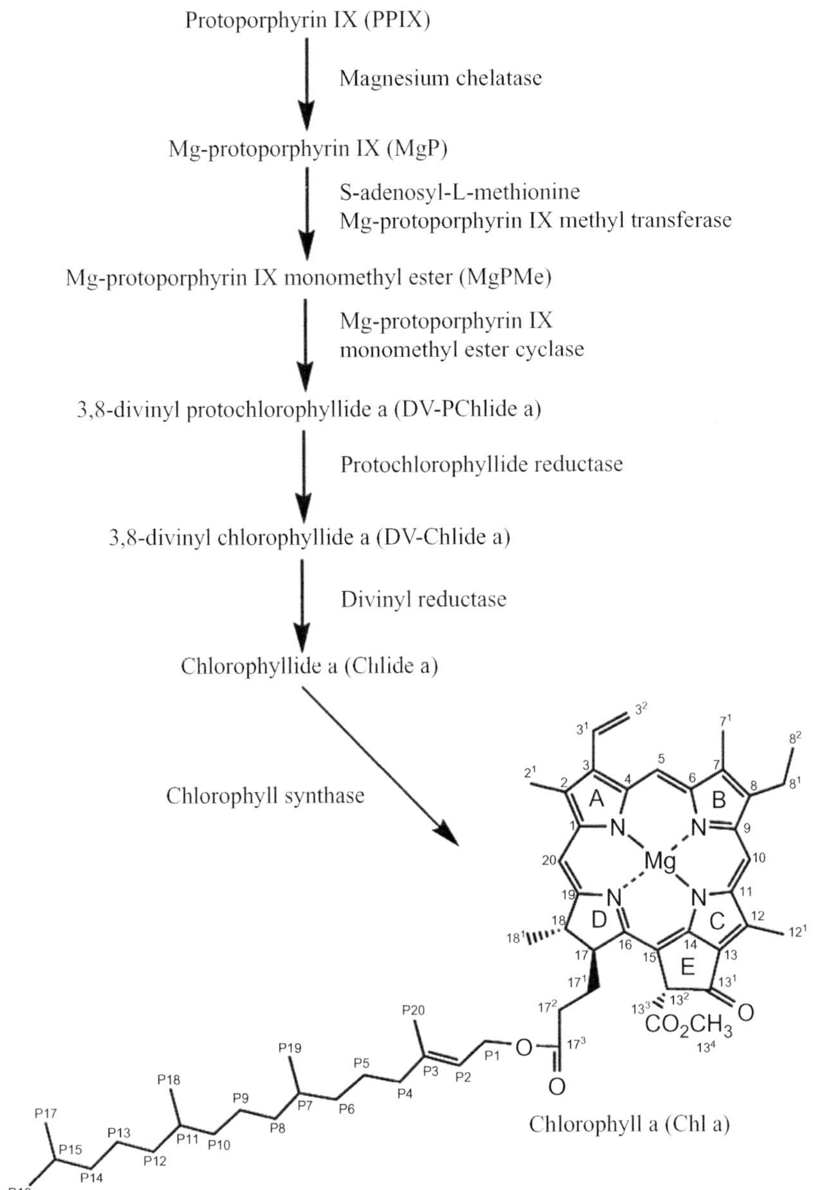

Fig. 1 Synthesis of chlorophyll *a* from protoporphyrin IX.

Willows, 1997; Willows & Hansson, 2003). Magnesium insertion into PPIX is relatively difficult to achieve chemically, requiring heating overnight at $>100\,^{\circ}C$ in dry pyridine together with magnesium acetate or magnesium perchlorate, compared to insertion of other metal ions such as Fe^{2+}, Zn^{2+}

and Ni^{2+} that can be inserted in aqueous conditions (Buchler, 1975; Hambright, 1975). Other tetrapyrrole cofactors with metal ions coordinated include vitamin B12 with inserted Co^{2+}, heme and siroheme with Fe^{2+}, the cofactor F430 with Ni^{2+}. The metal ions of these tetrapyrrole variants have key functions that allow different types of chemistry to be performed based on both the metal redox state and the number of coordination bonds. In addition, tetrapyrroles with other divalent metal ions have also been reported in various biological systems including those with Sn^{2+}, Cu^{2+} and Zn^{2+}, although this is usually a consequence of either chemical or enzymatic insertion into PPIX, the latter by ferrochelatase, which can in some cases use these alternative divalent metal ions as substrate.

2.1 Overview of magnesium chelatase structure and activity

Magnesium chelatase requires a minimum of three protein subunits for activity and requires ATP hydrolysis and the substrates Mg^{2+} and PPIX, as shown in Fig. 2. The protein subunits are generally known as **I**, **D** and **H** with the prefix "Bch" and "*bch*" to the proteins and encoding genes, respectively, from bacteriochlorophyll-synthesizing organisms and "Chl" and "*chl*" to the proteins and encoding genes, respectively, from chlorophyll-synthesizing organisms (Walker & Willows, 1997; Willows, 2003; Willows & Hansson, 2003). These proteins and genes also have numerous alternative names in plants and algae with the names corresponding to the mutants in these organisms with a brief summary shown in Table 1. An accessory protein, GUN4, is also important for

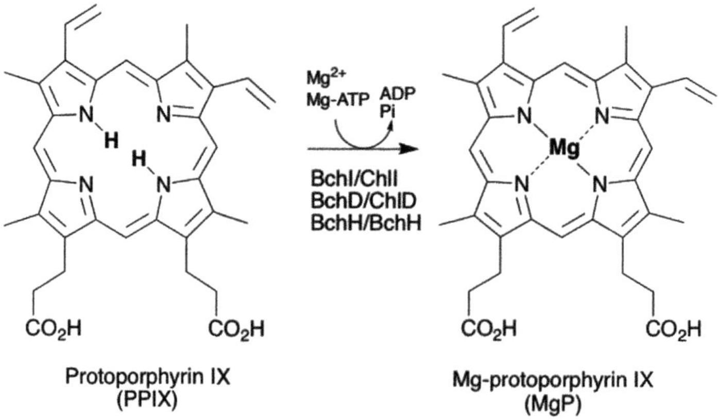

Fig. 2 Catalytic reaction of magnesium chelatase.

Table 1 Names of mutants in magnesium chelatase subunits.

Subunit	Organism	Gene or mutant name	References
I	*Arabidopsis thaliana*	*ch42*	Koncz et al. (1990)
I	*Zea mays*	*oy-1*	Sawers, Farmer, Moffett, and Brutnell (2006) and Sawers, Viney, et al. (2006)
I	*Hordeum vulgare* (barley)	*xantha-h*	Jensen, Willows, et al. (1996)
I	*Euglena gracilis*	*ccsA*	Orsat, Monfort, Chatellard, and Stutz (1992)
I	*Oryza sativa* (rice)	*chlorina-9*	Zhang et al. (2006)
D	*Hordeum vulgare* (barley)	*xantha-g*	Petersen, Moller, Jensen, and Henningsen (1999)
D	*Oryza sativa* (rice)	*chlorina-1*	Zhang et al. (2006)
H	*Antirhinnum majus* (snapdragon)	*olive*	Hudson, Carpenter, Doyle, and Coen (1993)
H	*Hordeum vulgare* (barley)	*xantha-f*	Jensen, Willows, et al. (1996)
H	*Arabidopsis thaliana*	*cch*	Wu et al. (2009)
H	*Arabidopsis thaliana*	*gun5*	Mochizuki, Brusslan, Larkin, Nagatani, and Chory (2001)

regulating magnesium chelatase activity in oxygenic photosynthetic organisms as discussed later.

The **I** protein sequences belong to a unique protein family (pfam), http://pfam.xfam.org/family/PF01078, with 41% sequence homology across the family and range in size from 38 to 45 kDa. In contrast, the **D** proteins are quite divergent in their sequences and they do not belong to a single pfam. The **D** proteins have an AAA+ (ATPases-associated with diverse cellular activities')-N-terminal domain, similar to the **I** protein, which is linked to an integrin-I/von Willibrand C-terminal domain by a polyproline sequence containing 8–10 prolines followed by a low complexity highly charged domain. The molecular weight of the **D** subunits ranges from 60 to 85 kDa. The **H** protein sequences belong to a unique pfam, http://pfam.xfam.org/family/PF02514, consisting of magnesium chelatase **H** and cobalt chelatase CobN subunits. These **H** proteins range in size from 130 to 155 kDa.

2.2 Assays for magnesium chelatase

Early *in organello* or permeabilized cell assays were performed using cucumber chloroplasts or with *Rhodobacter sphaeroides* cells (Ellsworth & Lawrence, 1973; Fuesler, Wong, & Castelfranco, 1984; Fuesler, Wright, & Castelfranco, 1981; Gorchein, 1972, 1973; Pardo, Chereskin, Castelfranco, Franceschi, & Wezelman, 1980; Smith & Rebeiz, 1977). These assays required addition of PPIX, magnesium and ATP or an ATP regenerating system and all activity appeared to be lost on lysis of the chloroplasts or cells. True *in vitro* activity from lysed cucumber chloroplasts was very low and was confirmed by using U-^{14}C PPIX and chromatographic separation and purification of the products and this activity had a similar requirement for ATP and magnesium to the intact chloroplasts (Richter & Rienits, 1980, 1982).

Several years later *in vitro* magnesium chelatase activity, similar to that observed in intact chloroplasts, was demonstrated with developing pea chloroplasts that were lysed by freeze thawing. This activity was separated into membrane and soluble fractions, but the most important finding here was that activity was protein concentration-dependent, implying in hindsight that a concentration-dependent protein complex had to be formed or reformed during catalysis (Walker, Hupp, & Weinstein, 1992; Walker & Weinstein, 1991a, 1991b, 1994, 1995). Recombinantly expressed proteins encoded by *bchI, bchD* and *bchH* genes from *Rhodobacter* were used to reconstitute magnesium chelatase activity *in vitro* (Gibson, Willows, Kannanagara, Wettstein, & Hunter, 1995; Willows & Beale, 1998; Willows, Gibson, Kanangara, Hunter, & von Wettstein, 1996). The homologous genes have been identified in plants, algae, cyanobacteria and other photosynthetic bacteria and the proteins have been similarly expressed and activity reconstituted with genes from *Synechocystis* and *Chlorobium* (Jensen, Gibson, Henningsen, & Hunter, 1996; Jensen, Gibson, & Hunter, 1998; Jensen, Stummann, Hansen, Karlebjerg, & Henningsen, 1996; Jensen, Willows, et al., 1996; Petersen et al., 1998). Much later, *in vitro* magnesium chelatase activity was demonstrated and characterized using a complete set of recombinantly expressed and purified plant and algal proteins (Muller et al., 2014; Sawicki, Zhou, Kwiatkowski, Luo, & Willows, 2017; Zhou, Sawicki, Willows, & Luo, 2012), although individual subunits had been expressed and analyzed prior to these reports as discussed below.

2.3 Regulation of enzyme activity and kinetic parameters

Recombinantly expressed and reconstituted *Rhodobacter* (Gibson, Jensen, & Hunter, 1999; Gibson et al., 1995; Hansson & Kannangara,

1997; Kannangara, Vothknecht, Hansson, & von Wettstein, 1997; Lake, Olsson, Willows, & Hansson, 2004; Sawicki & Willows, 2008, 2010; Willows & Beale, 1998; Willows et al., 1996), *Synechocystis* (Gibson et al., 1999; Jensen, Gibson, et al., 1996; Jensen et al., 1998; Jensen, Gibson, & Hunter, 1999; Karger, Reid, & Hunter, 2001; Reid, Siebert, Bullough, & Hunter, 2003; Sawers, Viney, et al., 2006) and *Thermosynechocystis* (Adams, Brindley, Hunter, & Reid, 2016; Adams, Marklew, Brindley, Hunter, & Reid, 2014; Brindley, Adams, Hunter, & Reid, 2015) magnesium chelatase enzymes have been studied in detail *in vitro*. The recent studies of the *Thermosynechocystis* enzymes revealed the thermodynamics of magnesium insertion with a $\Delta G°$ value of 25–33kJ/mol which explains the ATP hydrolysis requirement for this process (Adams, Brindley, et al., 2016). The K_m's for the enzyme's three main substrates Mg^{2+}, Mg-ATP and either PPIX or the more soluble deuteroporphyrin are within the same order of magnitude between the applied organismic systems. The reported K_m's being 2–5mM for Mg^{2+}, 0.09–0.5mM for Mg-ATP and with the porphyrin substrates being in the sub-μM to μM range.

The kinetic parameters for the porphyrin substrates are complicated because the **H** subunit behaves kinetically as a substrate while binding both PPIX and MgPPIX (Sawicki & Willows, 2008; Willows & Beale, 1998; Willows et al., 1996). BchH binds deuteroporphyrin with a K_D of 0.5μM, while ChlH binds with a lower affinity (K_D 1.2μM) that is further reduced in the presence of Mg-ATP (Karger et al., 2001). The accessory protein GUN4 both stimulates the activity and also binds the PPIX and MgPPIX with K_D's of 0.1–0.5μM and ~0.8μM, respectively, from both *Synechocystis* and rice (Verdecia et al., 2005; Zhou et al., 2012). In contrast, the relative affinities are reversed when using deuteroporphyrin and Mg-deuteroporphyrin with K_D's of ~2 and ~0.3μM for both the *Synechocystis* and *Thermosynechocytis* GUN4s (Davison et al., 2005). Given the differences in K_D's for porphyrin binding between GUN4 and ChlH subunits, it seems reasonable to suggest that GUN4 stimulation of activity may in some way be due to delivery and/or removal of the PPIX/MgP from the ChlH subunit. However, given that the relative binding affinity of deuteroporphyrin and Mg-deuteroporphyrin to ChlH is inverted compared to PPIX and MgP, it raises questions about the relevance of binding measurements using non-natural substrates and whether GUN4 is involved in substrate delivery or product removal. For example, the binding affinities of the non-natural substrates and products suggest ordered trafficking and transfer of substrate from GUN4 to ChlH followed by product back to GUN4. However, the tighter binding of PPIX than MgP to GUN4 does not allow this. Thus, the molecular mechanism for GUN4 stimulation of magnesium chelatase activity remains unclear.

Although the anoxygenic bacteria do not generally have GUN4, a stimulatory effect on magnesium chelatase activity has been reported with Tween 80, or the proteins BchM or BchJ. On further investigation when BchH and porphyrin are present at stoichiometric concentrations and the reaction was allowed to proceed to equilibrium only 71% of the PPIX was converted to MgP in the concentration range 10–125 nM. However, when Tween 80, BchM or BchJ were added the equilibrium shifted to 100% product formation with the MgP bound to the BchM or BchJ proteins or in the Tween 80 micelles, indicating that the stimulation of activity is most likely due to MgP removal from BchH (Sawicki & Willows, 2010).

GUN4's stimulatory effect in plant, algal and cyanobacterial magnesium chelatase is unlikely to be related to shifting the equilibrium as observed for *Rhodobacter* for a number of reasons. First, GUN4 specifically interacts with ChlH (Adhikari, Orler, Chory, Froehlich, & Larkin, 2009; Brzezowski et al., 2014; Davison et al., 2005; Larkin, Alonso, Ecker, & Chory, 2003; Peter & Grimm, 2009; Sobotka et al., 2008; Wilde, Mikolajczyk, Alawady, Lokstein, & Grimm, 2004), and the order of binding of PPIX in kinetic assays suggests that GUN4 stimulation is not due to delivery of PPIX (Zhou et al., 2012). Other evidence that substrate delivery or product removal is unlikely to be responsible for stimulation of magnesium chelatase activity by GUN4 are experiments conducted with site directed mutants of ChlH from *Synechocystis*, corresponding to those observed in *cch* and *gun5-1* ChlH mutants of *A. thaliana*. These ChlH mutants abolish Mg-chelatase activity measured without GUN4 present in the assay and form a ChlI-ChlD-ChlH complex. On addition of GUN4 to this ChlI-ChlD-ChlH complex, magnesium chelatase activity is restored indicating GUN4 has a direct role in activating and stimulating the catalytic cycle (Davison & Hunter, 2011).

In addition to these observations, the eukaryotic GUN4s have a C-terminal extension relative to the cyanobacterial GUN4s and the extension is highly conserved in higher plants. This C-terminal extension when removed prevents stimulation of magnesium chelatase but the truncated GUN4 still binds both MgP and PPIX (Zhou et al., 2012). The C-terminal extension is highly conserved in higher plants and it contains a conserved serine which is phosphorylated in *A. thaliana* (Reiland et al., 2009). When the serine residue is phosphorylated, GUN4 is unable to stimulate magnesium chelatase activity (Richter, Hochheuser, et al., 2016).

The most likely explanation, which fits the available data, is that the phosphorylation prevents GUN4 from interacting with ChlH, while other physiological implications on GUN4 phosphorylation were also suggested (Richter, Hochheuser, et al., 2016).

The BchH/ChlH behaves kinetically as a substrate and displays positive cooperativity (Muller, Sawicki, Tabrizi, et al., 2014; Sawicki & Willows, 2008) in most systems, but the BchI/ChlI can also be treated kinetically as a substrate for the enzyme. As such, they also have $S_{0.5}$ or K_m values reported of 100–300 nM for BchH/ChlH and 20 nM for BchI (Adams, Brindley, et al., 2016; Sawicki & Willows, 2008). However, the relevance of these Michaelis-Menton kinetic parameters may depend on the protein concentrations of the **I** and **D** used in the assays. A number of authors have assumed the **D** subunit is the enzymatic subunit for these types of calculations. In *Rhodobacter,* the k_{cat} was 5 min^{-1}, while in *Synechocystis,* it had a slower turnover of 0.16 min^{-1}. Given these measurements are *in vitro*, and that structural and interaction studies of the enzyme subunits and complexes are conflicting, as discussed below, the relevance of these values remains an open question as *in vivo* activity may depend on other factors.

Numerous plants have multiple ChlI subunit genes. In *A. thaliana* there are two ChlI proteins with ChlI1 required for activity, although the ChlI2 can substitute for ChlI1 and allow some chlorophyll synthesis (Kobayashi et al., 2008; Rissler, Collakova, DellaPenna, Whelan, & Pogson, 2002). This contrasts with the two ChlI proteins in *Chlamydomonas* with the ChlI2 unable to substitute for ChlI1 (Brzezowski et al., 2016). This *Chlamydomonas* ChlI2 has a C-terminal extension and is unable to form large ATP-dependent complexes as the ChlI/BchI proteins normally do. However, this ChlI2 stimulates magnesium chelatase activity in the presence of ChlI1 (Sawicki et al., 2017). The ChlI2 has a histidine kinase activity that specifically phosphorylates the ChlD subunit at the conserved H641 and it also prevents aggregation of ChlI1, suggesting that it also interacts with ChlI1. The ChlI1s of *Chlamydomonas* and rice also display histidine kinase activity, but at a lower level than ChlI2. The BchI subunit of *Rhodobacter* does not display this histidine kinase activity. The conserved histidine at position 641 (H641) is not absolutely required for activity because the phosphomimicking H641E mutant ChlD is still active and the mutant H641E ChlD is not phosphorylated. It should be noted that the H641 in ChlD is not conserved in BchD proteins. The stimulation of *in vitro* activity by ChlI2 may therefore be due to a combination of the stabilization of ChlI1 and phosphorylation of ChlD and the phosphorylation may also have a regulatory function. In addition to phosphorylation of ChlD by ChlI in plants and algae, nonequilibrium isotope experiments also indicate that an enzyme-phosphate complex exists in the *Synechocystis* magnesium chelatase (Adams & Reid, 2012) suggesting that phosphorylation of ChlD may occur in all oxygenic organisms.

The redox state of the chloroplast is another regulatory mechanism controlling magnesium chelatase activity. ChlI in *A. thaliana* was identified as a target of thioredoxin (TRX) (Balmer et al., 2003), and the ATPase activity of both ChlI1 and ChlI2 was inhibited by oxidation and restored by TRX-dependent reduction (Ikegami et al., 2007; Kobayashi et al., 2008). A cysteine residue on the C-terminal helical domain of these proteins was identified as being oxidation sensitive and its oxidation state *in vivo* was light-dependent (Ikegami et al., 2007; Kobayashi et al., 2008). Similarly the pea ChlI was also found to be a TRX target and the TRX-F and TRX-M isoforms were involved in restoring activity (Luo et al., 2012). NADPH-dependent thioredoxin reductase C (NTRC) mutants of *A. thaliana* are perturbed in chlorophyll synthesis (Stenbaek et al., 2008). These mutants accumulate intermediates from PPIX to MgPME and the NTRC-2-cysteine-peroxiredoxin system has been shown to be involved in regulating the redox status and activity of the first three enzymes shown in Fig. 1 (Da et al., 2017; Stenbaek et al., 2008). The redox state of these enzymes is modulated by TRX and NTRC and is, thus, important in regulating chlorophyll synthesis.

2.4 Structure activity relationships of magnesium chelatase subunits

BchI from *R. capsulatus* was crystallized as hexagonal rods (Willows, Hansson, Beale, Laurberg, & al-Karadaghi, S., 1999). The 2.1 Å X-ray structure revealed that BchI belonged to the AAA+ class of proteins with a novel inverted domain structure linked by a long kinked alpha helix (Fodje et al., 2001). The packing interactions in the crystal structure created a hexameric helical arrangement of the subunits. The arrangement was such that two neighboring subunits came together to form the ATP binding site. The Walker A and Walker B type ATP binding site on one subunit is positioned near a conserved arginine finger on a neighboring subunit that is usually provided within the same subunit of AAA+ proteins and is thought to mediate ATP hydrolysis. The same paper presented a structural model of the BchD which contains an N-terminal BchI-like domain linked to an integrin-I domain by a poly-proline region of ~10 amino acids followed by 30–50 mainly charged amino acids. The integrin-I domain contains a metal ion-dependent adhesion site (MIDAS) motif which is thought to allow interaction with RG(D/E) motifs in other proteins. All BchD and ChlD proteins have a similar domain architecture and all BchI/ChlI proteins contain a conserved RG(D/E) motif suggesting this interaction between the I and D subunits probably occurs through the same mechanism.

Single amino acid mutations in ChlI of barley, also called XANH, gave a semi-dominant phenotype and the homologous mutation in BchI of *Rhodobacter* replicated *in vitro* this semidominant behavior suggesting that the higher order hexameric structures observed by electron microscopy (EM) are important for full activity of the magnesium chelatase (Hansson, Willows, Roberts, & Hansson, 2002).

EM-single particle analysis showed that BchI formed ATP-dependent hexamers (Willows, Hansson, Birch, & al-Karadaghi, & Hansson, 2004). Cryo-EM of BchI/BchD complexes in the presence and absence of ATP and ATP-analogues revealed a double hexameric structure, as shown in Fig. 3, with BchI forming a trimer of dimers with a hexamer of BchD sitting on top of this trimer (Elmlund et al., 2008; Lundqvist et al., 2010). Large conformational differences are observed in the BchD hexamer when ATP, ADP or non-hydrolyzable ATP analogues were bound (Lundqvist et al., 2010). The BchD-BchD interactions are stable up to 4M urea maintaining a high molecular weight complex. Inactive point mutants of BchD based on identified barley XANG mutations can only be added to WT BchD proteins after denaturation in 6M urea and refolding. By doing this with different ratios of wild-type and mutant BchD it appears that either only fully wild-type BchD complexes were active or that complexes containing mutated BchD had reduced magnesium chelatase activity (Axelsson et al., 2006) similar to the experiments with XANH/BchI mentioned earlier (Hansson et al., 2002). Resolving this question with BchI was possible because the subunits exchange more readily and it appears that the BchI disassembles or exchanges from the complex in each catalytic cycle. This experiment indicated that all three pairs of dimers in the hexameric BchI/ChlI ring must be capable of ATP hydrolysis for magnesium chelatase activity (Lundqvist, Braumann, Kurowska, Muller, & Hansson, 2013).

Binding experiments using atomic force microscopy with a combination of tethered *Synechocystis* complete ChlI and ChlD as well as ChlD subdomains show that interaction between the ChlI and ChlD is *via* the AAA+ domain of ChlD, provided the polyproline linker is present, and the interaction requires Mg-ADP. This Mg-ADP-dependent interaction

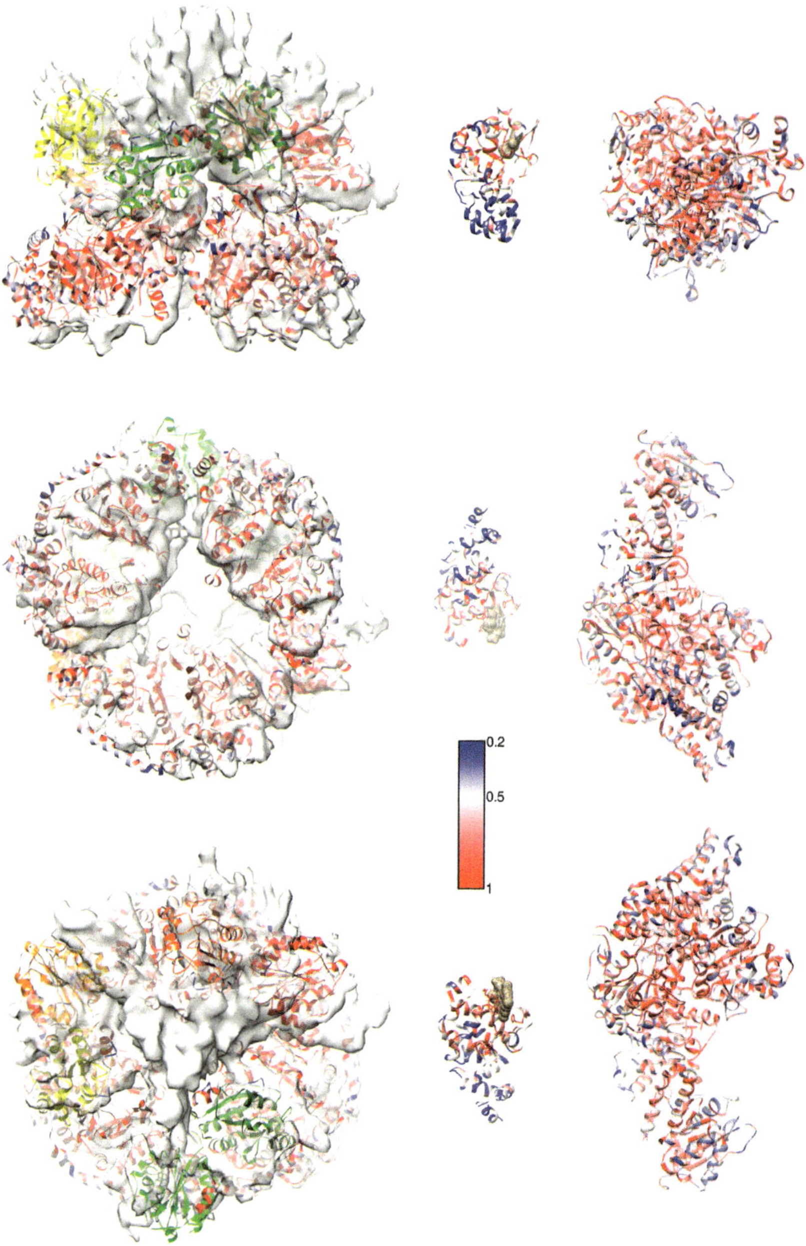

Fig. 3 Structure of magnesium chelatase subunits and complexes. Middle structures are rotated 90° to view the bottom of the upper models, and the lower structure is rotated by 180° to view the rear of the middle structures. *Left*: BchID complex (BchI/BchD: 2 × 31) with the space filling model being the electron microscopy structure with ATP bound.

was supported by microscale thermophoresis data (Adams, Vasilev, Brindley, & Hunter, 2016). Given this interaction was Mg-ADP-dependent it is interesting to speculate that this may be an intermediate in recycling or turnover/exchange of ChlI subunits discussed previously in the catalytic cycle.

Low resolution EM structures of BchH with and without PPIX bound showed a three lobed structure with two of the lobes coming together on PPIX binding. One of these two movable lobes was identified as the C-terminal domain (Sirijovski et al., 2008). It appears that BchH has an iron sulfur cluster within the central region of the protein which is not conserved in ChlH sequences. The function of this cluster is unclear but its oxidation state is not important for activity, suggesting a regulatory function possibly in turnover of the subunit (Sirijovski, Mamedov, Olsson, Styring, & Hansson, 2007). The EM structure of *Synechocystis* ChlH was also determined and it appeared to have an additional lobe with a structure more similar to BchH with PPIX bound (Qian et al., 2012). The high resolution X-ray structure of *Synechocystis* ChlH without PPIX bound revealed six different domains in the protein (Chen, Pu, Fang, et al., 2015) with the most central part of the protein being the most conserved between species (as shown red-colored in Fig. 3). The red region visible in the bottom view of Fig. 3 is thought to be the PPIX binding site as it contains a hydrophobic pocket (Chen, Pu, Fang, et al., 2015). This region may also be where GUN4 binds to activate the catalytic cycle. The large conformational change occurring on PPIX binding observed in the BchH structural may be related to the activation step observed kinetically between ChlH and PPIX (Zhou et al., 2012).

Crystal structures of GUN4 have been determined from two cyanobacteria (Chen, Pu, Wang, et al., 2015; Davison et al., 2005; Verdecia et al., 2005) and from *Chlamydomonas* (Tarahi Tabrizi, Sawicki, Zhou,

Center: Synechocystis GUN4 complexed with PPIX as a space filling model (GUN4: 1Y6I). *Right*: Synechocystis ChlH (ChlH: 4ZHJ). With the exception of BchD subunits in the model, coloring of the subunits, as shown in the scale bar, is based on conservation of residues as per the color scale with red being 100% to blue being non-conserved between species. Conservation of residue comparisons is with orthologous protein sequences from *Arabidopsis thaliana*, *Chlamydomonas reinhardtii*, *Orysa sativa* and *Synechocystis*. The BchI comparison in BchID and the ChlH comparison includes the *Rhodobacter capsulatus* subunits. The sequence alignments were performed with ClustalW and figure created in Chimera and all subunits are at the same scale (Pettersen et al., 2004).

Luo, & Willows, 2016). The core structures are similar as shown in red in Fig. 3, and the *Synechocystis* structure has been solved with PPIX bound in a hydrophobic pocket that is partly covered by a loop when PPIX is not bound. The PPIX propanoate groups are exposed with approximately two-thirds of the macrocycle bound in the pocket (Chen, Pu, Wang, et al., 2015).

Despite the progress in structural and mechanistic studies on the magnesium chelatase, the precise mechanism by which Mg^{2+} is inserted by this complex enzyme remains unclear. The most widely posited theory is that the **ID** complex with ATP bound binds the **H** subunit with PPIX bound. ATP hydrolysis then catalyzes a conformational change in the **H** subunit which bends the porphyrin to both expose and deprotonate the pyrrole nitrogens while allowing a bound and dehydrated Mg^{2+} ion to exchange into the pyrrole ring. As multiple ATP hydrolyses are required for catalytic insertion, the ATP may be hydrolyzed in concert by the BchI/ChlI trimer of dimers or it may occur in sequence around the ring. Future structural studies of whole Mg chelatase complexes and intermediate states will be required to prove this mechanism.

3. *S*-Adenosyl-L-methionine magnesium protoporphyrin IX *O*-methyl transferase

The second committed step in chlorophyll synthesis is the synthesis of MgPME catalyzed by *S*-adenosyl-L-methionine magnesium PPIX *O*-methyl transferase (MgPMTase). This enzyme catalyzes the transfer of the methyl group from *S*-adenosyl-L-methionine (SAM) to the carboxyl group on the C-13 propionate side chain of MgP as shown in Fig. 4. The literature on the MgPMTase was extensively reviewed by Bollivar (2003).

Early examples of *in vitro* assays used chromatophores from *Rhodobacter* (cited in Bollivar, 2003) and required separation of the reactants from the products. These studies identified the requirement for SAM using ^{14}C-methyl SAM as a substrate. MgPMTase activity was assayed and characterized from *Rhodobacter sphaeoroides*, wheat, *Zea mays* and *Euglena gracilis*. The *Rhodobacter* enzyme could utilize MgPPIX, Mg-deuteroporphyrin, Ca(II)-PPIX, and Zn-PPIX as substrates and was inhibited by SAM and *S*-adenosyl-L-homocysteine (SAH), see Bollivar (2003) for details. It is worth noting that early magnesium chelatase assays with *Rhodobacter* cells produced MgPMe suggesting that the chelatase and methyltransferase are

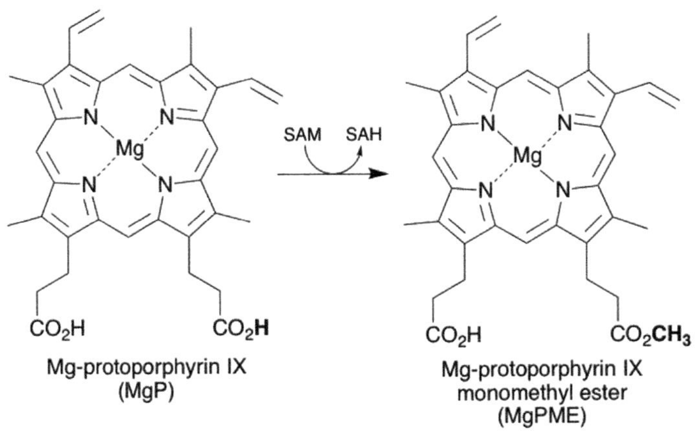

Fig. 4 Catalytic reaction of *S*-adenosyl-L-methionine magnesium protoporphyrin IX monomethylester transferase.

tightly coupled (Gorchein, 1972, 1973). This observation is consistent with the effect of BchM on magnesium chelatase activity mentioned early.

Rhodobacter bchM gene/s were confirmed to code for MgPMTase and *E. coli*-produced recombinant protein shows BchM activity (Bollivar, Jiang, Bauer, & Beale, 1994; Gibson & Hunter, 1994). As the enzyme activity was membrane-associated, it took some time before the *E. coli*-expressed *R. capsulatus* BchM was purified to homogeneity and characterized kinetically. This enzyme displayed a random sequential mechanism, with Co(II)- and Mn(III)-PPIX acting as inhibitors, while Zn(II)-PPIX and deuteroporphyrin were substrates in addition to MgPPIX. This specificity indicated a requirement for five-coordinate square-pyramidal metalloporphyrins as substrates (Sawicki & Willows, 2007).

ChlM is the orthologous protein to BchM in chlorophyll-synthesizing organisms. The *Synechocystis chlM* gene was shown to complement a *bchM Rhodobacter capsulatus* mutant (Smith, Suzuki, & Bauer, 1996). Expression and purification of *Synechocystis* ChlM from *E. coli* confirmed that ChlM possesses the MgPMTase activity (Shepherd, Reid, & Hunter, 2003). Further analysis of the *Synechocystis* ChlM revealed an unidentified intermediate in the reaction. Quenched flow analysis of this intermediate with the magnesium chelatase ChlH subunit suggests that ChlH is involved in the reaction pathway, although it did not affect steady-state kinetics of the reaction (Shepherd, McLean, & Hunter, 2005). An interaction between ChlM and the magnesium chelatase ChlH in tobacco was shown (Alawady, Reski, Yaronskaya, & Grimm, 2005) confirming that this coupling of magnesium chelatase to MgPMTase extends to plants. A novel continuous coupled spectrophotometric assay was developed to measure ChlM activity and this assay confirmed the stimulatory effect of ChlH on ChlM activity (McLean & Hunter, 2009).

The ChlM of *A. thaliana* was shown to be sensitive to oxidation. The cysteine-177 displayed a redox-dependent activation presumably dependent on NTRC and TRX (Richter, Wang, & Grimm, 2016).

The X-ray crystal structure of *Synechocystis* ChlM without substrate and with both SAM and SAH has been determined (Chen et al., 2014). The structure with SAM bound is shown in Fig. 5. The main area of conservation between species is in the active site with bound SAM, as indicated by the red shading. In contrast the residues away from the active site are not highly conserved. The differences in the structures reveal two "arm" regions that may

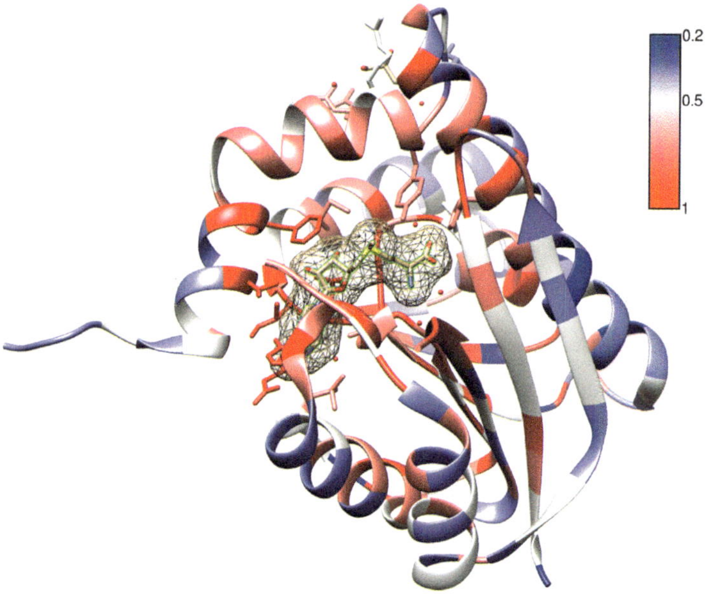

Fig. 5 Structure of SAM-magnesium protoporphyrin IX *O*-methyl transferase from *Synechocystis* with SAM bound as a space filling model (ChlM: 4QDJ). Coloring of protein backbone, as shown in the scale bar, is based on conservation of residues as per the color scale with red being 100% conserved to blue being non-conserved between species. Conservation of residue comparisons is made with ClustalW using aligned ChlM protein sequences from *Arabidopsis thaliana* (OAO99572), *Chlamydomonas reinhardtii* (XP_001702380), *Oryza sativa* (AAO33146), *Synechocystis* (BAA10812) and BchM protein sequence from *Rhodobacter capsulatus* (ADE84426). This figure was created in Chimera (Pettersen et al., 2004).

modulate binding and release of substrates and products, while tyrosine-28 and histidine-139 appear to play a role in methyl transfer reaction, but are not involved in binding of substrates or products.

4. Magnesium protoporphyrin IX monomethyl ester oxidative cyclase

The formation of the fifth isocyclic ring in chlorophyll *a*nd bacteriochlorophyll requires a number of oxidation steps on the 13-C methyl propanoate of MgPME and is catalyzed by the MgPME oxidative cyclase, to form 3,8 divinyl-protochlorophyllide (Pchlide). In oxygenic organisms the oxygen in the isocyclic ring is derived from molecular oxygen (Walker, Mansfield, Smith, & Castelfranco, 1989). In most anoxygenic bacteria that synthesize bacteriochlorophyll the oxygen is derived from water (Porra et al., 1996; Porra, Schafer, Katheder, & Scheer, Gad'on, et al., 1995; Porra, Schafer, Katheder, & Scheer, 1995) provided by a proposed hydratase. Thus, there appears to be two different pathways and mechanisms for formation of the isocyclic ring, a molecular oxygen requiring aerobic cyclase and an anaerobic cyclase.

Based on the biochemical evidence and proposed intermediates, the formation of the isocyclic ring utilizing molecular oxygen consists of three sequential steps in the following biosynthetic scheme involving: (a) hydroxylation of the ring C to methylpropionate, (b) oxidation to ketopropionate, (c) ring closure *via* carbon-carbon bond formation of an activated methylene group to the 15-*meso* carbon of the porphyrin ring (Wong, Castelfranco, Goff, & Smith, 1985) as shown in Fig. 6. The methylation of the propanoate is required to stabilize and prevent decarboxylation which can and does occur if the methyl group is removed after the ring forms.

4.1 Biochemical assays

MgPME oxidative cyclase activity was first assayed in the 1980s using cucumber chloroplasts (Chereskin & Castelfranco, 1982; Chereskin, Castelfranco, Dallas, & Straub, 1983; Chereskin, Wong, & Castelfranco, 1982; Fuesler et al., 1984) and wheat etioplasts (Nasrulhaq-Boyce, Griffiths, & Jones, 1987). The common features of these assay systems were an absolute requirement for NADPH and molecular oxygen. The activity was inhibited by lipid soluble metal ion chelators such as 2,2,-dipyridyl *in vivo*, which remove the iron. Additionally, the wheat system was able

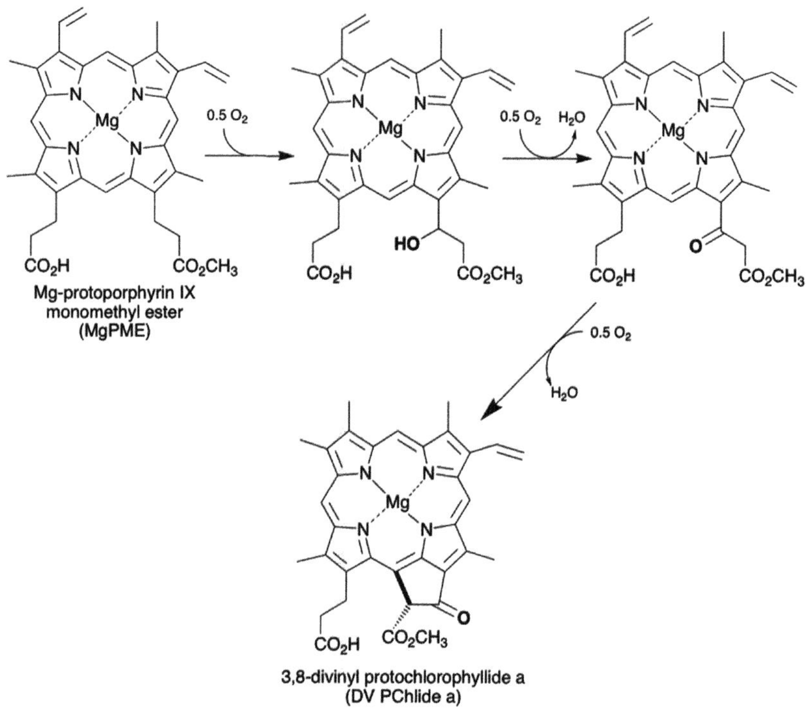

Fig. 6 Magnesium protoporphyrin IX monomethyl ester oxidative cyclase reaction intermediates. Reaction catalyzed by the AcsF-type cyclase using molecular oxygen and NADPH as a secondary reductant (Hollingshead et al., 2012; Wong et al., 1985).

to utilize Zn-protoporphyrin IX-monomethylester in addition to MgPME as substrate.

In vitro MgPME oxidative cyclase activity, as opposed to *in organello* activity, was demonstrated using extracts from *Chlamydomonas reinhardtii* and *Synechocystis* PCC 6803 (Bollivar & Beale, 1996), cucumber (Walker, Castelfranco, & Whyte, 1991) and barley (Rzeznicka et al., 2005) and was resolved into membrane and soluble components for these assays.

Further biochemical characterization also provided evidence for the incorporation of atmospheric oxygen into Pchlide in plants (Walker et al., 1989). This contrasts with a MgPME oxidative cyclase from bacteriochlorophyll-synthesizing anaerobic bacteria that has no requirement for molecular oxygen. The *bchE*-encoded gene product of *Rhodobacter* species was identified as a component of the MgPME oxidative cyclase through mutation analysis in *Rhodobacter capsulatus* (Bollivar, Suzuki, Beatty, Dobrowolski, & Bauer, 1994; Taylor, Cohen, Clark, & Marrs, 1983) and

Rhodobacter sphaeroides (Coomber, Chaudhri, Connor, Britton, & Hunter, 1990; Hunter & Coomber, 1988). This anaerobic cyclase requires an adenosyl-cobalamin cofactor, suggesting a free radical hydratase mechanism (Gough, Petersen, & Duus, 2000), which means that the protein differs both structurally and mechanistically from the oxygen requiring aerobic cyclase required by most chlorophyll-synthesizing organisms. Cyanobacteria such as *Cyanothece* and *Synechocystis* PCC 6803 have *bchE* orthologues (Minamizaki, Mizoguchi, Goto, Tamiaki, & Fujita, 2008; Yamanashi, Minamizaki, & Fujita, 2015). The *Cyanothece bchE* orthologue functionally complements a *Rhodobacter bchE* mutation and restores bacteriochlorophyll synthesis. However, the *bchE* genes are not required in these cyanobacteria indicating that additional aerobic cyclase is present.

An aerobic cyclase-encoding gene was identified in *Rubrivax gelatinosus* which contributed to synthesis of bacteriochlorophyll aerobically (Pinta, Picaud, Reiss-Husson, & Astier, 2002). The gene encoding this cyclase was called *acsF* standing for aerobic cyclization system Fe-containing subunit and subsequently *acsF* homologs have been identified in cyanobacteria (Minamizaki et al., 2008; Peter et al., 2009), plants (Tottey et al., 2003) and algae (Allen, Kropat, & Merchant, 2008; Moseley, Quinn, Eriksson, & Merchant, 2000; Quinn, Eriksson, Moseley, & Merchant, 2002).

AcsF contains a putative diiron site and, thus, is viewed as the catalytic subunit of MgPME cyclase (Pinta et al., 2002). As the reaction uses at least a soluble component and a membrane component and given the available biochemical data including inhibition by lipid soluble Fe^{2+} chelators, it is expected that the AcsF is the main membrane component of the MgPME cyclase. The membrane component appears to be formed from at least two different proteins; the first one is the homolog of AcsF, and the second one is the protein deficient in the viridis-k mutant of barley (Rzeznicka et al., 2005).

Synechocystis has two orthologues of the *acsF*-type cyclase, called *chlA* or *cycI* or sll1214 and *chlB* or *cycII* or sll1874. The chlB is specifically expressed under micro-oxic conditions, and its level is probably minimal under atmospheric oxygen (Minamizaki et al., 2008; Peter et al., 2009). Similarly, *Chlamydomonas* has two *acsF*-type cyclase genes called *CRD1* and *CTH1*, with the *CRD1* expressed under micro-oxic and/or under copper limitation conditions while *CTHI* is expressed under normal aerobic and nutritional conditions (Allen et al., 2008; Moseley et al., 2000; Quinn et al., 2002). *Arabidopsis* also has an *acsF*-type gene encoded by *CHL27* which is required for cyclase activity (Tottey et al., 2003).

4.2 Other cyclase components

A FLAG-tagged chlA/sll1214/cycI expressed in *Synechocytsis* PCC6803 was used to immune-precipitate this AcsF membrane component in order to identify any co-purifying proteins. Ycf54 was co-purified in these pull down experiments and was identified as an important component of the oxidative cyclase (Hollingshead et al., 2012). At around the same time the tobacco *ycf54*, also called LCAA, was suggested to be a component of the cyclase through its interaction with the AcsF-type membrane-localized CHL27 in *Arabidopsis thaliana* (Albus et al., 2012). The complete deletion of *ycf54* in *Synechocystis* was subsequently shown to still allow 13% of normal chlorophyll synthesis (Hollingshead et al., 2016), although these bacteria accumulated large amounts of MgPME indicating its conversion to Pchlide is very poor in the absence of Ycf54. Ycf54 is a soluble protein and the X-ray crystal structure of wild type (PDB ID: 5M2P) and mutant proteins (PDB ID: 5M2U and 5MRR) have been determined (Hollingshead, Bliss, Baker, & Neil Hunter, 2017). The protein appears to be rather hydrophilic and it was suggested that it may be the soluble component of the cyclase required in biochemical assays (Albus et al., 2012). However, it was demonstrated in barley that the Ycf54 is membrane-associated together with the AcsF-type homolog designated XanL and that it is not the soluble component required for *in vitro* activity (Bollivar, Braumann, Berendt, Gough, & Hansson, 2014).

Three different classes of AcsF have been recently identified that have different requirements for Ycf54 as an accessory protein in the cyclase reaction. There is a class of AcsF, which does not require an additional protein such as Ycf54, a second class of AcsF requiring Ycf54 and a third new class of AcsF that requires a protein called BciE instead of Ycf54 (Chen, Canniffe, & Hunter, 2017).

A possible candidate for the soluble component and an electron donor for the cyclase reaction has recently been identified as ferredoxin-NADP-reductase (FNR). FNR was found to be an interacting partner with Ycf54 in *Arabidopsis thaliana* and it is suggested that Ycf54 acts as a scaffolding factor between AcsF and other components involved in chlorophyll synthesis including FNR (Herbst, Girke, Hajirezaei, Hanke, & Grimm, 2018).

Significant work is still required on this enzyme to clarify how it is regulated and what factors are absolutely required for activity. A reproducible enzyme assay or expression system would go some way to addressing these questions. The reported *E. coli* system that is capable of synthesizing chlorophyll may be of benefit for *in vitro* assays in the future (Chen et al., 2018).

5. Protochlorophyllide oxidoreductase

One of the most well studied reactions in chlorophyll biosynthesis is the reduction of ring D of Pchlide to form chlorophyllide as shown in Fig. 7. There are two different enzymes for catalyzing this reaction and they have different substrate requirements, structure and properties and both enzymes have been reviewed a number of times over the past 15 years (Fujita & Bauer, 2003; Gabruk & Mysliwa-Kurdziel, 2015; Layer, Krausze, & Moser, 2017; Reinbothe et al., 2010; Rüdiger, 2003). The first enzyme to be discussed is known as the light-dependent protochlorophyllide oxidoreductase, or LPOR. The substrates for LPOR are Pchlide *a*, NADPH and light, and this Pchlide reduction is one of the few light-dependent enzyme-catalyzed reactions known. The second enzyme is light-independent and is often referred to as dark operative protochlorophyllide oxidoreductase or DPOR. The substrates for DPOR are Pchlide *a*, ATP and reduced ferredoxin.

5.1 Properties of LPOR

LPOR is encoded by the *por/POR* gene and is found in all oxygenic photosynthetic organisms that make chlorophyll. Until recently it was thought to be restricted to cyanobacteria and eukaryotic phototrophs. But a functional LPOR has recently been found in an anoxygenic phototroph, *Dinoroseobacter shibae* (Kaschner et al., 2014). An extensive review of LPOR has been recently published in 2015 (Gabruk & Mysliwa-Kurdziel, 2015), and additional six reviews were published in 2010–2011 on aspects of LPOR

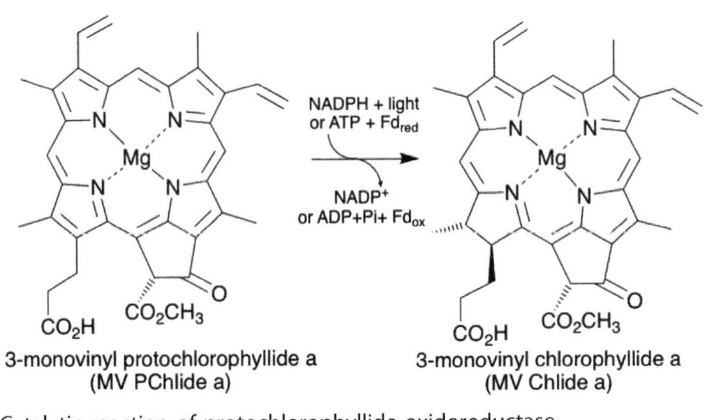

Fig. 7 Catalytic reaction of protochlorophyllide oxidoreductase.

(Amirjani, 2010; Belyaeva & Litvin, 2011; Chekunova, 2010; Fujita & Kurisu, 2011; Mysliwa-Kurdziel & Strzalka, 2010; Reinbothe et al., 2010). Despite the interest and work on LPOR there are no structures published. However, detailed spectroscopy of the reaction has been performed and spectral intermediates determined. Numerous studies on mutants and variants have also contributed to the understanding of the reaction mechanism.

LPOR belongs to the class of proteins known as short chain alcohol dehydrogenases and appears to have originally evolved in cyanobacteria due to strong evolutionary pressure because of the high oxygen sensitivity of DPOR (Yang & Cheng, 2004). Angiosperms lack DPOR and during germination in darkness they assemble Pchlide:LPOR:NADPH complexes inside the etioplasts, which are poised for photoreduction of Pchlide to chlorophyllide. These complexes are embedded in large membrane structures known as prolamellar bodies (PLBs). PLBs are visible in electron micrographs of transmission EM as semicrystalline membrane structures which have been shown by atomic force microscopy to consist of a branched tubular network (Grzyb, Solymosi, Strzalka, & Mysliwa-Kurdziel, 2013). These PLBs can be isolated and consist of both lipids and proteins (Blomqvist, Ryberg, & Sundqvist, 2008). After illumination, PLBs gradually disperse (Henningsen, Boynton, & Von, 1993), and given that photoreduction of Pchlide accompanies illumination, it is commonly accepted that PLB dispersion and formation of primary thylakoid lamella is induced by this photoreduction. The light-harvesting-like protein 3 (LIL3) has been shown to bind to LPOR (Hey et al., 2017; Zhou et al., 2017), as well as other later enzymes of chlorophyll synthesis. This suggests that chlorophyll synthesis and assembly of photosystems and light harvesting complexes are probably the driving force for PLB dispersal as discussed in detail in chapters "Chlorophyll-binding subunits of photosystem I and II: Biosynthesis, chlorophyll incorporation and assembly" by Sobotka and Komenda, Vol. 91 and "Posttranslational control of tetrapyrrole biosynthesis: Interacting proteins, chaperones, auxiliary factors" by Herbst et al., Vol. 91.

The three different isoforms of LPOR from *A. thaliana* were all shown to form Pchlide-LPOR-NADPH complexes with similar spectral and kinetic properties (Gabruk et al., 2015). The spectral properties of these complexes were shown to be similar to the complexes observed in PLB (Gabruk et al., 2015). Supramolecular aggregates of Pchlide-LPOR-NADPH complexes containing Pchlide *b* were reported in etioplasts of barley, which allow more efficient photoreduction under low light conditions. These Pchlide *b* complexes transferring energy to Pchlide *a* complexes have been proposed by Reinbothe et al. (2010). Although this is controversial and has been disputed due to lack of evidence for Pchlide *b* (Armstrong, Apel, & Rudiger, 2000) it has been suggested that it may be due to differences in barley LPOR (Yuan et al., 2012).

5.1.1 Regulation of LPOR activity

In addition to the obvious regulation of LPOR by light, LPOR is also regulated by lipids with allosteric regulation of NADPH-binding mediated by phosphatidyl glycerol and sulfoquinovosyl diacylglycerol (Gabruk, Mysliwa-Kurdziel, & Kruk, 2017). No other direct regulatory factors for LPOR have been identified. More LPOR interacting factors are reported in chapter "Posttranslational control of tetrapyrrole biosynthesis: Interacting proteins, chaperones, auxiliary factors" by Herbst et al., Vol. 91.

5.1.2 Enzyme activity, substrate specificity and mechanism

The penta-coordinate Zn^{2+} derivative of Pchlide can also function as substrate for LPOR, while the square-pyrimidal Co^{2+}, Ni^{2+} and Cu^{2+} derivatives are not substrates (Griffiths, 1980). Pchlide derivatives with substituents on rings A and B are all substrates for LPOR while substituents and modifications of ring E or esters on the C17 propanoic acid are not substrates (see Rüdiger, 2003 and cited references).

Precautions need to be taken to exclude light when setting up assays for LPOR enzyme activity since light is a substrate for the reaction. Only single turnovers of purified LPOR could be achieved until Griffith discovered that a firmly bound NADPH to a Pchlide:LPOR:NADPH complex was responsible for photoreduction and that flashes of light were required to allow substrate exchange (Griffiths, 1978). Numerous spectral shifts and forms have been reported for pigment complexes and reaction intermediates for LPOR pigment complexes. Sequential spectral forms of pigments and LPOR-pigment complexes were proposed by Griffiths and later by Rüdiger (2003) and cited references. These intermediates being: 1. Free Pchlide, 630 nm; 2. Pchlide:LPOR:NADPH, 638 nm; 3. Pchlide:LPOR:NADPH, 650 nm; 4. Chlide:LPOR:$NADP^+$, 674 nm; 5. Chlide:LPOR:NADPH, 684 nm; 6. Chlide, 672 nm. The shift from 684 to 672 nm, known as the Shibata shift, is observable *in vivo* and can be inhibited by fluoride and non-hydrolyzable ATP and promoted by ATP (Wiktorsson, Ryberg, & Sundqvist, 1996). The authors suggested protein phosphorylation as a mechanism, but there is no evidence for direct phosphorylation of LPOR.

The isoforms of LPOR from barley and *A. thaliana* have been studied *in vitro*, with the barley LPOR-B displaying a higher catalytic efficiency than LPOR-A (Garrone, Dietzek, Archipowa, Zipfel, & Hermann, 2015). This contrasts with the three *A. thaliana* isoforms that had no difference in kinetic parameters. However, when these different isozymes were incubated for extended periods in the dark and then illuminated to begin the assay, LPOR-C displayed lower rates than the LPOR-A and LPOR-B (Gabruk et al., 2015).

While there are no structures solved for LPOR, a number of structural models have been generated and numerous studies performed with site directed mutants on the residues suspected as important for catalytic activity. In addition time resolved spectroscopic measurements using both absorbance and fluorescence have helped to identify the reaction intermediates in photoreduction, from steps 3 to 4 in the scheme described above. Upon light absorption by the bound Pchlide, hydride transfer occurs from the pro-S face of the nicotinamide ring of NADPH to C17 of the Pchlide molecule, followed by proton transfer probably from tyrosine-280 of LPOR to C18 of Pchlide (Heyes, Martin, Reid, Hunter, & Wilks, 2000; Heyes, Ruban, Wilks, & Hunter, 2002; Wilks & Timko, 1995).

The order in this photoreaction sequence was confirmed by time resolved spectroscopy in which excited-state interactions occur between LPOR and the carboxyl group on the Pchlide molecule. This resulted in a polarized and highly reactive double bond allowing hydride transfer from NADPH to C18 of the Pchlide (Heyes et al., 2015). Additionally, it was recently found that the keto group on ring E is required for excited state charge separation, which drives the photochemistry in the photoreduction (Heyes et al., 2017).

5.2 Properties of DPOR

The three proteins ChlL, ChlN and ChlB constitute the DPOR activity in oxygenic phototrophs such as cyanobacteria, green algae and gymnosperms. The orthologous proteins BchL, BchN and BchB constitute DPOR activity in anoxygenic bacteria, and these proteins are also structurally similar to the BchX, BchY and BchZ proteins which catalyze a similar reduction on ring B in bacteriochlorophyll synthesis. The *chlL*, *chlN* and *chlB* genes encoding ChlL, ChlN and ChlB, when present, are always found on the chloroplast genomes. The genes encoding DPOR subunits are absent in Euglenophytes, Cryptophytes, Rhodophytes as well as in the flowering plants using LPOR as the only enzyme for Pchlide reduction. The ChlL, ChlN, ChlB subunits are homologous to the NifH, NifD and NifK proteins of nitrogenase, respectively. Similar to nitrogenase, DPOR is oxygen sensitive and requires a source of reducing power, supplied by reduced ferredoxin, and ATP hydrolysis (reviewed in Fujita & Bauer, 2003).

5.2.1 Structure and enzyme mechanism of DPOR

The first complete DPOR structure was characterized from *Rhodobacter capsulatus* (Muraki et al., 2010). The BchN/BchB complex in this structure has an iron-sulfur cluster and is structurally similar to the nitrogenase NifD/NifK proteins. Each of the BchN/BchB dimers, in the BchN/BchB tetramer, has an iron sulfur cluster and a bound Pchlide. The Pchlide occupies the same position as the molybdenum-iron complex found in nitrogenase. A BchL dimer containing a second iron sulfur cluster sits on top of the BchN/BchB dimer positioning the clusters to allow electron transfer to the Pchlide.

Structures of the ChlN/ChlB cyanobacterial component from *Thermosynechococcus elongatus* were solved at the same time and has a similar heterotetrameric structure 2ChlN/2ChlB with each ChlN/ChlB dimer containing an iron sulfur complex and a Pchlide binding site (Brocker, Schomburg, et al., 2010), as for the *Rhodobacter* enzyme. Studies on the *Prochlorococcus marina* DPOR show that the ChlL dimer interacts transiently with the ChlN/ChlB tetrameric complex (Brocker, Watzlich, et al., 2010). A stable octameric complex is formed with 2× ChlL-dimers/2× ChlN/ChlB heterodimers upon applied non-hydrolyzable ATP analogues (Moser & Brocker, 2011) enabling crystallization. The X-ray structure of this octameric *Prochlorococcus marina* complex with Pchlide bound was determined, as shown in Fig. 8 (Moser et al., 2013). Given that the ChlN/ChlB tetramer spans the membrane and that ChlL is only expected in the stromal or cytosolic compartment in cells, a hexameric complex with a single ChlL dimer bound to the ChlN/ChlB tetramer is likely the form to exist *in vivo*. This also suggests that only one ChlN/ChlB dimer within the membrane-spanning tetramer is functional *in vivo*.

ATP bound to ChlL is required for binding to the ChlN/ChlB complex and ATP hydrolysis is required to initiate electron transfer to the ChlN/ChlB complex which contains the Pchlide bound (Brocker, Watzlich, et al., 2010). *In vitro* the reductant for ChlL can be either dithionite or reduced ferredoxin (cited in Fujita & Bauer, 2003). The dissociation of ChlL dimers may be required to re-reduce the ChlL iron sulfur cluster during the catalytic cycle.

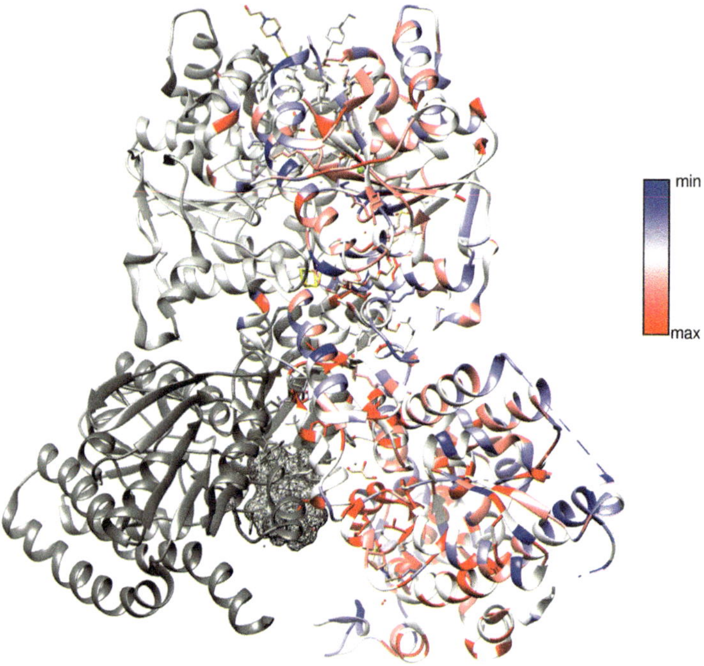

Fig. 8 Crystal structure of ChlL, ChlN and ChlB complex from *Prochlorococcus marinus* with substrate bound as space filling model (PDB: 2YNM). The structure shown is half of the octameric crystal complex containing two of the four ChlL subunits and a single ChlN and ChlB subunit. The ChlN/ChlB is at the bottom with a dimer of ChlL subunits on the top. The light gray is ChlL with the second ChlL colored red and blue at the top. The bottom half is ChlN and ChlB. ChlN is colored in dark gray and ChlB in red and blue, with Pchlide, as space filling model sandwiched between these two subunits. Blue and red coloring of the ChlL and ChlB subunits, as shown in the scale bar, is based on conservation of residues as per the color scale with red being 100% conserved to blue being non-conserved between species. The conservation of residue comparisons is performed with Clustal W aligned ChlL and ChlB protein sequences from *Prochlorococcus marina* (as per PDB: 2YNM), *Pinus sylvestris* (YP_009388294, YP_009388232), *Chlamydomonas reinhardtii* (ASF83651, AAA16321), Synechocystis (BAA18745, BAA10114) and BchL and BchB protein sequences from *Rhodobacter sphaeroides* (AAF24272, AAF24274). This figure was created with Chimera (Pettersen et al., 2004).

Yellow in the dark *y*-mutants of *C. reinhardtii*, at 10 loci *y-1* to *y-10*, has been identified which lack DPOR activity. But are able to synthesize chlorophyll in the light using LPOR. These *y*-mutants have nuclear mutations, which may posttranslationally prevent correct processing of the chloroplast-encoded ChlL (Cahoon & Timko, 2000). The mechanism by which ChlL translation is prevented appears to be related to translation initiation or elongation of only the ChlL since ChlN and ChlB translation is unaffected. However, the identity and characteristics of the proteins encoded at these *y* loci have not been further characterized.

As DPOR is oxygen sensitive it is assumed that in oxygenic phototrophs the enzyme will only be active during the dark or under low light conditions when oxygen tensions are low. As such, when oxygen tensions rise due to photosynthetic activity, LPOR will be the active reductase. It has been hypothesized that the loss of DPOR genes from chloroplasts has occurred multiple times from eukaryotic phototrophs as DPOR appears to be dispensable when LPOR is present (see Fujita & Bauer, 2003).

6. 3,8-Divinyl-chlorophyllide *a* reductase

The chlorophylls present in most oxygenic organism have an 8-ethyl group. Thus, the 8-ethylene group needs to be reduced to an 8-ethyl group at some stage of chlorophyll synthesis. *In vitro* assays for this conversion utilized an enzyme extract from plastid membranes to catalyze the NADPH dependent conversion of 3,8-divinyl-protochlorophyllide producing 3-vinyl 8-ethyl-protochlorophyllide (Parham & Rebeiz, 1992; Whyte & Griffiths, 1993). The gene encoding this enzyme in *A. thaliana* was identified and the mutants accumulate divinyl chlorophyll *a* and *b* (Nagata, Tanaka, Satoh, & Tanaka, 2005a, 2005b; Nakanishi et al., 2005). The enzyme displays some promiscuity for alternate substrates and, at least in *A. thaliana*, the preferred substrate *in vivo* is 3,8-divinyl-chlorophyllide *a* (Nagata, Tanaka, & Tanaka, 2007) with NADPH as the reducing cofactor as shown in Fig. 9. Interestingly, this NADPH-dependent divinyl reductase in higher plants is related to the hydroxymethyl chlorophyll *a* reductase (Ito & Tanaka, 2014) which is involved in the chlorophyll cycle described in chapter "The biochemistry, physiology, and evolution of the chlorophyll cycle" of this volume by Tanaka and Tanaka.

An NADPH-dependent divinyl reductase that is homologous to that found in plants is also found in green sulfur bacteria and is encoded by *bciA* (Chew & Bryant, 2007a). In contrast to higher plants *Synechocystis* has a ferredoxin-dependent divinyl reductase (Islam et al., 2008; Ito, Yokono, Tanaka, & Tanaka, 2008) which is not homologous to the NADPH dependent *bciA*. The *bciB* gene of green sulfur bacteria is homologous to the Synechocystis gene encoding the ferredoxin-dependent divinyl reductase and was shown to encode a functional ferredoxin-dependent divinyl reductase (Liu & Bryant, 2011; Saunders, Golbeck, & Bryant, 2013).

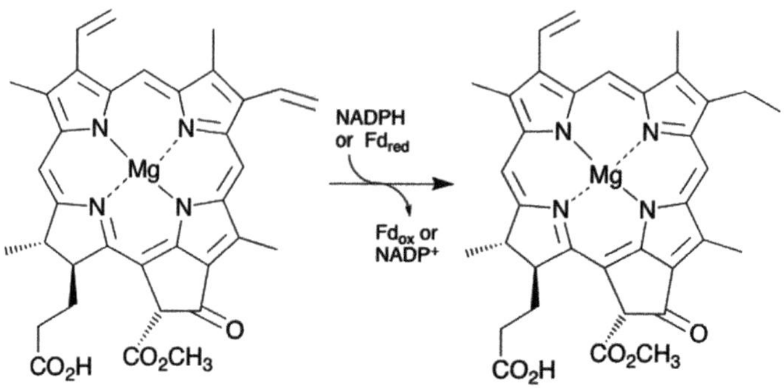

3,8-divinyl chlorophyllide a
(DV Chlide a)

3-monovinyl chlorophyllide a
(Chlide a)

Fig. 9 Catalytic reaction of 3,8-divinyl-chlorophyllide *a* divinyl reductase.

Subsequently, it appears that the cyanobacteria *Acaryochloris marina* as well as anoxygenic photosynthetic bacteria have *bciA* and *bciB* homologs which encode both an NADPH- and ferredoxin-dependent divinyl reductase.

7. Chlorophyll synthase

Chlorophyll synthase is the final enzyme in the synthesis of chlorophyll *a*. The enzyme catalyzes the trans-esterification of phytyl-pyrophosphate onto the C-17 propanoate of Chlide *a* to make chlorophyll *a* and pyrophosphate as products, as shown in Fig. 10. Chlorophyll synthase is encoded by *chlG/CHLG* and homologous to *bchG* which catalyzes a similar reaction in bacteriochlorophyll synthesis. The enzyme displays some substrate promiscuity as it can utilize chlorophyllide *b* as substrate to make chlorophyll *b* (Rudiger, Benz, & Guthoff, 1980) and also geranylgeranyl-pyrophosphate (GG-PP) to make geranylgeraniol-esterified chlorophylls (Rudiger et al., 1980; Rudiger, Hedden, Kost, & Chapman, 1977). In contrast, Pchlide, chlorophyllide *a′*, and chlorophyllide derivatives lacking a carboxymethyl group at C13-2 are not substrates (Helfrich, Schoch, Lempert, Cmiel, & Rudiger, 1994; Rüdiger, 2003). This indicates that both reduction of the 17,18 double bond and the presence and conformation of the C13-2 carboxymethyl group are essential for esterification. The metal ion coordinated in the chlorophyll is also relevant for catalysis. Chlorophyllides with alternative metal ions indicate that chlorophyllides containing pentacoordinate metals such as Mg^{2+} and Zn^{2+} are substrates, while the square-pyrimidal Co^{2+}-, Ni^{2+}- and Cu^{2+}- complexes are neither substrates or inhibitors (Helfrich & Ruediger, 1992).

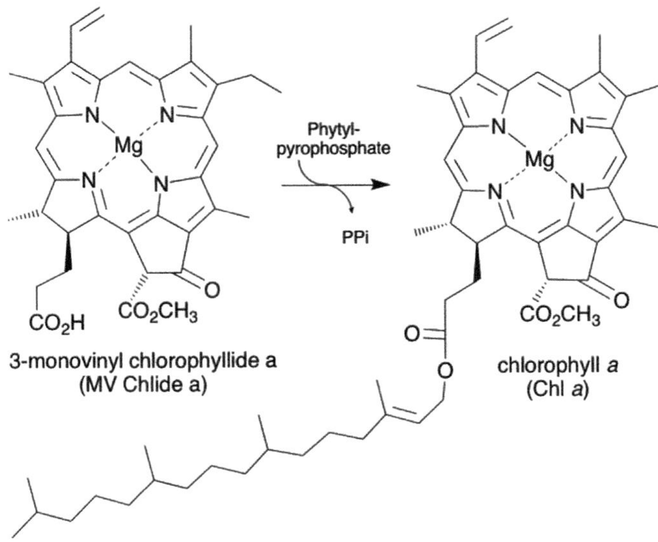

Fig. 10 Catalytic reaction of chlorophyll synthase.

8. Geranylgeranyl reductase

The enzyme that converts GG-PP to phytyl-pyrophosphate is the NADPH-GGPP reductase (ChlP) encoded by the *chlP/CHLP* gene. It is still unclear if geranylgeraniol-esterified chlorophylls are converted to chlorophyll, but this cannot be excluded. In *A. thaliana* the light-harvesting-like protein 3 (LIL3) has been shown to bind to LPOR and ChlP (Hey et al., 2017). In rice a recruiting protein, called geranylgeranyl-diphosphate synthase recruiting protein (GRP), forms a heterodimer with geranylgeranyl-diphosphate synthase. This dimer is directed to the ChlP-LIL3-ChlG complex in the thylakoid membrane (Zhou et al., 2017). This suggests that an *in vivo* formed complex allows substrate channeling from GG-PP and Chlide *a* to chlorophyll *a* synthesis and subsequent assembly of chlorophyll into light-harvesting complexes and reaction centers. This hypothesis is consistent with the observation that ChlG is bound to PLBs and redistributed on light induced dispersion of the PLBs (Lindsten, Wiktorsson, Ryberg, & Sundqvist, 1993).

9. Concluding remarks and future directions

While MgPMTase and DPOR 3D-structure have been well characterized with high resolution enabling a reasonably detailed knowledge of their molecular catalytic mechanism, there are many questions regarding the structure and precise molecular mechanisms of the other enzymes. Solving the structures of the complete magnesium chelatase complex, the oxidative cyclase complex, LPOR, ChlG and the divinyl reductases, will go some way toward our understanding of these reactions and help confirm the molecular mechanisms proposed. Even when high resolution structures have become available of components of the large complexes, such as the magnesium chelatase subunits, this does not necessarily give us a clear picture of the molecular mechanism. However, high resolution structures will provide a basis for identifying key residues that are involved in binding substrates and potentially identify important residues such as those red highlighted residues in Figs. 5 and 8.

The advent of rapid DNA synthesis and synthetic biology methods have also provide tools to study these reactions at the whole pathway level in organisms like *E. coli*. The use of these tools has allowed chlorophyll to be synthesized in *E. coli* (Chen et al., 2018) and this success will provide a way to study things like substrate channeling and regulation of the pathway in a background devoid of the normal regulatory mechanisms.

References

Adams, N. B., Brindley, A. A., Hunter, C. N., & Reid, J. D. (2016). The catalytic power of magnesium chelatase: A benchmark for the AAA(+) ATPases. *FEBS Letters*, *590*(12), 1687–1693. https://doi.org/10.1002/1873-3468.12214.

Adams, N. B., Marklew, C. J., Brindley, A. A., Hunter, C. N., & Reid, J. D. (2014). Characterization of the magnesium chelatase from Thermosynechococcus elongatus. *The Biochemical Journal*, *457*(1), 163–170. https://doi.org/10.1042/BJ20130834.

Adams, N. B., & Reid, J. D. (2012). Nonequilibrium isotope exchange reveals a catalytically significant enzyme-phosphate complex in the ATP hydrolysis pathway of the AAA(+) ATPase magnesium chelatase. *Biochemistry*, *51*(10), 2029–2031. https://doi.org/10.1021/bi300149z.

Adams, N. B., Vasilev, C., Brindley, A. A., & Hunter, C. N. (2016). Nanomechanical and thermophoretic analyses of the nucleotide-dependent interactions between the AAA(+) subunits of magnesium chelatase. *Journal of the American Chemical Society*, *138*(20), 6591–6597. https://doi.org/10.1021/jacs.6b02827.

Adhikari, N. D., Orler, R., Chory, J., Froehlich, J. E., & Larkin, R. M. (2009). Porphyrins promote the association of GENOMES UNCOUPLED 4 and a Mg-chelatase subunit with chloroplast membranes. *The Journal of Biological Chemistry*, *284*(37), 24783–24796. https://doi.org/10.1074/jbc.M109.025205.

Alawady, A., Reski, R., Yaronskaya, E., & Grimm, B. (2005). Cloning and expression of the tobacco CHLM sequence encoding Mg protoporphyrin IX methyltransferase and its interaction with Mg chelatase. *Plant Molecular Biology*, *57*(5), 679–691.

Albus, C. A., Salinas, A., Czarnecki, O., Kahlau, S., Rothbart, M., Thiele, W., et al. (2012). LCAA, a novel factor required for magnesium protoporphyrin monomethylester cyclase accumulation and feedback control of aminolevulinic acid biosynthesis in tobacco. *Plant Physiology*, *160*(4), 1923–1939.

Allen, M. D., Kropat, J., & Merchant, S. S. (2008). Regulation and localization of isoforms of the aerobic oxidative cyclase in Chlamydomonas reinhardtii. *Photochemistry and Photobiology, 84*(6), 1336–1342. https://doi.org/10.1111/j.1751-1097.2008.00440.x.

Amirjani, M. R. (2010). Protochlorophyllide spectral forms. *Pakistan Journal of Biological Sciences, 13*(12), 563–576. https://doi.org/10.3923/pjbs.2010.563.576.

Armstrong, G., Apel, K., & Rudiger, W. (2000). Does a light-harvesting protochlorophyllide a/b-binding protein complex exist? *Trends in Plant Science, 5*(1), 40–44.

Axelsson, E., Lundqvist, J., Sawicki, A., Nilsson, S., Schroder, I., Al-Karadaghi, S., et al. (2006). Recessiveness and dominance in barley mutants deficient in Mg-chelatase subunit D, an AAA protein involved in chlorophyll biosynthesis. *Plant Cell, 18*(12), 3606–3616.

Balmer, Y., Koller, A., del Val, G., Manieri, W., Schurmann, P., & Buchanan, B. B. (2003). Proteomics gives insight into the regulatory function of chloroplast thioredoxins. *Proceedings of the National Academy of Sciences of the United States of America, 100*(1), 370–375. https://doi.org/10.1073/pnas.232703799.

Belyaeva, O. B., & Litvin, F. F. (2011). Advances in understanding of the primary reactions of protochlorophyll(ide) photoreduction in cells and model systems. *Journal of Biophysical Chemistry, 2*(1), 1–9. https://doi.org/10.4236/jbpc.2011.21001.

Blomqvist, L. A., Ryberg, M., & Sundqvist, C. (2008). Proteomic analysis of highly purified prolamellar bodies reveals their significance in chloroplast development. *Photosynthesis Research, 96*(1), 37–50. https://doi.org/10.1007/s11120-007-9281-y.

Bollivar, D. W. (2003). *Intermediate steps in chlorophyll biosynthesis: Methylation and cyclization. 13* (pp. 49–69). Porphyrin Handbook.

Bollivar, D., & Beale, S. (1996). The chlorophyll biosynthetic enzyme Mg-protoporphyrin IX monomethyl ester (oxidative) cyclase. Characterization and partial purification from Chlamydomonas reinhardtii and Synechocystis sp. PCC 6803. *Plant Physiology, 112*(1), 105–114.

Bollivar, D., Braumann, I., Berendt, K., Gough, S. P., & Hansson, M. (2014). The Ycf54 protein is part of the membrane component of Mg-protoporphyrin IX monomethyl ester cyclase from barley (Hordeum vulgare L.). *The FEBS Journal, 281*(10), 2377–2386.

Bollivar, D., Jiang, Z., Bauer, C., & Beale, S. (1994). Heterologous expression of the bchM gene product from Rhodobacter capsulatus and demonstration that it encodes S-adenosyl-L-methionine:Mg-protoporphyrin IX methyltransferase. *Journal of Bacteriology, 176*(17), 5290–5296.

Bollivar, D., Suzuki, J., Beatty, J., Dobrowolski, J., & Bauer, C. (1994). Directed mutational analysis of bacteriochlorophyll a biosynthesis in Rhodobacter capsulatus. *Journal of Molecular Biology, 237*(5), 622–640.

Brindley, A. A., Adams, N. B., Hunter, C. N., & Reid, J. D. (2015). Five glutamic acid residues in the C-terminal domain of the ChlD subunit play a major role in conferring Mg(2 +) cooperativity upon magnesium chelatase. *Biochemistry, 54*(44), 6659–6662. https://doi.org/10.1021/acs.biochem.5b01080.

Brocker, M. J., Schomburg, S., Heinz, D. W., Jahn, D., Schubert, W.-D., & Moser, J. (2010). Crystal structure of the nitrogenase-like dark operative protochlorophyllide oxidoreductase catalytic complex (ChlN/ChlB)2. *The Journal of Biological Chemistry, 285*(35), 27336–27345.

Brocker, M. J., Watzlich, D., Saggu, M., Lendzian, F., Moser, J., & Jahn, D. (2010). Biosynthesis of (bacterio)chlorophylls: ATP-dependent transient subunit interaction and electron transfer of dark operative protochlorophyllide oxidoreductase. *The Journal of Biological Chemistry, 285*(11), 8268–8277.

Brzezowski, P., Morhard, M. K., Grimm, B., Sharifi, M. N., Dent, R. M., & Niyogi, K. K. (2016). Mg chelatase in chlorophyll synthesis and retrograde signaling in Chlamydomonas reinhardtii: CHLI2 cannot substitute for CHLI1. *Journal of Experimental Botany, 67*(13), 3925–3938.

Brzezowski, P., Schlicke, H., Richter, A., Dent, R. M., Niyogi, K. K., & Grimm, B. (2014). The GUN4 protein plays a regulatory role in tetrapyrrole biosynthesis and chloroplast-to-nucleus signalling in Chlamydomonas reinhardtii. *Plant Journal, 79*(2), 285–298. https://doi.org/10.1111/tpj.12560.

Buchler, J. W. (1975). Static coordination chemistry of metalloporphyrins. In K. M. Smith (Ed.), *Porphyrins and metalloporphyrins*. Amsterdam: Elsevier Scientific Pub. Co., pp. 157–232.

Cahoon, A., & Timko, M. (2000). Yellow-in-the-dark mutants of Chlamydomonas lack the CHLL subunit of light-independent protochlorophyllide reductase. *Plant Cell, 12*(4), 559–568.

Chekunova, E. M. (2010). The genetics of chlorophyll biosynthesis: Light-independent and light-dependent pathways. *Ecological Genetics*, 8(3), 38–51. https://doi.org/10.17816/ecogen8338-51.

Chen, G. E., Canniffe, D. P., Barnett, S. F. H., Hollingshead, S., Brindley, A. A., Vasilev, C., et al. (2018). Complete enzyme set for chlorophyll biosynthesis in Escherichia coli. *Science Advances*, *4*(1), eaaq1407.

Chen, G. E., Canniffe, D. P., & Hunter, C. N. (2017). Three classes of oxygen-dependent cyclase involved in chlorophyll and bacteriochlorophyll biosynthesis. *Proceedings of the National Academy of Sciences of the United States of America, 114*(24), 6280–6285.

Chen, X., Pu, H., Fang, Y., Wang, X., Zhao, S., Dai, H.-E., et al. (2015). Crystal structure of the catalytic subunit of magnesium chelatase. *Nature Plants, 1*, 15125.

Chen, X., Pu, H., Wang, X., Long, W., Lin, R., & Liu, L. (2015). Crystal structures of GUN4 in complex with porphyrins. *Molecular Plant, 8*(7), 1125–1127. https://doi.org/10.1016/j.molp.2015.04.013.

Chen, X., Wang, X., Feng, J., Fang, Y., Zhao, S., Zhao, A., et al. (2014). Structural insights into the catalytic mechanism of Synechocystis magnesium protoporphyrin IX O-methyltransferase (ChlM). *The Journal of Biological Chemistry, 289*(37), 25690–25698.

Chereskin, B., & Castelfranco, P. (1982). Effects of iron and oxygen on chlorophyll biosynthesis. II. Observations on the biosynthetic pathway in isolated etiochloroplasts. *Plant Physiology, 69*(1), 112–116.

Chereskin, B., Castelfranco, P., Dallas, J., & Straub, K. (1983). Magnesium-2,4-divinyl pheoporphyrin a5: The product of a reaction catalyzed in vitro by developing chloroplasts. *Archives of Biochemistry and Biophysics, 226*(1), 10–18.

Chereskin, B., Wong, Y., & Castelfranco, P. (1982). In vitro synthesis of the chlorophyll isocyclic ring. Transformation of magnesium-protoporphyrin IX and magnesium-protoporphyrin IX monomethyl ester into magnesium-2,4-divinyl pheoporphyrin A5. *Plant Physiology, 70*(4), 987–993.

Chew, A. G., & Bryant, D. A. (2007a). Characterization of a plant-like protochlorophyllide a divinyl reductase in green sulfur bacteria. *The Journal of Biological Chemistry, 282*(5), 2967–2975. https://doi.org/10.1074/jbc.M609730200.

Chew, A. G., & Bryant, D. A. (2007b). Chlorophyll biosynthesis in bacteria: The origins of structural and functional diversity. *Annual Review of Microbiology, 61*, 113–129. https://doi.org/10.1146/annurev.micro.61.080706.093242.

Coomber, S., Chaudhri, M., Connor, A., Britton, G., & Hunter, C. (1990). Localized transposon Tn5 mutagenesis of the photosynthetic gene cluster of Rhodobacter sphaeroides. *Molecular Microbiology, 4*(6), 977–989.

Da, Q., Wang, M., Sun, T., Jin, H., Liu, B., Wang, J., et al. (2017). Thioredoxin and NADPH-dependent thioredoxin reductase C regulation of tetrapyrrole biosynthesis. *Plant Physiology, 175*(2), 652–666.

Davison, P. A., & Hunter, C. N. (2011). Abolition of magnesium chelatase activity by the gun5 mutation and reversal by Gun4. *FEBS Letters, 585*(1), 183–186. https://doi.org/10.1016/j.febslet.2010.11.037.

Davison, P. A., Schubert, H. L., Reid, J. D., Iorg, C. D., Heroux, A., Hill, C. P., et al. (2005). Structural and biochemical characterization of Gun4 suggests a mechanism for its role in chlorophyll biosynthesis. *Biochemistry*, *44*(21), 7603–7612.

Ellsworth, R., & Lawrence, G. (1973). Synthesis of magnesium-protoporphyrin IX in vitro. *Photosynthetica*, 7(1), 73–86.

Elmlund, H., Lundqvist, J., Al-Karadaghi, S., Hansson, M., Hebert, H., & Lindahl, M. (2008). A new cryo-EM single-particle ab initio reconstruction method visualizes secondary structure elements in an ATP-fueled AAA+ motor. *Journal of Molecular Biology*, *375*(4), 934–947.

Fodje, M. N., Hansson, A., Hansson, M., Olsen, J. G., Gough, S., Willows, R. D., et al. (2001). Interplay between an AAA module and an integrin I domain may regulate the function of magnesium Chelatase. *Journal of Molecular Biology*, *311*(1), 111–122.

Fuesler, T. P., Wong, Y.-S., & Castelfranco, P. A. (1984). Localisation of Mg-Chelatase and Mg-protoporphyrin IX monomethyl ester (oxidative) cyclase activities within isolated, developing cucumber chloroplasts. *Plant Physiology*, *75*, 662–664.

Fuesler, T. P., Wright, L. A., & Castelfranco, P. A. (1981). Properties of magnesium chelatase in greening etioplasts. *Plant Physiology*, *67*, 246–249.

Fujita, Y., & Bauer, C. E. (2003). The light-independent protochlorophyllide reductase: A nitrogenase-like enzyme catalysing a key reaction for greening in the dark. In K. M. Kadish, K. M. Smith, & R. Guilard (Eds.), *Vol. 13. The porphyrin handbook* (pp. 109–156). Amsterdam. Academic Press.

Fujita, Y., & Kurisu, G. (2011). Two protochlorophyllide reductases to make chlorophyll green: Structural and evolutionary aspects. *Seibutsu Butsuri*, *51*(2), 66–71. https://doi.org/10.2142/biophys.51.066.

Gabruk, M., & Mysliwa-Kurdziel, B. (2015). Light-dependent protochlorophyllide oxidoreductase: Phylogeny, regulation, and catalytic properties. *Biochemistry*, *54*(34), 5255–5262. https://doi.org/10.1021/acs.biochem.5b00704.

Gabruk, M., Mysliwa-Kurdziel, B., & Kruk, J. (2017). MGDG, PG and SQDG regulate the activity of light-dependent protochlorophyllide oxidoreductase. *The Biochemical Journal*, *474*(7), 1307–1320. https://doi.org/10.1042/BCJ20170047.

Gabruk, M., Stecka, A., Strzalka, W., Kruk, J., Strzalka, K., & Mysliwa-Kurdziel, B. (2015). Photoactive protochlorophyllide-enzyme complexes reconstituted with PORA, PORB and PORC proteins of A. thaliana: Fluorescence and catalytic properties. *PLoS ONE*, 10(2), e0116990. https://doi.org/10.1371/journal.pone.0116990.

Garrone, A., Dietzek, B., Archipowa, N., Zipfel, P. F., & Hermann, G. (2015). Plant protochlorophyllide oxidoreductases A and B: Catalytic efficiency and initial reaction steps. *The Journal of Biological Chemistry*, *290*(47), 28530–28539.

Gibson, L., & Hunter, C. (1994). The bacteriochlorophyll biosynthesis gene, bchM, of Rhodobacter sphaeroides encodes S-adenosyl-L-methionine: Mg protoporphyrin IX methyltransferase. *FEBS Letters*, *352*(2), 127–130.

Gibson, L. C., Jensen, P. E., & Hunter, C. N. (1999). Magnesium chelatase from Rhodobacter sphaeroides: Initial characterization of the enzyme using purified subunits and evidence for a BchI-BchD complex. *Biochemical Journal*, *337*(Pt. 2), 243–251.

Gibson, L. C. D., Willows, R. D., Kannanagara, C. G., Wettstein, D. v., & Hunter, C. N. (1995). Magnesium-protoporphyrin chelatase of Rhodobacter sphaeroides: Reconstitution of activity by combining the products of the bchH, -I, and -D genes expressed in Escherichia coli. *Proceedings of the National Academy of Sciences of the United States of America*, *92*(6), 1941–1944.

Gorchein, A. (1972). Magnesium protoporphyrin chelatase activity in Rhodopseudomonas spheroides. Studies with whole cells. *The Biochemical Journal*, *127*(1), 97–106.

Gorchein, A. (1973). Control of magnesium-protoporphyrin chelatase activity in Rhodopseudomonas spheroides. Role of light, oxygen, and electron and energy transfer. *The Biochemical Journal, 134*(4), 833–845.

Gough, S., Petersen, B., & Duus, J. (2000). Anaerobic chlorophyll isocyclic ring formation in Rhodobacter capsulatus requires a cobalamin cofactor. *Proceedings of the National Academy of Sciences of the United States of America, 97*(12), 6908–6913.

Griffiths, W. T. (1978). Reconstitution of chlorophyllide formation by isolated etioplast membranes. *The Biochemical Journal, 174*(3), 681–692.

Griffiths, W. (1980). Substrate-specificity studies on protochlorophyllide reductase in barley (Hordeum vulgare) etioplast membranes. *The Biochemical Journal, 186*(1), 267–278.

Grzyb, J. M., Solymosi, K., Strzalka, K., & Mysliwa-Kurdziel, B. (2013). Visualization and characterization of prolamellar bodies with atomic force microscopy. *Journal of Plant Physiology, 170*(14), 1217–1227. https://doi.org/10.1016/j.jplph.2013.04.017.

Hambright, P. (1975). Dynamic coordination chemistry of metalloporphyrins. In K. M. Smith (Ed.), *Porphyrins and metalloporphyrins*. Amsterdam: Elsevier Scientific Pub. Co., pp. 233–278.

Hansson, M., & Kannangara, C. (1997). ATPases and phosphate exchange activities in magnesium chelatase subunits of Rhodobacter sphaeroides. *Proceedings of the National Academy of Sciences of the United States of America, 94*(24), 13351–13356.

Hansson, M., Lundqvist, J., Sirijovski, N., & Al-Karadaghi, S. (2014). Magnesium chelatase: The molecular motor of chlorophyll biosynthesis. In G. C. Ferreira (Ed.), *Handbook of porphyrin science*. World Scientific Publishing Co., *28*, 41–84.

Hansson, A., Willows, R. D., Roberts, T. H., & Hansson, M. (2002). Three semidominant barley mutants with single amino acid substitutions in the smallest magnesium chelatase subunit form defective AAA+ hexamers. *Proceedings of the National Academy of Sciences of the United States of America, 99*(21), 13944–13949.

Helfrich, M., & Ruediger, W. (1992). Various metallopheophorbides as substrates for chlorophyll synthetase. *Zeitschrift für Naturforschung Section C, 47*(3–4), 231–238.

Helfrich, M., Schoch, S., Lempert, U., Cmiel, E., & Rudiger, W. (1994). Chlorophyll synthetase cannot synthesize chlorophyll a'. *European Journal of Biochemistry, 219*(1–2), 267–275.

Henningsen, K., Boynton, J., & Von, W. (1993). *Mutants at xantha and albina loci in relation to chloroplast biogenesis in barley (Hordeum vulgare L.). 42* (pp. 1–348). Kongelige Danske Videnskabernes Selskab Biologiske Skrifter.

Herbst, J., Girke, A., Hajirezaei, M. R., Hanke, G., & Grimm, B. (2018). Potential roles of YCF54 and ferredoxin-NADPH Reductase for magnesium protoporphyrin monomethylester cyclase. *The Plant Journal, 94*, 485–496. https://doi.org/10.1111/tpj.13869.

Hey, D., Rothbart, M., Herbst, J., Wang, P., Mueller, J., Wittmann, D., et al. (2017). LIL3, a light-harvesting complex protein, links terpenoid and tetrapyrrole biosynthesis in Arabidopsis thaliana. *Plant Physiology, 174*(2), 1037–1050. https://doi.org/10.1104/pp.17.00505.

Heyes, D. J., Hardman, S. J., Hedison, T. M., Hoeven, R., Greetham, G. M., Towrie, M., et al. (2015). Excited-state charge separation in the photochemical mechanism of the light-driven enzyme protochlorophyllide oxidoreductase. *Angewandte Chemie (International Ed. in English), 54*(5), 1512–1515. https://doi.org/10.1002/anie.201409881.

Heyes, D. J., Hardman, S. J. O., Mansell, D., Ni, C. A., Gardiner, J. M., Johannissen, L. O., et al. (2017). Excited-state properties of protochlorophyllide analogues and implications for light-driven synthesis of chlorophyll. *The Journal of Physical Chemistry B, 121*(6), 1312–1320.

Heyes, D., Martin, G., Reid, R., Hunter, C., & Wilks, H. (2000). NADPH:protochlorophyllide oxidoreductase from Synechocystis: Overexpression, purification and preliminary characterization. *FEBS Letters*, *483*(1), 47–51.

Heyes, D., Ruban, A., Wilks, H., & Hunter, C. (2002). Enzymology below 200 K: The kinetics and thermodynamics of the photochemistry catalyzed by protochlorophyllide oxidoreductase. *Proceedings of the National Academy of Sciences of the United States of America*, *99*(17), 11145–11150.

Hollingshead, S., Bliss, S., Baker, P. J., & Neil Hunter, C. (2017). Conserved residues in Ycf54 are required for protochlorophyllide formation in Synechocystis sp. PCC 6803. *The Biochemical Journal*, *474*(5), 667–681. https://doi.org/10.1042/BCJ20161002.

Hollingshead, S., Kopecna, J., Armstrong, D. R., Chen, G. E., Williamson, M. P., Hunter, C. N., et al. (2016). Synthesis of chlorophyll-binding proteins in a fully segregated Δycf54 strain of the Cyanobacterium Synechocystis PCC 6803. *Frontiers in Plant Science*, 7, 292.

Hollingshead, S., Kopecna, J., Jackson, P. J., Canniffe, D. P., Davison, P. A., Dickman, M. J., et al. (2012). Conserved chloroplast open-reading frame ycf54 is required for activity of the magnesium protoporphyrin monomethylester oxidative cyclase in Synechocystis PCC 6803. *The Journal of Biological Chemistry*, *287*(33), 27823–27833.

Hudson, A., Carpenter, R., Doyle, S., & Coen, E. (1993). Olive: A key gene required for chlorophyll biosynthesis in Antirrhinum majus. *The EMBO Journal*, *12*(10), 3711–3719.

Hunter, C., & Coomber, S. (1988). Cloning and oxygen-regulated expression of the bacteriochlorophyll biosynthesis genes bch E, B, A and C of Rhodobacter sphaeroides. *Journal of General Microbiology*, *134*(6), 1491–1497.

Ikegami, A., Yoshimura, N., Motohashi, K., Takahashi, S., Romano, P. G., Hisabori, T., et al. (2007). The CHLI1 subunit of Arabidopsis thaliana magnesium chelatase is a target protein of the chloroplast thioredoxin. *The Journal of Biological Chemistry*, *282*(27), 19282–19291. https://doi.org/10.1074/jbc.M703324200.

Islam, M. R., Aikawa, S., Midorikawa, T., Kashino, Y., Satoh, K., & Koike, H. (2008). slr1923 of Synechocystis sp. PCC6803 is essential for conversion of 3,8-divinyl(proto)chlorophyll(ide) to 3-monovinyl(proto)chlorophyll(ide). *Plant Physiology*, *148*(2), 1068–1081. https://doi.org/10.1104/pp.108.123117.

Ito, H., & Tanaka, A. (2014). Evolution of a new chlorophyll metabolic pathway driven by the dynamic changes in enzyme promiscuous activity. *Plant & Cell Physiology*, *55*(3), 593–603.

Ito, H., Yokono, M., Tanaka, R., & Tanaka, A. (2008). Identification of a novel vinyl reductase gene essential for the biosynthesis of monovinyl chlorophyll in Synechocystis sp. PCC6803. *The Journal of Biological Chemistry*, *283*(14), 9002–9011. https://doi.org/10.1074/jbc.M708369200.

Jensen, P. E., Gibson, L. C. D., Henningsen, K. W., & Hunter, C. N. (1996). Expression of the chlI, chlD, and chlH genes from the Cyanobacterium Synechocystis PCC6803 in Escherichia coli and demonstration that the three cognate proteins are required for magnesium-protoporphyrin chelatase activity. *The Journal of Biological Chemistry*, *271*(28), 16662–16667.

Jensen, P. E., Gibson, L. C. D., & Hunter, C. N. (1998). Determinants of catalytic activity with the use of purified I, D and H subunits of the magnesium protoporphyrin IX chelatase from Synechocystis PCC6803. *Biochemical Journal*, *334*(2), 335–344.

Jensen, P. E., Gibson, L. C. D., & Hunter, C. N. (1999). ATPase activity associated with the magnesium-protoporphyrin IX chelatase enzyme of Synechocystis PCC6803: Evidence for ATP hydrolysis during Mg2+ insertion, and the MgATP-dependent interaction of the ChlI and ChlD subunits. *Biochemical Journal*, 339(Pt. 1), 127–134.

Jensen, P., Stummann, B., Hansen, N., Karlebjerg, K., & Henningsen, K. (1996). Characterization of a gene for a Mg-chelatase subunit from Synechocystis sp. PCC6803. *Plant Molecular Biology*, *30*, 1075–1076.

Jensen, P., Willows, R., Petersen, B., Vothknecht, U., Stummann, B., Kannangara, C., et al. (1996). Structural genes for Mg-chelatase subunits in barley: Xantha-f, −g and -h. *Molecular & General Genetics*, *250*(4), 383–394.

Kannangara, C. G., & von Wettstein, D. (2010). Magnesium chelatase. *Advances in Photosynthesis of Respiratory*, *31*(chloroplast), 79–88.

Kannangara, C., Vothknecht, U., Hansson, M., & von Wettstein, D. (1997). Magnesium chelatase: Association with ribosomes and mutant complementation studies identify barley subunit Xantha-G as a functional counterpart of Rhodobacter subunit BchD. *Molecular & General Genetics*, *254*(1), 85–92.

Karger, G. A., Reid, J. D., & Hunter, C. N. (2001). Characterization of the binding of deuteroporphyrin IX to the magnesium chelatase H subunit and spectroscopic properties of the complex. *Biochemistry*, *40*(31), 9291–9299.

Kaschner, M., Loeschcke, A., Krause, J., Minh, B. Q., Heck, A., Endres, S., et al. (2014). Discovery of the first light-dependent protochlorophyllide oxidoreductase in anoxygenic phototrophic bacteria. *Molecular Microbiology*, *93*(5), 1066–1078. https://doi.org/10.1111/mmi.12719.

Kobayashi, K., Mochizuki, N., Yoshimura, N., Motohashi, K., Hisabori, T., & Masuda, T. (2008). Functional analysis of Arabidopsis thaliana isoforms of the Mg-chelatase CHLI subunit. *Photochemical & Photobiological Sciences*, 7(10), 1188–1195. https://doi.org/10.1039/b802604c.

Koncz, C., Mayerhofer, R., Koncz-Kalman, Z., Nawrath, C., Redei, G. P., & Schell, J. (1990). Isolation of a gene encoding a novel chloroplast protein by T-DNA tagging in Arabidopsis thaliana. *The EMBO Journal*, *9*(5), 1337–1346.

Lake, V., Olsson, U., Willows, R. D., & Hansson, M. (2004). ATPase activity of magnesium chelatase subunit I is required to maintain subunit D in vivo. *European Journal of Biochemistry*, *271*(11), 2182–2188.

Larkin, R. M., Alonso, J. M., Ecker, J. R., & Chory, J. (2003). GUN4, a regulator of chlorophyll synthesis and intracellular signaling. *Science*, *299*(5608), 902–906. https://doi.org/10.1126/science.1079978.

Layer, G., Krausze, J., & Moser, J. (2017). Reduction of chemically stable multibonds: Nitrogenase-like biosynthesis of tetrapyrroles. *Advances in Experimental Medicine and Biology*, *925*, 147–161.

Lindsten, A., Wiktorsson, B., Ryberg, M., & Sundqvist, C. (1993). Chlorophyll synthetase activity is relocated from transforming prolamellar bodies to developing thylakoids during irradiation of dark-grown wheat. *Physiologia Plantarum*, *88*(1), 29–36.

Liu, Z., & Bryant, D. A. (2011). Multiple types of 8-vinyl reductases for (bacterio)chlorophyll biosynthesis occur in many green sulfur bacteria. *Journal of Bacteriology*, *193*(18), 4996–4998. https://doi.org/10.1128/JB.05520-11.

Lundqvist, J., Braumann, I., Kurowska, M., Muller, A. H., & Hansson, M. (2013). Catalytic turnover triggers exchange of subunits of the magnesium chelatase AAA+ motor unit. *The Journal of Biological Chemistry*, *288*(33), 24012–24019. https://doi.org/10.1074/jbc.M113.480012.

Lundqvist, J., Elmlund, H., Wulff, R. P., Berglund, L., Elmlund, D., Emanuelsson, C., et al. (2010). ATP-induced conformational dynamics in the AAA+ motor unit of magnesium chelatase. *Structure*, *18*, 354–365.

Luo, T., Fan, T., Liu, Y., Rothbart, M., Yu, J., Zhou, S., et al. (2012). Thioredoxin redox regulates ATPase activity of magnesium chelatase CHLI subunit and modulates redox-mediated signaling in tetrapyrrole biosynthesis and homeostasis of reactive oxygen species in pea plants. *Plant Physiology*, *159*(1), 118–130. https://doi.org/10.1104/pp.112.195446.

McLean, S., & Hunter, C. N. (2009). An enzyme-coupled continuous spectrophotometric assay for magnesium protoporphyrin IX methyltransferases. *Analytical Biochemistry, 394*(2), 223–228. https://doi.org/10.1016/j.ab.2009.07.036.

Minamizaki, K., Mizoguchi, T., Goto, T., Tamiaki, H., & Fujita, Y. (2008). Identification of two homologous genes, chlAI and chlAII, that are differentially involved in isocyclic ring formation of chlorophyll a in the cyanobacterium Synechocystis sp. PCC 6803. *The Journal of Biological Chemistry, 283*(5), 2684–2692. https://doi.org/10.1074/jbc.M708954200.

Mochizuki, N., Brusslan, J., Larkin, R., Nagatani, A., & Chory, J. (2001). Arabidopsis genomes uncoupled 5 (GUN5) mutant reveals the involvement of Mg-chelatase H subunit in plastid-to-nucleus signal transduction. *Proceedings of the National Academy of Sciences of the United States of America, 98*(4), 2053–2058.

Moseley, J., Quinn, J., Eriksson, M., & Merchant, S. (2000). The Crd1 gene encodes a putative di-iron enzyme required for photosystem I accumulation in copper deficiency and hypoxia in Chlamydomonas reinhardtii. *The EMBO Journal, 19*(10), 2139–2151.

Moser, J., & Brocker, M. J. (2011). Enzymatic systems with homology to nitrogenase. *Methods in Molecular Biology, 766*, 67–77.

Moser, J., Lange, C., Krausze, J., Rebelein, J., Schubert, W.-D., Ribbe, M. W., et al. (2013). Structure of ADP-aluminium fluoride-stabilized protochlorophyllide oxidoreductase complex. *Proceedings of the National Academy of Sciences of the United States of America, 110*(6), 2094–2098.

Muller, A. H., Sawicki, A., Tabrizi, S. T., Zhou, S., Luo, M., Hansson, M., et al. (2014). Inducing the oxidative stress response in Escherichia coli improves the quality of a recombinant protein: Magnesium chelatase ChlH. *Protein Expression and Purification, 101*, 61–67.

Muraki, N., Nomata, J., Ebata, K., Mizoguchi, T., Shiba, T., Tamiaki, H., et al. (2010). X-ray crystal structure of the light-independent protochlorophyllide reductase. *Nature, 465*(7294), 110–114. https://doi.org/10.1038/nature08950.

Mysliwa-Kurdziel, B., & Strzalka, K. (2010). Light-induced reduction of protochlorophyllide in angiosperms and chloroplast development. *Postepy Biochemii, 56*(4), 418–426.

Nagata, N., Tanaka, R., Satoh, S., & Tanaka, A. (2005a). Identification of a vinyl reductase gene for chlorophyll synthesis in Arabidopsis thaliana and implications for the evolution of Prochlorococcus species. *Plant Cell, 17*(1), 233–240.

Nagata, N., Tanaka, R., Satoh, S., & Tanaka, A. (2005b). Identification of a vinyl reductase gene for chlorophyll synthesis in Arabidopsis thaliana. *Plant and Cell Physiology, 46*, S89.

Nagata, N., Tanaka, R., & Tanaka, A. (2007). The major route for chlorophyll synthesis includes [3,8-divinyl]-chlorophyllide a reduction in Arabidopsis thaliana. *Plant & Cell Physiology, 48*(12), 1803–1808. https://doi.org/10.1093/pcp/pcm153.

Nakanishi, H., Nozue, H., Suzuki, K., Kaneko, Y., Taguchi, G., & Hayashida, N. (2005). Characterization of the Arabidopsis thaliana mutant pcb2 which accumulates divinyl chlorophylls. *Plant & Cell Physiology, 46*(3), 467–473.

Nasrulhaq-Boyce, A., Griffiths, W., & Jones, O. (1987). The use of continuous assays to characterize the oxidative cyclase that synthesizes the chlorophyll isocyclic ring. *The Biochemical Journal, 243*(1), 23–29.

Orsat, B., Monfort, A., Chatellard, P., & Stutz, E. (1992). Mapping and sequencing of an actively transcribed Euglena gracilis chloroplast gene (ccsA) homologous to the Arabidopsis thaliana nuclear gene cs(ch-42). *FEBS Letters, 303*(2–3), 181–184.

Pardo, A., Chereskin, B., Castelfranco, P., Franceschi, V., & Wezelman, B. (1980). ATP requirement for magnesium chelatase in developing chloroplasts. *Plant Physiology, 65*(5), 956–960.

Parham, R., & Rebeiz, C. A. (1992). Chloroplast biogenesis: [4-vinyl] Chlorophyllide a reductase is a divinyl chlorophyllide a-specific, NADPH-dependent enzyme. *Biochemistry, 31*(36), 8460–8464.

Peter, E., & Grimm, B. (2009). GUN4 is required for posttranslational control of plant tetrapyrrole biosynthesis. *Molecular Plant*, *2*(6), 1198–1210. https://doi.org/10.1093/mp/ssp072.

Peter, E., Salinas, A., Wallner, T., Jeske, D., Dienst, D., Wilde, A., et al. (2009). Differential requirement of two homologous proteins encoded by sll1214 and sll1874 for the reaction of Mg protoporphyrin monomethylester oxidative cyclase under aerobic and micro-oxic growth conditions. *Biochimica et Biophysica Acta*, *1787*(12), 1458–1467. https://doi.org/10.1016/j.bbabio.2009.06.006.

Petersen, B. L., Jensen, P. E., Gibson, L. C., Stummann, B. M., Hunter, C. N., & Henningsen, K. W. (1998). Reconstitution of an active magnesium chelatase enzyme complex from the bchI, −D, and -H gene products of the green sulfur bacterium Chlorobium vibrioforme expressed in Escherichia coli. *Journal of Bacteriology*, *180*(3), 699–704.

Petersen, B., Moller, M., Jensen, P., & Henningsen, K. (1999). Identification of the Xan-g gene and expression of the Mg-chelatase encoding genes Xan-f, −g and -h in mutant and wild type barley (Hordeum vulgare L.). *Hereditas*, *131*(2), 165–170.

Pettersen, E., Goddard, T., Huang, C., Couch, G., Greenblatt, D., Meng, E., et al. (2004). UCSF chimera—A visualization system for exploratory research and analysis. *Journal of Computational Chemistry*, *25*(13), 1605–1612.

Pinta, V., Picaud, M., Reiss-Husson, F., & Astier, C. (2002). Rubrivivax gelatinosus acsF (previously orf358) codes for a conserved, putative binuclear-iron-cluster-containing protein involved in aerobic oxidative cyclization of Mg-protoporphyrin IX monomethylester. *Journal of Bacteriology*, *184*(3), 746–753.

Porra, R., Schaefer, W., Gad'on, N., Katheder, I., Drews, G., & Scheer, H. (1996). Origin of the two carbonyl oxygens of bacteriochlorophylla. Demonstration of two different pathways for the formation of ring E in Rhodobacter sphaeroides and Roseobacter denitrificans, and a common hydratase mechanism for 3-acetyl group formation. *European Journal of Biochemistry*, *239*(1), 85–92.

Porra, R., Schaefer, W., Katheder, I., Scheer, H., Gad'on, N., & Drews, G. (1995). Evidence for two different pathways for the formation of isocyclic ring E of bacteriochlorophyll a in Rhodobacter sphaeroides and Roseobacter denitrificans using 18O-labeling and mass spectrometry. In *Photosynthesis: From light to biosphere, Proceedings of the International Photosynthesis Congress, 10th, Montpellier, Fr., Aug. 20-25, 1995, 3* (pp. 881–885).

Porra, R., Schafer, W., Katheder, I., & Scheer, H. (1995). The derivation of the oxygen atoms of the 13(1)-oxo and 3-acetyl groups of bacteriochlorophyll a from water in Rhodobacter sphaeroides cells adapting from respiratory to photosynthetic conditions: Evidence for an anaerobic pathway for the formation of isocyclic ring E. *FEBS Letters*, *371*(1), 21–24.

Qian, P., Marklew, C. J., Viney, J., Davison, P. A., Brindley, A. A., Soederberg, C., et al. (2012). Structure of the cyanobacterial magnesium chelatase H subunit determined by single particle reconstruction and small-angle x-ray scattering. *The Journal of Biological Chemistry*, *287*(7), 4946–4956. https://doi.org/10.1074/jbc.M111.308239.

Quinn, J. M., Eriksson, M., Moseley, J. L., & Merchant, S. (2002). Oxygen deficiency responsive gene expression in Chlamydomonas reinhardtii through a copper-sensing signal transduction pathway. *Plant Physiology*, *128*(2), 463–471. https://doi.org/10.1104/pp.010694.

Reid, J. D., Siebert, C. A., Bullough, P. A., & Hunter, C. N. (2003). The ATPase activity of the ChlI subunit of magnesium chelatase and formation of a heptameric AAA+ ring. *Biochemistry*, *42*(22), 6912–6920.

Reiland, S., Messerli, G., Baerenfaller, K., Gerrits, B., Endler, A., Grossmann, J., et al. (2009). Large-scale Arabidopsis phosphoproteome profiling reveals novel chloroplast kinase substrates and phosphorylation networks. *Plant Physiology*, *150*(2), 889–903. https://doi.org/10.1104/pp.109.138677.

Reinbothe, C., El Bakkouri, M., Buhr, F., Muraki, N., Nomata, J., Kurisu, G., et al. (2010). Chlorophyll biosynthesis: Spotlight on protochlorophyllide reduction. *Trends in Plant Science*, *15*(11), 614–624. https://doi.org/10.1016/j.tplants.2010.07.002.

Richter, A. S., Hochheuser, C., Fufezan, C., Heinze, L., Kuhnert, F., & Grimm, B. (2016). Phosphorylation of GENOMES UNCOUPLED 4 alters stimulation of Mg chelatase activity in angiosperms. *Plant Physiology*, *172*(3), 1578–1595. https://doi.org/10.1104/pp.16.01036.

Richter, M. L., & Rienits, K. G. (1980). The synthesis of magnesium and zinc protoporphyrin IX and their monomethyl esters in etioplast preparations studied by high pressure liquid chromatography. *FEBS Letters*, *116*(2), 211–216.

Richter, M. L., & Rienits, K. G. (1982). The synthesis of magnesium-protoporphyrin IX by etiochloroplast membrane preparations. *Biochimica et Biophysica Acta*, *717*, 255–264.

Richter, A. S., Wang, P., & Grimm, B. (2016). Arabidopsis Mg-protoporphyrin IX methyltransferase activity and redox regulation depend on conserved cysteines. *Plant & Cell Physiology*, *57*(3), 519–527. https://doi.org/10.1093/pcp/pcw007.

Rissler, H., Collakova, E., DellaPenna, D., Whelan, J., & Pogson, B. (2002). Chlorophyll biosynthesis. Expression of a second chl I gene of magnesium chelatase in Arabidopsis supports only limited chlorophyll synthesis. *Plant Physiology*, *128*(2), 770–779.

Rüdiger, W. (2003). The last steps in chlorophyll synthesis. In K. M. Kadish, K. M. Smith, & R. Guilard (Eds.), *Vol. 13. The porphyrin handbook* (pp. 71–108). Amsterdam: Academic Press.

Rudiger, W., Benz, J., & Guthoff, C. (1980). Detection and partial characterization of activity of chlorophyll synthetase in etioplast membranes. *European Journal of Biochemistry*, *109*(1), 193–200.

Rudiger, W., Hedden, P., Kost, H., & Chapman, D. (1977). Esterification of chlorophyllide by geranylgeranyl pyrophosphate in a cell-free system from maize shoots. *Biochemical and Biophysical Research Communications*, *74*(3), 1268–1272.

Rzeznicka, K., Walker, C. J., Westergren, T., Kannangara, C. G., von Wettstein, D., Merchant, S., et al. (2005). Xantha-l encodes a membrane subunit of the aerobic Mg-protoporphyrin IX monomethyl ester cyclase involved in chlorophyll biosynthesis. *Proceedings of the National Academy of Sciences of the United States of America*, *102*(16), 5886–5891.

Saunders, A. H., Golbeck, J. H., & Bryant, D. A. (2013). Characterization of BciB: A ferredoxin-dependent 8-vinyl-protochlorophyllide reductase from the green sulfur bacterium Chloroherpeton thalassium. *Biochemistry*, *52*(47), 8442–8451. https://doi.org/10.1021/bi401172b.

Sawers, R., Farmer, P., Moffett, P., & Brutnell, T. (2006). In planta transient expression as a system for genetic and biochemical analyses of chlorophyll biosynthesis. *Plant Methods*, *2*, 15.

Sawers, R., Viney, J., Farmer, P., Bussey, R., Olsefski, G., Anufrikova, K., et al. (2006). The maize oil yellow1 (Oy1) gene encodes the I subunit of magnesium chelatase. *Plant Molecular Biology*, *60*(1), 95–106.

Sawicki, A., & Willows, R. D. (2007). S-adenosyl-L-methionine:magnesium-protoporphyrin IX O-methyltransferase from Rhodobacter capsulatus: Mechanistic insights and stimulation with phospholipids. *The Biochemical Journal*, *406*(3), 469–478. https://doi.org/10.1042/BJ20070284.

Sawicki, A., & Willows, R. D. (2008). Kinetic analyses of the magnesium chelatase provide insights into the mechanism, structure, and formation of the complex. *The Journal of Biological Chemistry*, *283*(46), 31294–31302. https://doi.org/10.1074/jbc.M805792200.

Sawicki, A., & Willows, R. D. (2010). BchJ and BchM interact in a 1:1 ratio with the magnesium chelatase BchH subunit of Rhodobacter capsulatus. *The FEBS Journal*, *277*(22), 4709–4721. https://doi.org/10.1111/j1742-4658.2010.07877.x.

Sawicki, A., Zhou, S., Kwiatkowski, K., Luo, M., & Willows, R. D. (2017). 1-N-histidine phosphorylation of ChlD by the AAA(+) ChlI2 stimulates magnesium chelatase activity in chlorophyll synthesis. *The Biochemical Journal*, *474*(12), 2095–2105. https://doi.org/10.1042/BCJ20161094.

Shepherd, M., McLean, S., & Hunter, C. (2005). Kinetic basis for linking the first two enzymes of chlorophyll biosynthesis. *The FEBS Journal*, *272*(17), 4532–4539.

Shepherd, M., Reid, J. D., & Hunter, C. N. (2003). Purification and kinetic characterization of the magnesium protoporphyrin IX methyltransferase from Synechocystis PCC6803. *The Biochemical Journal*, *371*(Pt. 2), 351–360.

Sirijovski, N., Lundqvist, J., Rosenback, M., Elmlund, H., Al-Karadaghi, S., Willows, R. D., et al. (2008). Substrate-binding model of the chlorophyll biosynthetic magnesium chelatase BchH subunit. *The Journal of Biological Chemistry*, *283*(17), 11652–11660.

Sirijovski, N., Mamedov, F., Olsson, U., Styring, S., & Hansson, M. (2007). Rhodobacter capsulatus magnesium chelatase subunit BchH contains an oxygen sensitive iron-sulfur cluster. *Archives of Microbiology*, *188*(6), 599–608.

Smith, B., & Rebeiz, C. (1977). Chloroplast biogenesis: Detection of Mg-protoporphyrin chelatase in vitro. *Archives of Biochemistry and Biophysics*, *180*(1), 178–185.

Smith, C. A., Suzuki, J. Y., & Bauer, C. E. (1996). Cloning and characterization of the chlorophyll biosynthesis gene chlM from Synechocystis PCC 6803 by complementation of a bacteriochlorophyll biosynthesis mutant of Rhodobacter capsulatus. *Plant Molecular Biology*, *30*(6), 1307–1314.

Sobotka, R., Duhring, U., Komenda, J., Peter, E., Gardian, Z., Tichy, M., et al. (2008). Importance of the cyanobacterial Gun4 protein for chlorophyll metabolism and assembly of photosynthetic complexes. *The Journal of Biological Chemistry*, *283*(38), 25794–25802. https://doi.org/10.1074/jbc.M803787200.

Stenbaek, A., Hansson, A., Wulff, R. P., Hansson, M., Dietz, K. J., & Jensen, P. E. (2008). NADPH-dependent thioredoxin reductase and 2-Cys peroxiredoxins are needed for the protection of Mg-protoporphyrin monomethyl ester cyclase. *FEBS Letters*, *582*(18), 2773–2778. https://doi.org/10.1016/j.febslet.2008.07.006.

Tanaka, R., & Tanaka, A. (2007). Tetrapyrrole biosynthesis in higher plants. *Annual Review of Plant Biology*, *58*, 321–346. https://doi.org/10.1146/annurev.arplant.57.032905.105448.

Tarahi Tabrizi, S., Sawicki, A., Zhou, S., Luo, M., & Willows, R. D. (2016). GUN4-protoporphyrin IX is a singlet oxygen generator with consequences for plastid retrograde signalling. *The Journal of Biological Chemistry*, *291*, 8978–8984. https://doi.org/10.1074/jbc.C116.719989.

Taylor, D. P., Cohen, S. N., Clark, W. G., & Marrs, B. L. (1983). Alignment of genetic and restriction maps of the photosynthesis region of the Rhodopseudomonas capsulatus chromosome by a conjugation-mediated marker rescue technique. *Journal of Bacteriology*, *154*(2), 580–590.

Tottey, S., Block, M., Allen, M., Westergren, T., Albrieux, C., Scheller, H., et al. (2003). Arabidopsis CHL27, located in both envelope and thylakoid membranes, is required for the synthesis of protochlorophyllide. *PNAS*, *100*(26), 16119–16124.

Tripathy, B. C., & Pattanayak, G. K. (2012). Chlorophyll biosynthesis in higher plants. *Advances in Photosynthesis and Respiration*, 34, 63–94. https://doi.org/10.1007/978-94-007-1579-0_3. (Photosynthesis).

Verdecia, M. A., Larkin, R. M., Ferrer, J.-L., Riek, R., Chory, J., & Noel, J. P. (2005). Structure of the Mg-chelatase cofactor GUN4 reveals a novel hand-shaped fold for porphyrin binding. *PLoS Biology*, *3*(5), 12. https://doi.org/10.1371/journal.pbio.0030151.t001.

Walker, C. J., Castelfranco, P. A., & Whyte, B. J. (1991). Synthesis of divinyl protochlorophyllide. Enzymological properties of the Mg-protoporphyrin IX monomethyl ester oxidative cyclase system. *The Biochemical Journal*, *276*(Pt. 3), 691–697.

Walker, C. J., Hupp, L. R., & Weinstein, J. D. (1992). Activation and stabilization of Mg-chelatase activity by ATP as revealed by a novel in vitro continuous assay. *Plant Physiology and Biochemistry, 30*(3), 263–269.

Walker, C. J., Mansfield, K. E., Smith, K. M., & Castelfranco, P. A. (1989). Incorporation of atmospheric oxygen into the carbonyl functionality of the protochlorophyllide isocyclic ring. *The Biochemical Journal, 257*(2), 599–602.

Walker, C. J., & Weinstein, J. D. (1991a). Further characterization of the magnesium chelatase in isolated developing cucumber chloroplasts: Substrate specificity, regulation, intactness and ATP requirements. *Plant Physiology (Bethesda), 95*(4), 1189–1196.

Walker, C. J., & Weinstein, J. D. (1991b). In vitro assay of the chlorophyll biosynthetic enzyme magnesium chelatase: Resolution of the activity into soluble and membrane bound fractions. *Proceedings of the National Academy of Sciences of the United States of America, 88*(13), 5789–5793.

Walker, C. J., & Weinstein, J. D. (1994). The magnesium-insertion step of chlorophyll biosynthesis is a two-stage reaction. *Biochemical Journal, 299*(1), 277–284.

Walker, C. J., & Weinstein, J. D. (1995). Re-examination of the localization of Mg-chelatase within the chloroplast. *Physiologia Plantarum, 94*(3), 419–424.

Walker, C. J., & Willows, R. D. (1997). Mechanism and regulation of Mg-chelatase. *Biochemical Journal*, 327(Pt. 2), 321–333.

Whyte, B. J., & Griffiths, W. T. (1993). 8-vinyl reduction and chlorophyll a biosynthesis in higher plants. *The Biochemical Journal*, 291(Pt. 3), 939–944.

Wiktorsson, B., Ryberg, M., & Sundqvist, C. (1996). Aggregation of NADPH-protochlorophyllide oxidoreductase-pigment complexes is favored by protein phosphorylation. *Plant Physiology and Biochemistry (Paris), 34*(1), 23–34.

Wilde, A., Mikolajczyk, S., Alawady, A., Lokstein, H., & Grimm, B. (2004). The gun4 gene is essential for cyanobacterial porphyrin metabolism. *FEBS Letters, 571*(1–3), 119–123. https://doi.org/10.1016/j.febslet.2004.06.063.

Wilks, H. M., & Timko, M. P. (1995). A light-dependent complementation system for analysis of NADPH:protochlorophyllide oxidoreductase: Identification and mutagenesis of two conserved residues that are essential for enzyme activity. *Proceedings of the National Academy of Sciences of the United States of America, 92*(3), 724–728.

Willows, R. D. (2003). Biosynthesis of chlorophylls from protoporphyrin IX. *Natural Product Reports, 20*(3), 327–341.

Willows, R. D., & Beale, S. I. (1998). Heterologous expression of the Rhodobacter capsulatus BchI, -D, and -H genes that encode magnesium chelatase subunits and characterization of the reconstituted enzyme. *The Journal of Biological Chemistry, 273*(51), 34206–34213.

Willows, R. D., Gibson, L. C. D., Kanangara, C. G., Hunter, C. N., & von Wettstein, D. (1996). Three separate proteins constitute the magnesium chelatase of Rhodobacter sphaeroides. *European Journal of Biochemistry, 235*(1/2), 438–443.

Willows, R. D., & Hansson, M. (2003). Mechanism, structure and regulation of magnesium chelatase. In K. M. Kadish, K. Smith, & R. Guilard (Eds.), *Vol. 13. The porphyrin handbook II* (pp. 1–48). Academic Press.

Willows, R. D., Hansson, M., Beale, S. I., Laurberg, M., & al-Karadaghi, S. (1999). Crystallization and preliminary X-ray analysis of the Rhodobacter capsulatus magnesium chelatase BchI subunit. *Acta Crystallographica Section D, Biological Crystallography, 55*(Pt. 3), 689–690.

Willows, R. D., Hansson, A., Birch, D., al-Karadaghi, S., & Hansson, M. (2004). EM single particle analysis of the ATP-dependent BchI complex of magnesium chelatase: An AAA + hexamer. *Journal of Structural Biology, 146*(1-2), 227–233.

Wong, Y. S., Castelfranco, P. A., Goff, D. A., & Smith, K. M. (1985). Intermediates in the formation of the chlorophyll isocyclic ring. *Plant Physiology, 79*(3), 725–729.

Wu, F. Q., Xin, Q., Cao, Z., Liu, Z. Q., Du, S. Y., Mei, C., et al. (2009). The magnesium-chelatase H subunit binds abscisic acid and functions in abscisic acid signaling: New evidence in Arabidopsis. *Plant Physiology*, *150*(4), 1940–1954. https://doi.org/10.1104/pp.109.140731.

Yamanashi, K., Minamizaki, K., & Fujita, Y. (2015). Identification of the chlE gene encoding oxygen-independent Mg-protoporphyrin IX monomethyl ester cyclase in cyanobacteria. *Biochemical and Biophysical Research Communications*, *463*(4), 1328–1333.

Yang, J., & Cheng, Q. (2004). Origin and evolution of the light-dependent protochlorophyllide oxidoreductase (LPOR) genes. *Plant Biology (Stuttgart, Germany)*, *6*(5), 537–544. https://doi.org/10.1055/s-2004-821270.

Yuan, M., Zhang, D. W., Zhang, Z. W., Chen, Y. E., Yuan, S., Guo, Y. R., et al. (2012). Assembly of NADPH: Protochlorophyllide oxidoreductase complex is needed for effective greening of barley seedlings. *Journal of Plant Physiology*, *169*(13), 1311–1316. https://doi.org/10.1016/j.jplph.2012.05.010.

Zhang, H., Li, J., Yoo, J., Yoo, S., Cho, S., Koh, H., et al. (2006). Rice Chlorina-1 and Chlorina-9 encode ChlD and ChlI subunits of Mg-chelatase, a key enzyme for chlorophyll synthesis and chloroplast development. *Plant Molecular Biology*, *62*(3), 325–337.

Zhou, S., Sawicki, A., Willows, R. D., & Luo, M. (2012). C-terminal residues of Oryza sativa GUN4 are required for the activation of the ChlH subunit of magnesium chelatase in chlorophyll synthesis. *FEBS Letters*, *586*(3), 205–210. https://doi.org/10.1016/j.febslet.2011.12.026.

Zhou, F., Wang, C.-Y., Gutensohn, M., Jiang, L., Zhang, P., Zhang, D., et al. (2017). A recruiting protein of geranylgeranyl diphosphate synthase controls metabolic flux toward chlorophyll biosynthesis in rice. *Proceedings of the National Academy of Sciences of the United States of America*, *114*(26), 6866–6871. https://doi.org/10.1073/pnas.1705689114.

CHAPTER SIX

The biochemistry, physiology, and evolution of the chlorophyll cycle

Ayumi Tanaka*, Ryouichi Tanaka
Institute of Low Temperature Science, Hokkaido University, Sapporo, Japan
*Corresponding author: e-mail address: ayumi@pop.lowtem.hokudai.ac.jp

Contents

Abstract

The chlorophyll cycle is a metabolic pathway in plants and green algae that interconverts chlorophyll *a* and chlorophyll *b*. The chlorophyll *a*-to-*b* conversion is catalyzed by a Rieske-type oxygenase, chlorophyll(ide) *a* oxygenase, while the chlorophyll *b*-to-*a* conversion is catalyzed in two steps by chlorophyll *b* reductase (a short-chain dehydrogenase) and 7-hydroxymethyl chlorophyll *a* reductase (a flavoprotein). The level of chlorophyll(ide) *a* oxygenase is regulated in a feedback loop by the accumulation of

Advances in Botanical Research, Volume 90
ISSN 0065-2296
https://doi.org/10.1016/bs.abr.2019.03.005

chlorophyll *b* in plants. In contrast, chlorophyll *b* reductase levels are regulated by a pool of light-harvesting complexes (LHC) that are energetically uncoupled to the core complexes of photosystems. Experimental evidence suggests that LHC levels in plants are regulated by chlorophyll *b* biosynthesis. LHC levels, however, appear to be independent of chlorophyll *b* in green algae. Instead, Prasinophytes, a group of green algae, have a chlorophyllide *a* oxygenase protein possessing structural alterations that differ dramatically from the rest of the green algae, which appear to boost chlorophyll *b* biosynthesis in these organisms. Hypothetical scenarios for the evolution of the chlorophyll cycle are presented.

Abbreviations

ABA abscisic acid
CAO chlorophyllide *a* oxygenase
CBR chlorophyll *b* reductase
CMO choline monooxygenase
DMO 3,6-dichloro-2-methoxybenzoic acid monooxygenase
FAD flavin adenine dinucleotide
HCAR 7-hydroxymethyl chlorophyll *a* reductase
LHC light-harvesting complex
PIFs phytochrome-interacting factors
SAM *S*-adenosyl-L-methionine
WT wild type

1. Introduction: A brief history of the identification of the chlorophyll cycle

Green algae, land plants, and some cyanobacteria utilize the two chlorophyll species chlorophyll *a* and chlorophyll *b* (Fig. 1) for their photosynthesis (Tomitani et al., 1999). Chlorophyll *a* has a methyl group in the C7 position, while chlorophyll *b* has a formyl group in the same position. Chlorophyll *a* and chlorophyll *b* have distinct absorption spectra in the blue and red regions, which enable this combination of pigments to absorb wider ranges of light spectra (Chen, 2014). Chlorophyll *a* exists in all light-harvesting complexes (LHCs) of oxygenic photosynthetic organisms, whereas chlorophyll *b* principally resides in the peripheral antenna complexes (Green & Durnford, 1996; Neilson & Durnford, 2010).

Several biosynthetic pathways for chlorophyll *b* synthesis were proposed (Smith & French, 1963) prior to the actual identification of the pathway. Many observations indicated that chlorophyll *b* was synthesized from chlorophyll *a*. For example, chlorophyll *b* is synthesized during the greening phase of etiolated seedlings only after the accumulation of chlorophyll *a*

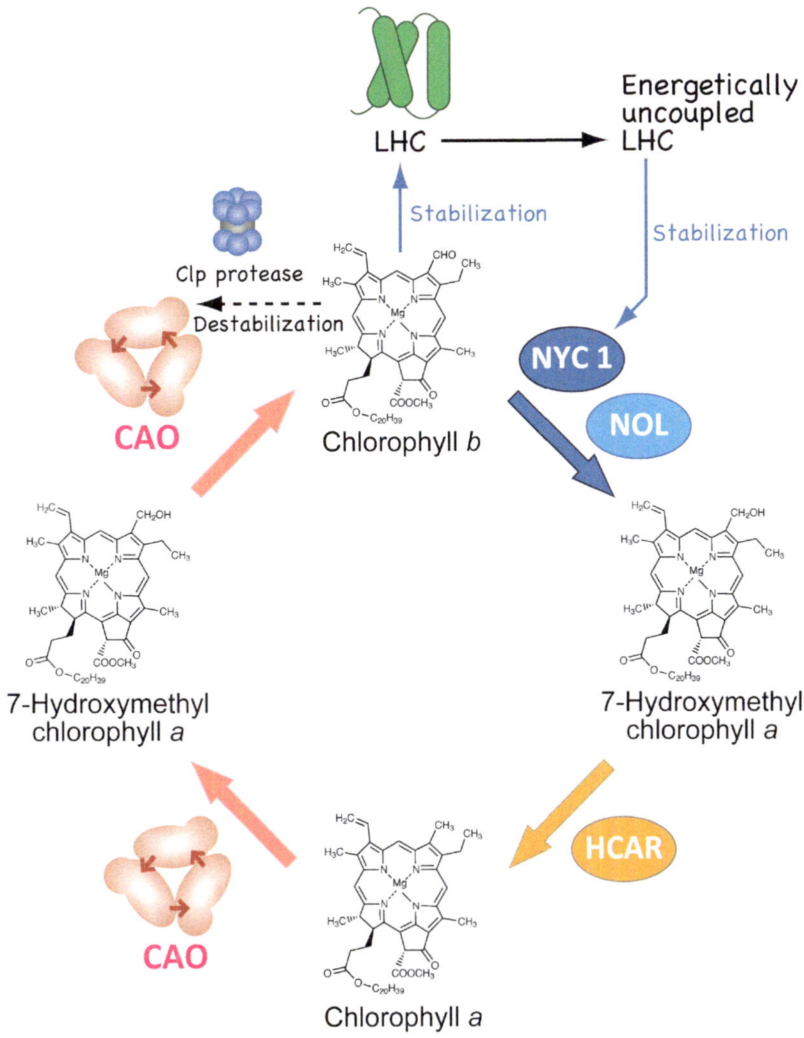

Fig. 1 Regulation of the enzymes of the chlorophyll cycle. Chlorophyll *a* is converted to chlorophyll *b* via 7-hydroxymethyl chlorophyll *a* by the action of chlorophyllide *a* oxygenase (CAO). This enzyme presumably consists of three identical subunits. Chlorophyll *b* triggers destabilization of CAO, a process involving stromal Clp protease. Chlorophyll *b* is necessary for the stabilization of LHC in plants. Chlorophyll *b* is converted to 7-hydroxymethyl chlorophyll *a* and to chlorophyll *a* by the action of chlorophyll *b* reductase (CBR) and 7-hydroxymethyl reductase (HCAR) during senescence or other conditions in which plants need to reduce PSII antenna size. There are two types of CBR in plants, one (NYC1) with and one (NOL) without transmembrane domains. The stabilization of NYC1 is presumably controlled by the level of a pool of LHC that is energetically uncoupled from photosystem II.

(Ohashi, Tanaka, & Tsuji, 1989). Chlorophyll *b* deficient mutants of barley and *Arabidopsis thaliana* (Arabidopsis) could still synthesize chlorophyll *a* (Hirono & Redei, 1963; Wettstein, Kahn, Nielsen, & Gough, 1974). Several isotopic labeling experiments in which ^{14}C-labeled precursors of chlorophyll biosynthesis were administered to plants or algae indicated that newly synthesized chlorophyll *a* is a substrate for chlorophyll *b* synthesis (see Shlyk, 1971 for the summary of these experiments). Further observations supported this pathway that an increase in chlorophyll *b* content in barley seedlings was concomitant with a decrease in chlorophyll *a* content in the presence of exogenous calcium, which was known to stabilize LHC by unknown mechanism while the *de novo* synthesis of chlorophyll does not occur (Tanaka & Tsuji, 1981). An experiment monitoring ^{18}O isotope incorporation into chlorophyll *b* subsequently demonstrated that the formyl group of chlorophyll *b* is derived from molecular oxygen, thus indicating that 7-hydroxymethyl chlorophyll *a* is an intermediate of chlorophyll *b* biosynthesis (Porra, Schafer, Cmiel, Katheder, & Scheer, 1994).

Although the reduction of a formyl group to a methyl group is a difficult chemical reaction, several experiments indicated that the conversion of chlorophyll *b* to chlorophyll *a* takes place *in vivo*. When greening tissues were incubated in the dark, chlorophyll *b* increased with a concomitant decrease in chlorophyll *a* (Kupke & Huntington, 1963; Tanaka & Tsuji, 1981). More direct evidence was derived from the observation that chlorophyll *a* accumulated when chlorophyllide *b* was infiltrated into etiolated tissues (Rudoi, Vezitskii, & Shlyk, 1982). These experiments indicated a conversion of chlorophyll(ide) *b* to chlorophyll *a*, because chlorophyll *b* accumulation was observed in darkness when chlorophyll synthesis was discontinued. The complete pathway was finally elucidated by several *in vitro* experiments. When chlorophyllide *b* was incubated with etioplasts (chlorophyll-free plastids present in etiolated seedlings) isolated from etiolated barley seedlings, the accumulation of 7-hydroxymethyl chlorophyllide *a* and chlorophyll *a* were observed. Additionally, the conversion of 7-hydroxymethyl chlorophyllide *a* to chlorophyll *a* in etioplasts was also demonstrated (Ito, Tanaka, Tsuji, & Tanaka, 1993; Ohtsuka, Ito, & Tanaka, 1997). Thus, the interconversion of chlorophyll *a* and chlorophyll *b* via 7-hydroxymethyl chlorophyll *a* was clearly elucidated and the pathway was named the chlorophyll cycle (Fig. 1, further information of chlorophyll *b* to chlorophyll *a* conversion is also found in "Chlorophyll Catabolites and the Biochemistry of Chlorophyll Breakdown" by B. Kräutler and S. Hörtensteiner).

As discussed later, the homologues of chlorophyll cycle enzymes are found in all of the sequenced land-plant genomes. As a result, it is likely that the chlorophyll cycle is present in all land plants. However, the presence of the chlorophyll cycle in chlorophyll-*b* synthesizing cyanobacteria is still not fully documented at the present time. The present review summarizes our current understanding of the enzymes involved in the chlorophyll cycle, its regulation, physiological functions, and evolution. Additionally, there is an attempt to highlight the remaining questions on this topic to encourage future scientific research. Readers may refer to previous review articles for wider background information (Tanaka & Tanaka, 2011; Tanaka, Kobayashi, & Masuda, 2011).

2. Genes and enzymes of the chlorophyll cycle

2.1 Chlorophyllide *a* oxygenase

2.1.1 Genes

The gene responsible for chlorophyll *b* synthesis was identified via the use of chlorophyll *b*-less mutants of *Chlamydomonas reinhardtii* (Tanaka et al., 1998). The identified gene was found to encode a Rieske mononuclear iron oxygenase, later named chlorophyllide *a* oxygenase (CAO) (Oster, Tanaka, Tanaka, & Ruedigger, 2000). Mutations in the *CAO* genes in *C. reinhardtii* and Arabidopsis result in a complete loss of chlorophyll *b*, indicating that CAO is solely responsible for chlorophyll *b* synthesis in these organisms (Espineda, Linford, Devine, & Brusslan, 1999; Tanaka et al., 1998). Subsequently, 14 previously isolated chlorophyll *b*-deficient barley mutants were all shown to contain mutations in the *CAO* gene (Mueller et al., 2012). *CAO* genes were also identified in chlorophyll *b*-synthesizing cyanobacteria, including Prochlorophytes (Satoh & Tanaka, 2006; Tomitani et al., 1999) and *Acaryochloris marina* (Partensky et al., 2018). In an *in vitro* assay, recombinant Arabidopsis CAO was shown to catalyze the conversion of the methyl group of chlorophyll *a* to a formyl group without any other protein (Oster et al., 2000). Thus, it appears, to the best of our knowledge, that CAO is the only enzyme responsible for chlorophyll *b* synthesis.

All green plants examined thus far, including Streptophyta and Chlorophyta, have only one copy of a *CAO* gene (Kunugi et al., 2016); with the exception of rice, which has two *CAO* genes (*OsCAO1* and *OsCAO2*) (Lee et al., 2005). Since the two rice genes are tandemly located in the genome, it was speculated that they were generated by a recent gene duplication event. The *OsCAO1* mutant completely lacks chlorophyll *b* indicating that OsCAO1 is responsible for chlorophyll *b* formation, while *OsCAOs2* may be a pseudogene or expressed only under specific conditions.

Green sulfur bacteria are strict anaerobes that possess bacteriochlorophyll *e*, which, like chlorophyll *b*, has a formyl group in the C7 position (Orf & Blankenship, 2013). This formyl group is formed from the methyl group of bacteriochlorophyll *d*, similar to the chlorophyll *b* biosynthesis pathway in oxygenic photosynthetic organisms. This reaction, however, is not catalyzed by a CAO homologue but rather by a radical *S*-adenosyl-L-methionine (SAM) enzyme BciD (Harada et al., 2013; Thweatt, Ferlez, Golbeck, & Bryant, 2017) which has no sequence similarity to CAO. It is reasonable to expect that strict anaerobes, such as green sulfur bacteria, would adopt radical SAM enzymes instead of oxygenases in their metabolism.

2.1.2 Enzymes

Recombinant CAO protein expressed in *E. coli* has been shown to convert chlorophyllide *a* to 7-hydroxymethyl chlorophyllide *a* (Oster et al., 2000), indicating that this enzyme catalyzes two successive oxygenation reactions. The first oxygenation reaction produces a hydroxyethyl group and the second oxygenation reaction yields an aldehyde hydrate which spontaneously loses water to form a formyl group. This research also demonstrates that CAO uses chlorophyllide *a* and not chlorophyll *a* as a substrate. CAO, however, has been hypothesized to utilize chlorophyll *a* as a substrate *in vivo* based on results of experiments in which transient overexpression of CAO in a chlorophyll *b*-less Arabidopsis mutant led to chlorophyll *b* accumulation even in the dark (Jia, Ito, & Tanaka, 2016). In the reported experiments, chlorophyllide *a* was not available for the CAO reaction because *de novo* synthesis of chlorophyllide *a* requires light. Thus, theoretically, chlorophyll *a* is the sole substrate available to CAO. These experiments also imply that CAO not only catalyzes free chlorophyll *a*, but also chlorophyll *a* bound to apoproteins since almost all chlorophyll molecules are bound to proteins *in vivo*.

In vitro experiments demonstrated that CAO requires reduced ferredoxin as a reductant (Oster et al., 2000). A detailed reaction mechanism for CAO, however, is not clear at the present time. CAO has sequence similarity to 3,6-dichloro-2-methoxybenzoic acid monooxygenase (DMO). Both CAO and DMO belong to the family of Rieske mononuclear iron oxygenases (Mason & Cammack, 1992). DMO catalyzes the hydroxylation of a methyl group of dicamba to a hydroxymethyl group, similar to the CAO reaction. Thus, it was hypothesized that the reaction mechanism of CAO is similar to that of DMO where an electron is transferred from ferredoxin to the Fe-S center, and to the mononuclear iron, which finally reduces O_2 to form O_2^-. The crystal structure of DMO indicates that it forms a trimer (Dumitru, Jiang, Weeks, & Wilson, 2009). Notably, inter-subunit electron transfer from a Rieske cofactor of one subunit to a mononuclear iron of another subunit takes place during its catalytic reaction indicating that trimer formation is a requisite for the reaction of DMO (Dumitru et al., 2009). It is tempting to speculate that CAO may have a similar organization of its subunits. In fact, recombinant Arabidopsis CAO protein appears to be present in oligomeric forms under non-denaturing conditions, which may single, double, or triple trimers (Kunugi, Takabayashi, & Tanaka, 2013, Fig. 2)

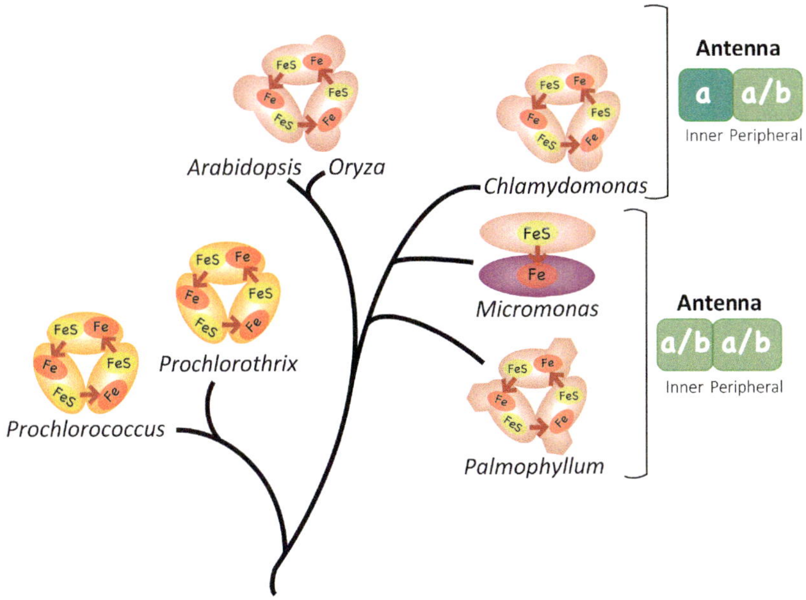

Fig. 2 A scenario for the evolution of chlorophyllide *a* oxygenase (CAO) and changes in pigment configuration of the photosynthetic antenna complexes. In cyanobacteria (*Prochlorococcus* and *Prochlorthrix*), CAO may form a trimer structure and does not have an N-terminal regulatory domain. At an early stage in evolution, CAO appears to have acquired an N-terminal extension involved in a chlorophyll *b*-dependent mechanism for controlling CAO stability. For adaptation to relatively weak light environments, however, *Palmophyllum* modified the control mechanism and evolved a high level of chlorophyll *b* in its antenna system. In *Micromonas*, however, the CAO sequence lost its N-terminal extension, and was further split into two peptides; which probably contributes to the enhanced accumulation of chlorophyll *b* in these algae by boosting chlorophyll *b* synthesizing capacity. The chlorophyll *b*-dependent feedback mechanism in *Chlamydomonas* and land plants was retained during evolution, which contributes to the controlled distribution of chlorophyll *b* in their antennae system.

though a possibility cannot be excluded that these oligomeric forms are simply aggregates due to expression in *E. coli*. In the end, the structure and the reaction mechanism of CAO remain to be further elucidated in future studies.

In most chlorophyll-*b* synthesizing organisms, CAO consists of a single polypeptide, while in Prasinophytes, CAO consists of two polypeptides. Then, CAO is encoded by two genes, which appear to be derived from a single ancestral *CAO* gene (Fig. 2). The first half of CAO in the Prasinophyte *Micromonaspusila* possesses a Rieske-binding motif encoded by *MpCAO2*, while the second half of CAO contains a mononuclear iron-binding motif encoded by *MpCAO1* (Kunugi et al., 2013). Simultaneous introduction of both *MpCAO1* and *MpCAO2* into a chlorophyll *b*-less Arabidopsis mutant compliments its chlorophyll *b* deficiency, while transformation with either *MpCAO1* or *MpCAO2* alone does not compliment the phenotype. Taken together, these data indicate that a combination of MpCAO1 or MpCAO2 polypeptides is necessary for CAO activity. In addition, blue-native PAGE analysis indicates that recombinant MpCAO1 and MpCAO2 form a heterodimer (Kunugi et al., 2013). It is reasonable to assume that electrons are transferred between the two subunits. The crystal structure of DMO indicates that the Rieske and catalytic domains (mononuclear iron-containing domain) are structurally and functionally independent, although the two domains are structurally connected. These structural characteristics are likely conserved in CAO, which would have made the splitting of the CAO protein feasible into two polypeptides encoded by two different genes.

To date, the localization of CAO in chloroplasts is not evident. It is because CAO is not detectable by immune and mass spectrometry analysis, due to its extremely low level in chloroplasts. In addition, no CAO assay is applicable with proteins extracts of leaves or chloroplasts. When truncated CAO is overexpressed, it is localized to both the chloroplast envelope and thylakoid membranes (Yamasato, Nagata, Tanaka, & Tanaka, 2005). However, it is possible that intact CAO may not be similarly localized.

2.2 Chlorophyll *b* reductase

2.2.1 Genes

Chlorophyll *b* reductase (CBR) catalyzes the conversion of chlorophyll *b* to 7-hydroxymethyl chlorophyll *a*. Two isoforms of CBR have been identified in plants. One isoform is encoded by *non-yellow coloring1* (*NYC1*)

(Kusaba et al., 2007) and the other by *NYC1-like* (*NOL*) (Sato et al., 2009). Both genes belong to the short-chain dehydrogenase/reductase family. Since both isoforms exist in most of green plants, it is likely that they play distinct roles. NYC1 possesses three transmembrane domains, while NOL has none. Species in the Chlorophyta and Streptophyta have both CBR isoforms, indicating that functional diversification of these isoforms occurred in a common ancestor of green plants.

In contrast, Prasinophytes, *Ostreococcus*, and *Micromonas* (Mamiellophyceae) only have one CBR gene, which does not encode a transmembrane domain. Thus, it is likely that Prasinophytes lost one of the CBR isoforms during the course of evolution. Considering that Prasinophytes accumulate a high level of chlorophyll *b* (nearly equal to levels of chlorophyll *a*), the alterations in CBR- and CAO-encoding genes noted above may play a role in maintaining the high level of chlorophyll *b* present in these organisms.

2.2.2 Enzymes

Presently, researchers have only been successful in measuring the enzymatic activity of NOL but not NYC1; although both isoforms have retained the conserved motifs of short-chain dehydrogenase/reductase (Horie, Ito, Kusaba, Tanaka, & Tanaka, 2009; Kusaba et al., 2007). Accordingly, our understanding of CBR enzymatic activity is limited to NOL. NOL can catalyze "free" chlorophyll *b*, chlorophyllide *b*, pheophorbide *b*, and pheophytin *b* that are unbound to LHCs into their respective hydroxymethyl molecules. These observations indicate that CBR has a wide substrate specificity and can reduce the formyl groups of all the chlorophyll *b* derivatives that potentially occur in chloroplasts (Shimoda, Ito, & Tanaka, 2012). This is in contrast to Mg-dechelatase, the first committed enzyme of chlorophyll degradation, which catalyzes the removal of Mg^{2+} only from chlorophyll(ide) *a* (Shimoda, Ito, & Tanaka, 2016). Due to the specificity of Mg-dechelatase, pheophytin *b* and pheophorbide *b* are not produced in chloroplasts *in planta*. Thus, the real substrates of CBR in chloroplasts *in vivo* should be chlorophyll *b* and chlorophyllide *b*.

CBR can catalyze both free chlorophyll *b* and chlorophyll *b* that is bound to LHC. When purified LHC trimers were incubated with NOL, almost all of the chlorophyll *b* molecules were converted to 7-hydroxymethyl chlorophyll *a* (Horie et al., 2009). Since six chlorophyll *b* molecules bind to an LHC protein and some of them are not located at the exterior of LHC (Standfuss, van Scheltinga, Lamborghini, & Kuhlbrandt, 2005), it is probably difficult for CBR to access all chlorophyll *b* molecules at once. Therefore, we speculate that NOL first catalyzes the chlorophyll *b* molecules at the exterior of LHC. This reaction may unfold the LHC structure, which would further expose the rest of chlorophyll molecules that are attached in the more internal binding sites of LHC to the enzymes. In brief, it is hypothesized that CBR activity triggers a substantial structural change in the LHC. This scenario is consistent with the observation that LHC is only degraded to a minor degree during senescence in the *nol/nyc1* double mutant. As the mutant is unable to unfold the structure of LHC, we assume that NOL/NYC action is essential prior to LHC degradation (Horie et al., 2009; Kusaba et al., 2007). Collectively, the data indicate the major substrate of CBR in chloroplasts is chlorophyll *b* bound to LHC.

Rice *nol* and *nyc1* mutants have both been reported to be defective in chlorophyll *b* degradation. Moreover, no additive phenotype was observed in the *nyc1/nol* double mutant (Sato et al., 2009). These results indicate that both NOL and NYC1 are required for chlorophyll *b* degradation in rice. An interaction between NOL and NYC1 has also been demonstrated using a yeast two-hybrid system. It was concluded that the formation of a NOL-NYC1 hetero complex is required for the function of CBR in rice. In contrast, the Arabidopsis *nol* mutant degrades chlorophyll *b* during leaf senescence in a manner similar to wild-type (WT) plants. Only the Arabidopsis *nyc1* mutant and the Arabidopsis *nyc1/nol* double mutant exhibits a defect in chlorophyll *b* degradation (Horie et al., 2009). These results indicate that the formation of a NOL-NYC1 heterodimer is not necessary for CBR function in Arabidopsis. Moreover, gene expression patterns for *NYC1* and *NOL* genes are different. In Arabidopsis, *NYC1* is mainly expressed during leaf senescence while *NOL* is constitutively expressed (Sakuraba, Schelbert, et al., 2012). The functional diversification of NYC1 and NOL still remains to be clarified.

Another complexity concerning CBR is its intra-organellar localization. CBR activity in senescing barley chloroplasts has been detected in thylakoid membranes but not in stroma or envelopes (Scheumann, Schoch, & Rudiger, 1998, 1999). In rice, both NYC1 and NOL localize to the stromal side of thylakoid membranes (Sato et al., 2009). Considering the complex formation of rice NYC1 and NOL, the localization of these isoforms in thylakoids is reasonable. It would be useful to determine if Arabidopsis NOL is also localized to thylakoids despite its lack of transmembrane domains and lack of interaction with NYC1. NOL localization in Arabidopsis chloroplasts, however, remains to be determined.

2.3 7-Hydroxymethyl chlorophyll *a* reductase

2.3.1 Genes

The second reaction in the chlorophyll *b*-to-*a* conversion is catalyzed by 7-hydroxymethyl chlorophyll *a* reductase (HCAR). The gene encoding this enzyme was identified by analyzing an Arabidopsis mutant that accumulates 7-hydroxymethyl chlorophyll *a* during dark-induced senescence (Meguro, Ito, Takabayashi, Tanaka, & Tanaka, 2011). To date, all sequenced genomes of plants contain a single copy of an *HCAR* gene, indicating that HCAR and the chlorophyll cycle are essential in plants.

2.3.2 Enzymes

HCAR catalyzes the complex reaction of converting a hydroxymethyl to a methyl group by adopting a mechanism of proton-coupled electron transfer, which is similar to the activity of ribonucleotide reductase (Cotruvo & Stubbe, 2011; Wang & Liu, 2016). The crystal structure of HCAR indicates that this enzyme contains two [4Fe-4S] clusters and flavin adenine dinucleotide (FAD). Ferredoxin donates an electron to one of the [4Fe-4S] cluster and then to another [4Fe-4S] cluster followed by the reduction of FAD. FAD donates an electron and an Asp residue also appears to be involved in proton coupling (Wang & Liu, 2016). In addition to the reduction of a hydroxymethyl group, HCAR also exhibits NADH dehydrogenase activity. It is uncertain whether the dehydrogenase activity is a promiscuous activity or has some physiological function (Ito & Tanaka, 2014). The crystal structure of HCAR indicates that this enzyme forms a homo trimer. It has also been proposed that the HCAR trimer interacts with an LHCII trimer (Wang & Liu, 2016), which may be important for the rapid degradation of chlorophyll *b* to chlorophyll *a*.

HCAR accepts 7-hydroxymethyl chlorophyll *a* and 7-hydroxymethyl chlorophyllide *a* as substrates, but cannot catalyze 7-hydroxymethyl pheophorbide *a* or 7-hydroxymethyl pheophytin *a in vitro* (Matsuda, Shimoda, Tanaka, & Ito, 2016). Therefore, the conversion of 7-hydroxymethyl chlorophyll *a* to chlorophyll *a* during leaf senescence is considered to precede the reaction of Mg dechelation (Matsuda et al., 2016; Shimoda et al., 2012). It is hypothesized that HCAR activity is constitutively high so that 7-hydroxymethyl chlorophyll *a* does not accumulate in leaves. In contrast, the Arabidopsis *hcar* mutant not only accumulates 7-hydroxymethyl chlorophyll *a* but also 7-hydroxymethyl pheophorbide *a* and 7-hydroxymethyl pheophytin *a*, both of which do not appear to be metabolized within chloroplasts (Shimoda et al., 2012).

3. Regulation

The chlorophyll cycle plays an essential role in many physiological processes in plants, such as the biogenesis of LHC during greening, antenna size regulation during acclimation to light intensity; as well as chlorophyll degradation during senescence and seed maturation. It is not surprising that the chlorophyll cycle is finely regulated at various levels, including gene expression, enzyme stability and activity. In this section, the regulation of the chlorophyll cycle is classified into three categories: developmental, environmental, and homeostatic regulation. The first category of regulation of the chlorophyll cycle is largely reflected in different developmental stages, such as germination, greening, senescence, and seed formation. Chlorophyll cycle activity is also differentially regulated in various organs. The chlorophyll cycle is very active in leaves but exhibits low activity in roots. This type of regulation is principally achieved by phytohormones and transcriptional factors as described below. The second category of regulation considers the response of the chlorophyll cycle to environmental changes. Generally speaking, plants construct their photosynthetic machinery based on prevailing environmental conditions that will presumably persist into the future. Photoreceptors and transcriptional factors are involved in this type of regulation (Cheminant et al., 2011; Waters et al., 2009). The third category of regulation is homeostatic control. Plants monitor the current internal status of the cell and make rapid adjustments. This is an immediate type of regulation and is mainly achieved by the regulation of enzyme activity by the redox state, feedforward/feedback mechanisms, and enzyme stability.

3.1 Regulation related to plant development and morphogenesis

Light signaling, phytohormones and activated transcription factors play a crucial role in the development and morphogenesis of plants (Bar & Ori, 2014). GOLDEN2-LIKE (GLK), which is under the control of cytokinin and auxin, is responsible for chloroplast development (Kobayashi et al., 2012) and induces the expression of chlorophyll synthesis-related genes, including *CAO* (Waters et al., 2009). Phytochrome-Interacting Factors (PIFs) also regulate *CAO* expression and chlorophyll synthesis (Cheminant et al., 2011) (further information in "Transcriptional control for the chlorophyll metabolism" by K. Kobayashi and T. Masuda).

NYC1 is expressed during seed maturation and leaf senescence. In Arabidopsis, *NYC1* expression is controlled by abscisic acid (ABA) signaling during the process of seed maturation (Nakajima, Ito, Tanaka, & Tanaka, 2012). In the cited study, defects in chlorophyll breakdown during seed maturation resulted in decreased germination rates. *NYC1* expression is controlled in a more complex manner during leaf senescence via multiple signaling pathways, including those involving phytohormones (Kuai, Chen, & Hortensteiner, 2018), and more specifically, jasmonate (Zhu et al., 2015) as well as ethylene and ABA (Qiu et al., 2015). In contrast, *HCAR* and *NOL* are stably expressed during development, and to the best of our knowledge, are not regulated by phytohormone or light signaling. Thus, the regulation of the chlorophyll *b*-to-*a* conversion appears focused on NYC1. NYC1 activity may represent a bottle neck in the chlorophyll *b* to *a* conversion during leaf senescence or seed maturation.

3.2 Regulation of the chlorophyll cycle in response to environmental changes

Although chlorophyll *b* biosynthesis is mainly regulated at the posttranslational level, transcriptional control of *CAO* responds to changes of varying light intensities (Tanaka et al., 2001). When plants are transferred to high-light conditions, the level of *CAO* mRNA is immediately decreased within several minutes, which may be achieved by both the degradation of *CAO* mRNA and the suppression of *CAO* expression. In contrast, when plants are transferred to from high-light to low-light conditions, the expression of the *CAO* gene is up-regulated (Tanaka & Tanaka, 2005). *CAO* expression in green algae is regulated in a light-intensity-dependent manner. Calmodulin and redox systems have been proposed to control *CAO* gene expression in the green alga, *Dunaliella salina* (Masuda, Tanaka, & Melis, 2003).When plants are exposed to abiotic stresses, such as light, osmotic, salt, and oxidative stresses, the expression of *NYC1* is up-regulated (Sakuraba et al., 2014).

3.3 Homeostatic control

3.3.1 CAO

In green plants, CAO consists of three domains which are termed A, B, and C. The C domain has a catalytic function and is conserved in nearly all CAO sequences, including those in cyanobacteria. The only exception, as described below, is the CAO sequences in the Prasinophytes. The A domain has a regulatory function, which will also be subsequently discussed. This domain is conserved in land plants and most green algae, but it is absent in cyanobacteria and the Mamiellales (a group of green algae) CAO sequences. The B domain appears to function as a linker and is less conserved even among land plants. As other chlorophyll metabolic enzymes in plants do not possess N-terminal extensions compared with corresponding prokaryotic sequences, the presence of a regulatory domain in its N-terminus seems to be unique to CAO (Nagata, Satoh, Tanaka, & Tanaka, 2004).

The role of regulatory domain in plant CAO sequences was deciphered when only the C domain was introduced to a chlorophyll *b*-less Arabidopsis mutant (*ch1-1*). The C domain accumulated to a substantial level in the transgenic plants and drastically increased chlorophyll *b* accumulation (Yamasato et al., 2005). In contrast, CAO accumulated below a detectable level when the full-length CAO was expressed in the *ch1-1* mutant, which only resulted in a moderate accumulation of chlorophyll *b*. These results indicate that the A and B domains prevent excessive accumulation of CAO protein. Interestingly, when the A domain is fused with GFP and introduced into wild-type (WT) Arabidopsis, the chimeric protein only accumulates to a low level, while this protein accumulates to a much higher level in the *ch1-1* background (Yamasato et al., 2005). Based on these experiments, we hypothesized that the A domain monitors the presence of (excessive) chlorophyll *b* and destabilizes CAO (or any A-domain containing protein). The C-domain is not involved in this regulatory process.

The next question to answer might be which protease is responsible for chlorophyll-*b* dependent degradation of CAO. To identify a protease which degrades CAO, an Arabidopsis mutant was isolated that accumulates a larger amount of full-length CAO than WT plants (Nakagawara, Sakuraba, Yamasato, Tanaka, & Tanaka, 2007). The mutant had a mutation in a gene encoding the ClpC subunit of the stromal Clp protease. These results indicated that Clp protease is responsible for the destabilization of CAO. In general, the Clp protease consists of a proteolytic core component and an ATP-dependent chaperone (Nishimura & van Wijk, 2015). ClpC1 is a chaperone of the stroma Clp machinery. The ATP-dependent chaperone of Clp protease was reported to recognize substrates containing specific degron sequences, and to denature and transfer them into a degradation chamber. Subsequently, a putative degron sequence was identified in the A domain of Arabidopsis CAO (Sakuraba, Tanaka, Yamasato, & Tanaka, 2009). Based on these findings, the following hypothetical mechanism was postulated. The degron sequence of the A domain may be buried within the domain. When the A domain recognizes the presence of chlorophyll *b*, however, the structure of the A domain changes to expose the degron to the exterior of the A domain. Then, the Clp chaperone recognizes the degron sequence and transfers CAO into the degradation chamber. Importantly, this scenario still remains to be confirmed in future studies. It is noteworthy that Clp protease is shown to regulate also another step of the chlorophyll metabolism, namely, 5-aminolevulinic acid synthesis, which represents a committed step of chlorophyll biosynthesis (Apitz et al., 2016, see chapters "Posttranslational control of 5-aminolevulinic acid synthesis" by Richter and Grimm, Vol. 91 and "Posttranslational control of tetrapyrrole biosynthesis: Interacting proteins, chaperones, auxiliary factors" by Herbst et al., Vol. 91).

An analysis of 14 alleles of barley chlorophyll *b*-less mutants revealed that a mutation in a conserved cysteine residue impairs chlorophyll *b* biosynthesis (Mueller et al., 2012). It is possible that this cysteine residue is involved in a redox-dependent regulation (Mueller et al., 2012). Interestingly, *Prochlorthrix* CAO does not contain this conserved cysteine. *Prochlorthrix* exhibits a low chlorophyll *a*/*b* ratio under high light conditions. Arabidopsis plants which overexpress *Prochlorthrix* CAO also exhibit a low chlorophyll *a*/*b* ratio under high light, which is in sharp contrast to WT Arabidopsis plants (Hirashima, Satoh, Tanaka, & Tanaka, 2006). Thus, it is tempting to speculate that plants have acquired a redox dependent system for regulating CAO.

3.3.2 NYC1

Based on an analysis of co-expressed genes in large-scale gene expression datasets (Obayashi, Aoki, Tadaka, Kagaya, & Kinoshita, 2017), *NYC1* expression is induced during leaf senescence, which is similar to other genes encoding chlorophyll catabolic enzymes, such as magnesium dechelatase (SGR) and pheophorbide *a* oxygenase. It is likely that these genes are under the control of the same signaling network that regulates senescence-related gene expression. An analysis of *NYC1* overexpressing plants also indicates that NYC1 protein levels are strictly regulated at the posttranslational level. For example, *NYC1* mRNA increases during senescence in the chlorophyll *b*-less mutant *ch1-1* background, while NYC1 never accumulates above a detectable level (Jia, Ito, Hu, & Tanaka, 2015), indicating that chlorophyll *b* is indispensable for the accumulation of NYC1 protein. In contrast, the level of NYC1 dramatically increased when the level of chlorophyll *b* was enhanced by the introduction of truncated CAO lacking a regulatory domain. The next question to ask about NYC1 regulation is whether the chlorophyll *b* level itself serves as a signal to modulate the level of NYC1, or do the levels of chlorophyll *b* binding-proteins regulate NYC1 protein levels? Since LHC is the main protein that binds chlorophyll *b*, it should be considered whether LHC accumulation is a signal that controls the level of NYC. It was shown, however, that LHC levels only slightly correlate with the level of NYC1. In contrast, the minimal chlorophyll fluorescence (Fo) level in dark-adapted leaves was highly correlated with the NYC1 levels as it was analyzed in transgenic plants accumulating various levels of chlorophyll *b* (Jia et al., 2015). Therefore, it is hypothesized that a population of LHC that is energetically uncoupled from photosystem cores (which is correlated with the Fo level) somehow functions as a signal to regulate the level of NYC1.

As previously mentioned, chlorophyll *b* levels are regulated by the chlorophyll cycle (Fig. 1). Furthermore, LHC levels are correlated with the accumulation of chlorophyll *b*. This implies that the chlorophyll cycle also regulates LHC levels. When the regulation of the chlorophyll cycle is taken into account, the regulation of LHC and chlorophyll *b* appears to be as follows (Fig. 1). First, upon low levels of chlorophyll *b*, CAO accumulates and actively synthesizes chlorophyll *b*. In contrast, NYC1 does not accumulate under this condition. When chlorophyll *b* accumulates to some extent, CAO is then degraded by Clp protease and chlorophyll *b* synthesis is stopped. Upon an excessive amount of LHC or a part of non-functional LHC (e.g., during senescence), NYC1 accumulates and degrades chlorophyll *b*. The regulation of CAO and NYC1 likely comprises a sophisticated system to optimize the level of LHC.

4. Physiological roles of the chlorophyll cycle

The chlorophyll cycle does not merely catalyze the interconversion of chlorophyll *a* and chlorophyll *b*, but, is also connected to various physiological processes (Tanaka & Tanaka, 2011; Voitsekhovskaja & Tyutereva, 2015), such as the regulation of chlorophyll synthesis, seed maturation, acclimation to light intensity, degradation of LHC, and senescence.

4.1 Regulation of chlorophyll synthesis by CAO

Deficiency in chlorophyll *b* synthesis adversely affects overall chlorophyll synthesis. On a fresh weight basis, chlorophyll content is low in chlorophyll *b*-less mutants of rice (Yang et al., 2016) and Arabidopsis. Similarly, chlorophyll content on a per cell basis is also decreased in a unicellular *Chlamydomonas* chlorophyll *b*-less mutant (Bujaldon et al., 2017).

Chlorophyll content increased almost by twofold, however, when the *CAO* gene was transiently induced in a chlorophyll *b*-less mutant (Jia et al., 2016). These data indicate that conversion of chlorophyll *a* to chlorophyll *b* potentially enhances chlorophyll synthesis. In contrast, chlorophyll levels in barley chlorophyll *b* deficient mutants are variable. Mueller and his colleagues examined 14 chlorophyll *b* deficient barley mutants, all of which contained a mutation in the *CAO* gene (Mueller et al., 2012). Chlorophyll content was low in most of the mutants. The authors found, however, that *clo-f2*.107 (chlorophyll *a*/*b* ratio is 5:7) accumulated chlorophyll *a* more than its parental cultivar, which has a mutation that results in an amino acid substitution in the C domain, while another chlorophyll *b*-less mutant (*clo-f2*.108) accumulated a level of chlorophyll similar to its parental cultivar. Therefore, it is possible that a regulatory mechanism exists to balance total chlorophyll synthesis and the ratio of chlorophyll *a*/*b*. In this regard, the regulatory mechanism controlling chlorophyll synthesis by CAO remains to be elucidated.

4.2 Distribution of chlorophyll *b*

Chlorophyll *a* and chlorophyll *b* are differentially distributed in various chlorophyll-protein complexes. For example, chlorophyll *b* exists in peripheral antenna complexes, whereas the level of chlorophyll *b* is extremely low in core antenna complexes, such as CP43, CP47, and the P700-chlorphyll complex. One possible mechanism explaining the different distribution of chlorophyll species is the different affinity of chlorophyll-binding sites for chlorophyll *a* and chlorophyll *b*. However, in an *in vitro* reconstitution experiment, it was demonstrated (Ikegami, Satoh, & Aoki, 2007) that the binding affinity of chlorophyll *b* for chlorophyll *a*-binding sites in the P700-chlorophyll *a*-protein complex is similar to that of chlorophyll *a*. This observation is supported by experiments in which the *CAO* gene from land plants was introduced into a cyanobacterial species that is naturally incapable of synthesizing chlorophyll *b*. This approach led to the production of chlorophyll *b* in this species and it was found that chlorophyll *b* was incorporated into the core antenna complexes and functioned as a light-harvesting pigment (Satoh, Ikeuchi, Mimuro, & Tanaka, 2001; Xu, Vavilin, & Vermaas, 2001). Interestingly, when the plant *CAO* gene was introduced into *Acaryochloris marina*, a chlorophyll *d* containing cyanobacterium, [7-formyl]-chlorophyll *d* was produced in the cell (Tsuchiya et al., 2012). Even though this chlorophyll moiety does not exist in this organism in nature, it was still incorporated into chlorophyll-protein complexes.

Collectively, these data indicate that chlorophyll-binding proteins potentially have broad binding affinity to various chlorophyll species.

These observations suggest that the differential distribution of chlorophylls among various complexes is not due to binding affinity alone, and that a mechanism likely exists to control the distribution of chlorophyll in photosynthetic machinery. While presently only limited information of such a mechanism is available, Hoober and Eggink (2001) proposed that the assembly of LHC apoprotein and chlorophyll *b* biosynthesis might occur in the envelope membranes and the assembling of the core complex subunits occurs in thylakoid membranes. This hypothesis, however, needs to be evaluated by experiments. Alternatively, a few lines of evidence exist which show that optimal *a*/*b* ratios are necessary for correct binding of pigments to LHC apoprotein. Overproduction of chlorophyll *b* has been observed to disturb the normal distribution of chlorophyll *b* into the photosynthetic machinery (Hirashima et al., 2006). The CAO protein sequence of *Prochlorthrix hollandica* (*PhCAO*) does not contain an N-terminal extension corresponding to the A domain of green plant CAO. When *PhCAO* was overexpressed in the Arabidopsis *ch1-1* mutant, chlorophyll *b* was overproduced and incorporated into core antenna complexes of both photosystems (Hirashima et al., 2006). Likewise, overexpression of a truncated Arabidopsis CAO lacking the A domain also resulted in a similar alteration in chlorophyll *b* distribution (Sakuraba, Yokono, Akimoto, Tanaka, & Tanaka, 2010). In nature, green algae that live in deep sea water have higher levels of chlorophyll *b* in their core antenna complexes. Interestingly, CAO from those species lacks the N-terminal regulatory domain (Kunugi et al., 2016). Thus, it appears that the function of the A domain to maintain chlorophyll *b* levels in an appropriate range is a requisite for the proper distribution of chlorophyll *b* in the photosynthetic machinery.

4.3 Impact of CAO on the accumulation of LHC

Chlorophyll *b* is known to be necessary for the accumulation of LHC in higher plants (Bellemare, Bartlett, & Chua, 1982; Thornber & Highkin, 1974). In the chlorophyll *b*-less Arabidopsis mutant, *ch1-1*, nine LHC subtypes (Lhcb1–4, Lhcb6, and Lhca1–4) are unstable (Kim et al., 2009; Takabayashi et al., 2011; Tanaka & Tanaka, 2015); with the exception of Lhcb5. It is likely that the LHC proteins (except for Lhcb5) are degraded by a specific protease if they do not bind chlorophyll *b*. It has been proposed that FtsH6 is responsible for the degradation of Lhcb3 in plants, but the deficiency of this protease only marginally affects the accumulation of LHC (Zelisko, Garcia-Lorenzo, Jackowski, Jansson, & Funk, 2005).

Based on the study of chlorophyll-*b*-less mutants, it has been hypothesized that chlorophyll *b* biosynthesis controls LHC accumulation. An alternative hypothesis, however, assumes that LHC levels are regulated by the transcription of *LHC* genes. Flachmann and Kuehlbrandt (1995) evaluated transgenic tobacco lines whether down-regulation of the *LHC* gene affects LHC protein levels. They found, although *LHC* gene expression was reduced to less than 5% of the WT level, LHC levels were not affected in these transgenic plants, indicating that the level of *LHC* transcription does not correlate with protein levels. Furthermore, Andersson et al. (2003) reported that a reduced level of LHC was observed only when *LHC* mRNA levels in WT plants are barely detectable. These results indicate that LHC levels are not predominantly controlled at the transcriptional level. The apoprotein of LHC is considered to be synthesized in excess but the LHC is only stabilized when both chlorophyll *a* and *b* are bound to it. Interestingly, among six LHC subunits that bind to PSII in plants, only Lhcb5 is not destabilized in the absence of chlorophyll *b*. The other LHC subunits are possibly degraded by a common protease in a chlorophyll *b* dependent manner, while Lhcb5 is not a target of this protease. Interestingly, LHC proteins are stable in the *Chlamydomonas reinhardtii* chlorophyll *b*-less mutant (BF3) indicating that this green alga does not possess a chlorophyll-*b* dependent degradation system for LHCs (Bujaldon et al., 2017). Instead, a double mutant (yid-BF3), which does not produce either chlorophyll, lacks LHC. Moreover, a triple mutant (yid-BF3-ftsh1), which does not synthesize chlorophyll but lacks the FtsH protease, retains LHCI apoproteins (Bujaldon et al., 2017). Taken together, these observations indicate that this alga has a mechanism to degrade chlorophyll-free LHCI apoprotein by FtsH.

Recently, another line of evidence was obtained to support chlorophyll-*b* dependent control of LHC in plants. Jia et al. (2016) introduced transient expression of the *CAO* gene in the *ch1-1* background. An increase in almost all of the LHC subunits was observed upon induced *CAO* expression. These results demonstrate that chlorophyll *b* biosynthesis is a limiting step in LHC accumulation.

4.4 LHC degradation is induced by the action of CBR

Degradation of LHC is a requisite for the completion of processes associated with seed maturation, light acclimation, and senescence. Whether the degradation of the protein moiety or that of pigments was the first to occur during LHC degradation was controversial (Dall'Osto, Bressan, & Bassi, 2015). Genetic and biochemical evidence, however, supports the latter process. The level of LHCII remained constant during senescence in the *nyc1*and the *nyc1/nol* double mutants (Horie et al., 2009; Kusaba et al., 2007), while other chlorophyll-protein complexes degraded in a manner similar to the WT. Collectively, these data indicate that NYC1 is responsible for the first step of LHC degradation. When purified LHCII trimers are incubated with recombinant CBR (NOL), all of the chlorophyll *b* is converted to 7-hydroxymethyl chlorophyll *a* and all of the pigment migrates as free chlorophyll after green native gel electrophoresis (Horie et al., 2009). Interestingly, LHCII apoproteins form a trimer after incubation without chlorophyll molecules. These results indicate that CBR acts on chlorophyll *b* while chlorophyll *b* is still bound to LHCII.

Immunoblotting analysis demonstrated that Lhcb1–6 remains in the *nyc1* and *nyc1/nol* mutants, while Lhca1–4 is degraded in a similar manner to WT Arabidopsis plants (Horie et al., 2009). These data indicate that chlorophyll *b*-to-*a* conversion is not necessary for the degradation of Lhca, which is consistent with the observation that Lhca stably accumulates in chlorophyll *b*-less plants. The degradation of Lhca may be controlled by another mechanism, which is not dependent on chlorophyll *b* binding.

4.5 Chlorophyll redistribution by the chlorophyll cycle

The turnover of chlorophyll molecules *in vivo* is much longer than that of chlorophyll-binding proteins (Vavilin, Yao, & Vermaas, 2007). This observation is explained by assuming that chlorophyll is "recycled." Experimental evidence indicates that chlorophyll is salvaged upon the degradation of the chlorophyll-protein complex and is re-used in the formation of a newly synthesized chlorophyll-protein complex. Chlorophyll *b* and LHCII were reported to be degraded with a concomitant increase in the core antenna of photosystem I (CP1) and photosystem II (CP43/47) when greening cucumber cotyledons are kept in the dark (Tanaka, Yamamoto, & Tsuji, 1991). Under these conditions, *de novo* chlorophyll biosynthesis should not have taken place because the penultimate step of chlorophyll biosynthesis requires light for the reaction to occur. Thus, chlorophyll molecules released from LHCII should have been re-incorporated into CP1 and CP43/CP47. Chlorophyll *b* was also demonstrated to be converted to chlorophyll *a* by chlorophyll cycle activity and used in the formation of core antenna complexes (Ohtsuka et al., 1997). This process is inhibited by chloramphenicol, an inhibitor of chloroplast translation. It is likely that chlorophyll *b*-to-*a* conversion is coupled with the translation of CP1, CP43 and CP47. In contrast, chlorophyll *a*-to-*b* conversion, and the formation of LHCII, is induced by treatment of greening cotyledons with calcium during dark incubation (Tanaka & Tsuji, 1982). In this process, chlorophyll *a* in the core complexes is converted to chlorophyll *b* and accompanied by the formation of LHC and degradation of core antenna protein. Although the mechanism underlying this observation is not clearly understood at this time, these collective observations indicate that chlorophyll can be redistributed between peripheral and core antenna complexes and that the chlorophyll cycle is responsible for these processes.

4.6 Acclimation to different light environments

The regulation of antenna size is one of the major strategies utilized by plants to acclimate to changing light environments (Murchie & Horton, 1997). In general, plants grown under low light conditions have larger antennae with a greater number of LHC proteins per core complex. Accordingly, the chlorophyll *a/b* ratios in leaves are lower. In contrast, plants grown under high light conditions have smaller antennae and higher chlorophyll *a/b* ratios.

During the long-term acclimation to light environments, transcriptional control of *CAO* gene expression is most likely the primary method of regulating antenna size. The level of *CAO* mRNA increased in the alga, *Dunaliella salina*, when it was transferred from high to low light. In contrast, *CAO* mRNA levels immediately decreased when low light-acclimated *D. salina* algae were transferred to high light conditions (Masuda et al., 2003). A strong correlation between *CAO* mRNA levels and chlorophyll *a/b* ratios was also observed in *Arabidopsis thaliana* during light acclimation (Harper et al., 2004; Tanaka & Tanaka, 2005). Chlorophyll *a/b* ratios remained low even under high light conditions in transgenic Arabidopsis plants overexpressing *CAO* (Tanaka & Tanaka, 2005). The accumulation of Lhcb1, Lhcb3 and Lhcb6 was enhanced, however, in *CAO*-overexpressing plants in high light, and was similar to the level in low light-grown WT plants. These data indicate that light intensity regulates antenna by controlling the expression of *CAO*. In contrast, the antenna size of photosystem I does not seem to be regulated by *CAO*, as Lhca1–4 accumulate in the chlorophyll *b*-less mutant to the same level as in the WT in both *Chlamydomonas* and Arabidopsis (Bujaldon et al., 2017; Tanaka et al., 2005). Instead, it is possible that CAO regulates PSI antenna size by regulating the synthesis of LHCII which is associated with PSI.

When plants are exposed to high light stress, the level of LHC often decreases. In *nyc1*, the reduction of LHC levels during a high light shift was suppressed (Sato, Ito, & Tanaka, 2015) resulting in a reduced maximum quantum efficiency of PSII (Sato et al., 2015). These observations suggest that NYC1 also plays an important role in the ability of plants to acclimate to high-light stress.

4.7 Stay green and biomass production

Genetically reduced antenna size in photosystems confers advantages in the mass cultures of microalgae by reducing self-shading effects and thus improves productivity (Melis, 2009). It has also been reported that reduced antenna size is beneficial for crop plants by alleviating light stress, resulting in increased plant canopy biomass accumulation (Kirst, Gabilly, Niyogi, Lemaux, & Melis, 2017). Reduction in the level of specific LHCs by mutation or RNAi technology may not always confer the same advantage as modulating chlorophyll *a*/*b* ratios because each LHC has a specific role, such as non-photochemical quenching. Reduction of CAO may represent a better compromise between a decrease in LHC levels and light-harvesting efficiency. By using RNAi strategies targeting *CAO* expression, partial suppression of chlorophyll *b* levels in *Chlamydomonas reinhardtii* reduced peripheral light-harvesting antennae size and increased photosynthetic efficiency, and thus, growth rate at saturating light intensities (Perrine, Negi, & Sayre, 2012).

Transgenic plants that overproduced chlorophyll *b* by overexpressing truncated CAO, exhibited delayed senescence, and, in addition, also had down-regulated expression of senescence-associated genes (Sakuraba, Balazadeh, Tanaka, Mueller-Roeber, & Tanaka, 2012). However, these transgenic plants exhibited many defects, such as low energy-transfer rates between photosynthetic pigments, photodamage, and yellow cotyledons. Further studies are needed to overcome problems associated with improving productivity. *CAO* overexpression in tobacco was reported to result in increased light-saturated photosynthetic carbon assimilation, starch content, and dry matter accumulation under both low and high light regimes (Biswal et al., 2012).

5. Evolution of the chlorophyll cycle

Molecular phylogenetic analysis clearly indicates that the chlorophyll cycle exists in the whole green lineage of photosynthetic organisms, including the Chlorophyta and Streptophyta. In addition, a *CAO* gene is present and chlorophyll *b* is produced in four cyanobacterial groups: Prochlorococcus, Prochloron, Prochlorthrix, and Acaryochloris. Existence of the chlorophyll *b*-to-*a* conversion in cyanobacteria, however, remains undocumented. Thus, it is not clear if the chlorophyll *b*-to-*a* conversion enzymes in green plants originated from cyanobacteria or if they arose independently during the evolution of green plants.

In the remaining section, the evolution of CAO and HCAR is discussed, but not CBR due to the absence of extensive phylogenetic analyses in the literature. CBR belongs to the short-chain dehydrogenases/reductases superfamily, which includes hundreds of divergent enzymes. Presently, the complexity of this family hampers discerning the origin and evolution of CBR. Understanding the evolution of this enzyme remains a future challenge.

5.1 Evolution of CAO

CAO belongs to the Rieske-type mononuclear non-heme iron oxygenase group of proteins (Gray et al., 2004). In the Arabidopsis genome, five genes in this family encode pheophorbide *a* oxygenase (*Pao*), choline monooxygenase (*Cmo*), flavone 8-hydroxylase (*PTC52*) (Berim, Park, & Gang, 2014), phyllobilin hydroxylase (*TIC55*), and *CAO*. CMO is usually involved in the synthesis of glycine betaine, but the function of Arabidopsis CMO is not clear, since Arabidopsis does not produce glycine betaine. Except for CMO, homologs of these genes are found in cyanobacteria. Among them, TIC55, Pao and CAO are involved in chlorophyll metabolism (Hauenstein, Christ, Das, Aubry, & Hörtensteiner, 2016) indicating that non-heme iron oxygenase diverged within the chlorophyll metabolic enzymes in the cyanobacterial lineage and green plants adopted them.

In green plants, the common ancestor of CAO may have obtained the A domain because it is common to both Chlorophyta and Streptophyta (Fig. 2). The structure of the catalytic and regulatory domain (A domain) has changed during the evolution of green plants (Kunugi et al., 2016). In *Micromonas*, CAO divided into two subunits which separately contain

Rieske and mononuclear iron-binding motifs, respectively. This specific CAO structure may be related to changes in the structure of LHCs in these organisms.

5.2 Evolution of HCAR

In general, enzymes have promiscuous activity in addition to their primary activity (Copley, 2017). The promiscuous activity has no physiological function and thus it is theoretically not subjected to selection pressure. It is hypothesized that the promiscuous activity of enzymes, however, becomes physiologically relevant under some environmental conditions. As a result, a promiscuous enzyme may eventually become a new enzyme when it comprises a functional part of a cellular metabolism. This is one of the scenarios of enzyme evolution for which HCAR serves as a good example. As previously mentioned, HCAR catalyzes the conversion of a hydroxymethyl group to a methyl group in 7-hydroxymethyl chlorophyll *a*. However, phylogenetic analysis indicates that HCAR also diverged off into a cluster of ferredoxin-dependent divinyl reductases (F-DVR) (Ito & Tanaka, 2014) that catalyze the conversion of a vinyl group in chlorophyllide (or protochlorophyllide) to an ethyl group (Ito, Yokono, Tanaka, & Tanaka, 2008). In present-day cyanobacteria, F-DVRs have been shown to still possess HCAR activity as a promiscuous activity, although cyanobacteria do not have 7-hydroxymethyl chlorophyll *a*, which is a substrate of HCAR. In contrast to cyanobacteria, plant HCAR does not have DVR activity. Thus, it is plausible that an ancestral F-DVR lost DVR activity during evolution, which was accompanied by an increase in HCAR activity.

6. Concluding remarks

Since the discovery of the chlorophyll cycle, our understanding of the genes and enzymes involved in this cycle, its regulation, physiological role, as well as its evolution, has greatly increased. Nevertheless, there are still unsolved biological questions that remain about this cycle. These include understanding the reaction mechanism of CAO, how chlorophyll *b* levels are sensed to regulate CAO stability, how NYC1 interacts with chlorophyll *b* on an LHC protein, and so on.

Acknowledgments

The authors are grateful for Ms. Junko Kishimoto for her help in illustration. This work was supported by Japan Society for the Promotion of Science, KAKENHI Grant Number 16H06554 and 17K07431 to R.T.

References

Andersson, J., Wentworth, M., Walters, R., Howard, C., Ruban, A., Horton, P., et al. (2003). Absence of the Lhcb1 and Lhcb2 proteins of the light-harvesting complex of photosystem II—Effects on photosynthesis, grana stacking and fitness. *The Plant Journal*, *35*, 350–361.

Apitz, J., Nishimura, K., Schmied, J., Wolf, A., Hedtke, B., van Wijk, K. J., et al. (2016). Posttranslational control of ALA synthesis includes GluTR degradation by Clp protease and stabilization by GluTR-binding protein. *Plant Physiology*, *170*, 2040–2051.

Bar, M., & Ori, N. (2014). Leaf development and morphogenesis. *Development*, *141*, 4219–4230.

Bellemare, G., Bartlett, S., & Chua, N. (1982). Biosynthesis of chlorophyll a/b-binding polypeptides in wild type and the chlorina *f2* mutant of barley. *The Journal of Biological Chemistry*, *257*, 7762–7767.

Berim, A., Park, J. J., & Gang, D. R. (2014). Unexpected roles for ancient proteins: Flavone 8-hydroxylase in sweet basil trichomes is a Rieske-type, PAO-family oxygenase. *The Plant Journal*, *80*, 385–395.

Biswal, A. K., Pattanayak, G. K., Pandey, S. S., Leelavathi, S., Reddy, V. S., Govindjee, et al. (2012). Light intensity-dependent modulation of chlorophyll b biosynthesis and photosynthesis by overexpression of chlorophyllide a oxygenase in tobacco. *Plant Physiology*, *159*, 433–449.

Bujaldon, S., Kodama, N., Rappaport, F., Subramanyam, R., de Vitry, C., Takahashi, Y., et al. (2017). Functional accumulation of antenna proteins in chlorophyll b-less mutants of *Chlamydomonas reinhardtii*. *Molecular Plant*, *10*, 115–130.

Cheminant, S., Wild, M., Bouvier, F., Pelletier, S., Renou, J. P., Erhardt, M., et al. (2011). DELLAs regulate chlorophyll and carotenoid biosynthesis to prevent photooxidative damage during seedling deetiolation in Arabidopsis. *The Plant Cell*, *23*, 1849–1860.

Chen, M. (2014). Chlorophyll modifications and their spectral extension in oxygenic photosynthesis. *Annual Review of Biochemistry*, *83*(83), 317–340.

Copley, S. D. (2017). Shining a light on enzyme promiscuity. *Current Opinion in Structural Biology*, *47*, 167–175.

Cotruvo, J. A., & Stubbe, J. (2011). Class I ribonucleotide reductases: Metallocofactor assembly and repair in vitro and in vivo. *Annual Review of Biochemistry*, *80*(80), 733–767.

Dall'Osto, L., Bressan, M., & Bassi, R. (2015). Biogenesis of light harvesting proteins. *Biochimica et Biophysica Acta-Bioenergetics*, *1847*, 861–871.

Dumitru, R., Jiang, W. Z., Weeks, D. P., & Wilson, M. A. (2009). Crystal structure of dicamba monooxygenase: A Rieske nonheme oxygenase that catalyzes oxidative demethylation. *Journal of Molecular Biology*, *392*, 498–510.

Espineda, C., Linford, A., Devine, D., & Brusslan, J. (1999). The AtCAO gene, encoding chlorophyll a oxygenase, is required for chlorophyll b synthesis in *Arabidopsis thaliana*. *Proceedings of the National Academy of Sciences of the United States of America*, *96*, 10507–10511.

Flachmann, R., & Kuehlbrandt, W. (1995). Accumulation of plant antenna complexes is regulated by post-transcriptional mechanisms in tobacco. *The Plant Cell*, 7, 149–160.

Gray, J., Wardzala, E., Yang, M. L., Reinbothe, S., Haller, S., & Pauli, F. (2004). A small family of LLS1-related non-heme oxygenases in plants with an origin amongst oxygenic photosynthesizers. *Plant Molecular Biology*, *54*, 39–54.

Green, B. R., & Durnford, D. G. (1996). The chlorophyll-carotenoid proteins of oxygenic photosynthesis. *Annual Review of Plant Physiology and Plant Molecular Biology*, *47*, 685–714.

Harada, J., Mizoguchi, T., Satoh, S., Tsukatani, Y., Yokono, M., Noguchi, M., et al. (2013). Specific gene bciD for C7-methyl oxidation in bacteriochlorophyll e biosynthesis of brown-colored green sulfur bacteria. *PLoS One*, *8*, e60026.

Harper, A. L., von Gesjen, S. E., Linford, A. S., Peterson, M. P., Faircloth, R. S., Thissen, M. M., et al. (2004). Chlorophyllide a oxygenase mRNA and protein levels correlate with the chlorophyll a/b ratio in *Arabidopsis thaliana*. *Photosynthesis Research*, *79*, 149–159.

Hauenstein, M., Christ, B., Das, A., Aubry, S., & Hörtensteiner, S. (2016). A role for TIC55 as a hydroxylase of phyllobilins, the products of chlorophyll breakdown during plant senescence. *The Plant Cell*, *28*, 2510–2527.

Hirashima, M., Satoh, S., Tanaka, R., & Tanaka, A. (2006). Pigment shuffling in antenna systems achieved by expressing prokaryotic chlorophyllide a oxygenase in Arabidopsis. *The Journal of Biological Chemistry*, *281*, 15385–15393.

Hirono, Y., & Redei, G. (1963). Multiple allelic control of chlorophyll b level in *Arabidopsis thaliana*. *Nature*, *197*, 1324–1325.

Hoober, J., & Eggink, L. (2001). A potential role of chlorophylls b and c in assembly of light-harvesting complexes. *FEBS Letters*, *489*, 1–3.

Horie, Y., Ito, H., Kusaba, M., Tanaka, R., & Tanaka, A. (2009). Participation of chlorophyll b reductase in the initial step of the degradation of light-harvesting chlorophyll a/b-protein complexes in Arabidopsis. *The Journal of Biological Chemistry*, *284*, 17449–17456.

Ikegami, I., Satoh, S., & Aoki, M. (2007). Binding affinity of Chl b for the Chl a-binding sites in PSI core complexes. *Plant and Cell Physiology*, *48*, 1092–1097.

Ito, H., & Tanaka, A. (2014). Evolution of a new chlorophyll metabolic pathway driven by the dynamic changes in enzyme promiscuous activity. *Plant and Cell Physiology*, *55*, 593–603.

Ito, H., Tanaka, Y., Tsuji, H., & Tanaka, A. (1993). Conversion of chlorophyll b to chlorophyll a by isolated cucumber etioplasts. *Archives of Biochemistry and Biophysics*, *306*, 148–151.

Ito, H., Yokono, M., Tanaka, R., & Tanaka, A. (2008). Identification of a novel vinyl reductase gene essential for the biosynthesis of monovinyl chlorophyll in *Synechocystis* sp PCC6803. *The Journal of Biological Chemistry*, *283*, 9002–9011.

Jia, T., Ito, H., Hu, X. Y., & Tanaka, A. (2015). Accumulation of the NON-YELLOW COLORING 1 protein of the chlorophyll cycle requires chlorophyll b in *Arabidopsis thaliana*. *The Plant Journal*, *81*, 586–596.

Jia, T., Ito, H., & Tanaka, A. (2016). Simultaneous regulation of antenna size and photosystem I/II stoichiometry in *Arabidopsis thaliana*. *Planta*, *244*, 1041–1053.

Kim, E.-H., Li, X.-P., Razeghifard, R., Anderson, J. M., Niyogi, K. K., Pogson, B. J., et al. (2009). The multiple roles of light-harvesting chlorophyll a/b-protein complexes define structure and optimize function of Arabidopsis chloroplasts: A study using two chlorophyll b-less mutants. *Biochimica et Biophysica Acta*, *1787*, 973–984.

Kirst, H., Gabilly, S. T., Niyogi, K. K., Lemaux, P. G., & Melis, A. (2017). Photosynthetic antenna engineering to improve crop yields. *Planta*, *245*, 1009–1020.

Kobayashi, K., Baba, S., Obayashi, T., Sato, M., Toyooka, K., Keranen, M., et al. (2012). Regulation of root greening by light and auxin/cytokinin signaling in Arabidopsis. *The Plant Cell*, *24*, 1081–1095.

Kuai, B., Chen, J. Y., & Hortensteiner, S. (2018). The biochemistry and molecular biology of chlorophyll breakdown. *Journal of Experimental Botany*, *69*, 751–767.

Kunugi, M., Satoh, S., Ihara, K., Shibata, K., Yamagishi, Y., Kogame, K., et al. (2016). Evolution of green plants accompanied changes in light-harvesting systems. *Plant & Cell Physiology*, *57*, 1231–1243.

Kunugi, M., Takabayashi, A., & Tanaka, A. (2013). Evolutionary changes in chlorophyllide a oxygenase (CAO) structure contribute to the acquisition of a new light-harvesting complex in micromonas. *The Journal of Biological Chemistry*, *288*, 19330–19341.

Kupke, D. W., & Huntington, J. L. (1963). Chlorophyll *a* appearance in dark in higher plants—Analytical notes. *Science*, *140*, 49.

Kusaba, M., Ito, H., Morita, R., Iida, S., Sato, Y., Fujimoto, M., et al. (2007). Rice NON-YELLOW COLORING1 is involved in light-harvesting complex II and grana degradation during leaf senescence. *The Plant Cell, 19*, 1362–1375.

Lee, S., Kim, J. H., Yoo, E., Lee, C. H., Hirochika, H., & An, G. H. (2005). Differential regulation of chlorophyll a oxygenase genes in rice. *Plant Molecular Biology, 57*, 805–818.

Mason, J. R., & Cammack, R. (1992). The electron-transport proteins of hydroxylating bacterial dioxygenases. *Annual Review of Microbiology, 46*, 277–305.

Masuda, T., Tanaka, A., & Melis, A. (2003). Chlorophyll antenna size adjustments by irradiance in *Dunaliella salina* involve coordinate regulation of chlorophyll a oxygenase (CAO) and Lhcb gene expression. *Plant Molecular Biology, 51*, 757–771.

Matsuda, K., Shimoda, Y., Tanaka, A., & Ito, H. (2016). Chlorophyll a is a favorable substrate for *Chlamydomonas* Mg-dechelatase encoded by STAY-GREEN. *Plant Physiology and Biochemistry, 109*, 365–373.

Meguro, M., Ito, H., Takabayashi, A., Tanaka, R., & Tanaka, A. (2011). Identification of the 7-hydroxymethyl chlorophyll a reductase of the chlorophyll cycle in Arabidopsis. *The Plant Cell, 23*, 3442–3453.

Melis, A. (2009). Solar energy conversion efficiencies in photosynthesis: Minimizing the chlorophyll antennae to maximize efficiency. *Plant Science, 177*, 272–280.

Mueller, A. H., Dockter, C., Gough, S. P., Lundqvist, U., von Wettstein, D., & Hansson, M. (2012). Characterization of mutations in barley fch2 encoding chlorophyllide a oxygenase. *Plant and Cell Physiology, 53*, 1232–1246.

Murchie, E., & Horton, P. (1997). Acclimation of photosynthesis to irradiance and spectral quality in British plant species: Chlorophyll content, photosynthetic capacity and habitat preference. *Plant, Cell & Environment, 20*, 438–448.

Nagata, N., Satoh, S., Tanaka, R., & Tanaka, A. (2004). Domain structures of chlorophyllide a oxygenase of green plants and *Prochlorothrix hollandica* in relation to catalytic functions. *Planta, 218*, 1019–1025.

Nakagawara, E., Sakuraba, Y., Yamasato, A., Tanaka, R., & Tanaka, A. (2007). Clp protease controls chlorophyll b synthesis by regulating the level of chlorophyllide a oxygenase. *The Plant Journal, 49*, 800–809.

Nakajima, S., Ito, H., Tanaka, R., & Tanaka, A. (2012). Chlorophyll b reductase plays an essential role in maturation and storability of Arabidopsis seeds. *Plant Physiology, 160*, 261–273.

Neilson, J. A., & Durnford, D. G. (2010). Structural and functional diversification of the light-harvesting complexes in photosynthetic eukaryotes. *Photosynthesis Research, 106*, 57–71.

Nishimura, K., & van Wijk, K. J. (2015). Organization, function and substrates of the essential Clp protease system in plastids. *Biochimica et Biophysica Acta-Bioenergetics, 1847*, 915–930.

Obayashi, T., Aoki, Y., Tadaka, S., Kagaya, Y., & Kinoshita, K. (2017). ATTED-II in 2018: A plant coexpression database based on investigation of the statistical property of the mutual rank index. *Plant & Cell Physiology, 59*, e3.

Ohashi, K., Tanaka, A., & Tsuji, H. (1989). Formation of the photosynthetic electron-transport system during the early phase of greening in barley leaves. *Plant Physiology, 91*, 409–414.

Ohtsuka, T., Ito, H., & Tanaka, A. (1997). Conversion of chlorophyll b to chlorophyll a and the assembly of chlorophyll with apoproteins by isolated chloroplasts. *Plant Physiology, 113*, 137–147.

Orf, G. S., & Blankenship, R. E. (2013). Chlorosome antenna complexes from green photosynthetic bacteria. *Photosynthesis Research, 116*, 315–331.

Oster, U., Tanaka, R., Tanaka, A., & Ruedigger, W. (2000). Cloning and functional expression of the gene encoding the key enzyme for chlorophyll b biosynthesis CAO from *Arabidopsis thaliana. The Plant Journal, 21*, 305–310.

Partensky, F., Six, C., Ratin, M., Garczarek, L., Vaulot, D., Probert, I., et al. (2018). A novel species of the marine cyanobacterium *Acaryochloris* with a unique pigment content and lifestyle. *Scientific Reports*, *8*, 9142.

Perrine, Z., Negi, S., & Sayre, R. T. (2012). Optimization of photosynthetic light energy utilization by microalgae. *Algal Research*, *1*, 134–142.

Porra, R. J., Schafer, W., Cmiel, E., Katheder, I., & Scheer, H. (1994). The derivation of the formyl-group oxygen of chlorophyll-B in higher-plants from molecular-oxygen—Achievement of high enrichment of the 7-formyl-group oxygen from O-18(2) greening maize leaves. *European Journal of Biochemistry*, *219*, 671–679.

Qiu, K., Li, Z. P., Yang, Z., Chen, J. Y., Wu, S. X., Zhu, X. Y., et al. (2015). EIN3 and ORE1 accelerate degreening during ethylene-mediated leaf senescence by directly activating chlorophyll catabolic genes in Arabidopsis. *PLoS Genetics*, *11*, 733–739.

Rudoi, A. B., Vezitskii, A. Y., & Shlyk, A. A. (1982). Study of the enzyme-system converting chlorophyllide to chlorophyll in etiolated leaves using exogenous substrates. *Biochemistry (Moscow)*, *47*, 615.

Sakuraba, Y., Balazadeh, S., Tanaka, R., Mueller-Roeber, B., & Tanaka, A. (2012). Overproduction of Chl b retards senescence through transcriptional reprogramming in Arabidopsis. *Plant and Cell Physiology*, *53*, 505–517.

Sakuraba, Y., Lee, S. H., Kim, Y. S., Park, O. K., Hortensteiner, S., & Paek, N. C. (2014). Delayed degradation of chlorophylls and photosynthetic proteins in Arabidopsis autophagy mutants during stress-induced leaf yellowing. *Journal of Experimental Botany*, *65*, 3915–3925.

Sakuraba, Y., Schelbert, S., Park, S. Y., Han, S. H., Lee, B. D., Andres, C. B., et al. (2012). STAY-GREEN and chlorophyll catabolic enzymes interact at light-harvesting complex II for chlorophyll detoxification during leaf senescence in Arabidopsis. *The Plant Cell*, *24*, 507–518.

Sakuraba, Y., Tanaka, R., Yamasato, A., & Tanaka, A. (2009). Determination of a chloroplast degron in the regulatory domain of chlorophyllide a oxygenase. *The Journal of Biological Chemistry*, *284*, 36689–36699.

Sakuraba, Y., Yokono, M., Akimoto, S., Tanaka, R., & Tanaka, A. (2010). Deregulated chlorophyll b synthesis reduces the energy transfer rate between photosynthetic pigments and induces photodamage in *Arabidopsis thaliana*. *Plant and Cell Physiology*, *51*, 1055–1065.

Sato, R., Ito, H., & Tanaka, A. (2015). Chlorophyll b degradation by chlorophyll b reductase under high-light conditions. *Photosynthesis Research*, *126*, 249–259.

Sato, Y., Morita, R., Katsuma, S., Nishimura, M., Tanaka, A., & Kusaba, M. (2009). Two short-chain dehydrogenase/reductases, NON-YELLOW COLORING 1 and NYC1-LIKE, are required for chlorophyll b and light-harvesting complex II degradation during senescence in rice. *The Plant Journal*, *57*, 120–131.

Satoh, S., Ikeuchi, M., Mimuro, M., & Tanaka, A. (2001). Chlorophyll b expressed in cyanobacteria functions as a light-harvesting antenna in photosystem I through flexibility of the proteins. *The Journal of Biological Chemistry*, *276*, 4293–4297.

Satoh, S., & Tanaka, A. (2006). Identification of chlorophyllide a oxygenase in the *Prochlorococcus* genome by a comparative genomic approach. *Plant and Cell Physiology*, *47*, 1622–1629.

Scheumann, V., Schoch, S., & Rudiger, W. (1998). Chlorophyll a formation in the chlorophyll b reductase reaction requires reduced ferredoxin. *The Journal of Biological Chemistry*, *273*, 35102–35108.

Scheumann, V., Schoch, S., & Rudiger, W. (1999). Chlorophyll b reduction during senescence of barley seedlings. *Planta*, *209*, 364–370.

Shimoda, Y., Ito, H., & Tanaka, A. (2012). Conversion of chlorophyll b to chlorophyll a precedes magnesium dechelation for protection against necrosis in Arabidopsis. *The Plant Journal*, *72*, 501–511.

Shimoda, Y., Ito, H., & Tanaka, A. (2016). Arabidopsis STAY-GREEN, Mendel's green cotyledon gene, encodes magnesium-dechelatase. *The Plant Cell, 28*, 2147–2160.

Shlyk, A. (1971). Biosynthesis of chlorophyll b. *Annual Review of Plant Physiology, 22*, 169–184.

Smith, J. H. C., & French, C. S. (1963). Major and accessory pigments in photosynthesis. *Annual Review of Plant Physiology and Plant Molecular Biology, 14*, 181–224.

Standfuss, J., van Scheltinga, A. C. T., Lamborghini, M., & Kuhlbrandt, W. (2005). Mechanisms of photoprotection and nonphotochemical quenching in pea light-harvesting complex at 2.5 A resolution. *The EMBO Journal, 24*, 919–928.

Takabayashi, A., Kurihara, K., Kuwano, M., Kasahara, Y., Tanaka, R., & Tanaka, A. (2011). The oligomeric states of the photosystems and the light-harvesting complexes in the Chl b-less mutant. *Plant & Cell Physiology, 52*, 2103–2114.

Tanaka, A., Ito, H., Tanaka, R., Tanaka, N. K., Yoshida, K., & Okada, K. (1998). Chlorophyll a oxygenase (CAO) is involved in chlorophyll b formation from chlorophyll a. *Proceedings of the National Academy of Sciences of the United States of America, 95*, 12719–12723.

Tanaka, R., Kobayashi, K., & Masuda, T. (2011). Tetrapyrrole metabolism in *Arabidopsis thaliana*. *The Arabidopsis Book, 6*, e0145.

Tanaka, R., Koshino, Y., Sawa, S., Ishiguro, S., Okada, K., & Tanaka, A. (2001). Overexpression of chlorophyllide a oxygenase (CAO) enlarges the antenna size of photosystem II in *Arabidopsis thaliana*. *The Plant Journal, 26*, 365–373.

Tanaka, A., & Tsuji, H. (1981). Changes in chlorophyll-a and chlorophyll-B content in dark-incubated cotyledons excised from Illuminated seedlings—The effect of calcium. *Plant Physiology, 68*, 567–570.

Tanaka, A., & Tsuji, H. (1982). Calcium-induced formation of chlorophyll-B and light-harvesting chlorophyll-a/B-protein complex in cucumber cotyledons in the dark. *Biochimica et Biophysica Acta, 680*, 265–270.

Tanaka, R., & Tanaka, A. (2005). Effects of chlorophyllide a oxygenase overexpression on light acclimation in *Arabidopsis thaliana*. *Photosynthesis Research, 85*, 327–340.

Tanaka, R., & Tanaka, A. (2011). Chlorophyll cycle regulates the construction and destruction of the light-harvesting complexes. *Biochimica et Biophysica Acta-Bioenergetics, 1807*, 968–976.

Tanaka, R., & Tanaka, A. (2015). Effects of chlorophyllide a oxygenase overexpression on light acclimation in Arabidopsis thaliana. *Photosynthesis Research, 85*, 327–340.

Tanaka, A., Yamamoto, Y., & Tsuji, H. (1991). Formation of chlorophyll-protein complexes during greening. 2. Redistribution of chlorophyll among apoproteins. *Plant and Cell Physiology, 32*, 195–204.

Thornber, J., & Highkin, H. (1974). Composition of the photosynthetic apparatus of normal barley leaves and a mutant lacking chlorophyll b. *European Journal of Biochemistry, 41*, 109–116.

Thweatt, J. L., Ferlez, B. H., Golbeck, J. H., & Bryant, D. A. (2017). BciD is a radical S-adenosyl-l-methionine (SAM) enzyme that completes bacteriochlorophyllide e biosynthesis by oxidizing a methyl group into a formyl group at C-7. *The Journal of Biological Chemistry, 292*, 1361–1373.

Tomitani, A., Okada, K., Miyashita, H., Matthijs, H., Ohno, T., & Tanaka, A. (1999). Chlorophyll b and phycobilins in the common ancestor of cyanobacteria and chloroplasts. *Nature, 400*, 159–162.

Tsuchiya, T., Akimoto, S., Mizoguchi, T., Watabe, K., Kindo, H., Tomo, T., et al. (2012). Artificially produced [7-formyl]-chlorophyll d functions as an antenna pigment in the photosystem II isolated from the chlorophyllide a oxygenase-expressing *Acaryochloris marina*. *Biochimica et Biophysica Acta-Bioenergetics, 1817*, 1285–1291.

Vavilin, D., Yao, D., & Vermaas, W. (2007). Small cab-like proteins retard degradation of photosystem II-associated chlorophyll in *Synechocystis* sp. PCC 6803: Kinetic analysis of pigment labeling with 15N and 13C. *The Journal of Biological Chemistry, 282*, 37660–37668.

Voitsekhovskaja, O. V., & Tyutereva, E. V. (2015). Chlorophyll b in angiosperms: Functions in photosynthesis, signaling and ontogenetic regulation. *Journal of Plant Physiology, 189*, 51–64.

Wang, X., & Liu, L. (2016). Crystal structure and catalytic mechanism of 7-hydroxymethyl chlorophyll a reductase. *The Journal of Biological Chemistry, 291*, 13349–13359.

Waters, M. T., Wang, P., Korkaric, M., Capper, R. G., Saunders, N. J., & Langdale, J. A. (2009). GLK transcription factors coordinate expression of the photosynthetic apparatus in Arabidopsis. *The Plant Cell, 21*, 1109–1128.

Wettstein, D. V., Kahn, A., Nielsen, O. F., & Gough, S. (1974). Genetic regulation of chlorophyll synthesis analyzed with mutants in barley. *Science, 184*, 800–802.

Xu, H., Vavilin, D., & Vermaas, W. (2001). Chlorophyll b can serve as the major pigment in functional photosystem II complexes of cyanobacteria. *Proceedings of the National Academy of Sciences of the United States of America, 98*, 14168–14173.

Yamasato, A., Nagata, N., Tanaka, R., & Tanaka, A. (2005). The N-terminal domain of chlorophyllide a oxygenase confers protein instability in response to chlorophyll B accumulation in Arabidopsis. *The Plant Cell, 17*, 1585–1597.

Yang, Y. L., Xu, J., Huang, L. C., Leng, Y. J., Dai, L. P., Rao, Y. C., et al. (2016). PGL, encoding chlorophyllide a oxygenase 1, impacts leaf senescence and indirectly affects grain yield and quality in rice. *Journal of Experimental Botany, 67*, 1297–1310.

Zelisko, A., Garcia-Lorenzo, M., Jackowski, G., Jansson, S., & Funk, C. (2005). AtFtsH6 is involved in the degradation of the light-harvesting complex II during high-light acclimation and senescence. *Proceedings of the National Academy of Sciences of the United States of America, 102*, 13699–13704.

Zhu, X. Y., Chen, J. Y., Xie, Z. K., Gao, J., Ren, G. D., Gao, S., et al. (2015). Jasmonic acid promotes degreening via MYC2/3/4-and ANAC019/055/072-mediated regulation of major chlorophyll catabolic genes. *The Plant Journal, 84*, 597–610.

CHAPTER SEVEN

Chlorophyll breakdown—Regulation, biochemistry and phyllobilins as its products

Stefan Hörtensteiner[a,*], Mareike Hauenstein[a], Bernhard Kräutler[b,*]

[a]Institute of Plant and Microbial Biology, University of Zurich, Zurich, Switzerland
[b]Institute of Organic Chemistry & Center for Molecular Biosciences (CMBI), University of Innsbruck, Innsbruck, Austria
*Corresponding authors: e-mail address: shorten@botinst.uzh.ch; bernhard.kraeutler@uibk.ac.at

Contents

Abstract

Chlorophyll breakdown occurs throughout the life cycle of a plant, but is most obvious during leaf senescence and fruit ripening when respective tissues visually degreen. Important milestones in elucidating the fate of chlorophyll, which was considered a "biological enigma" more than 30 years ago, were, first, the initial decryption of the structures of chlorophyll catabolites, now collectively termed phyllobilins, in 1991, and, second, the identification of pheophorbide *a* oxygenase (PAO), a key enzyme of

Advances in Botanical Research, Volume 90
ISSN 0065-2296
https://doi.org/10.1016/bs.abr.2019.03.004

breakdown that determines the structural backbone of the phyllobilins, in 2003. Hence, nowadays the biochemical pathway of chlorophyll breakdown is termed the "PAO/phyllobilin" pathway. This chapter summarizes the current knowledge on the diversity of phyllobilins with regard to structural characteristics and distribution within plant species, and presents biochemical, cellular and molecular, as well as regulatory and evolutionary aspects of the enzymes of the PAO/phyllobilin pathway.

Abbreviations

ABA abscisic acid
ABC ATP-binding cassette
ACD accelerated cell death
***bc*DFCC** bicyclo-DFCC
***bc*FCC** bicyclo-FCC
CBR Chl *b* reductase
CCE Chl catabolic enzyme
Chl chlorophyll
Chlide chlorophyllide
CYP cytochrome P450 mono-oxygenase
DCC dioxobilin-type (type-II) Chl catabolite
DFCC dioxobilin-type (type-II) fluorescent Chl catabolite
DNCC dioxobilin-type (type-II) nonfluorescent Chl catabolite
DPiCC dioxobilin-type (type-II) pink Chl catabolite
DYCC dioxobilin-type (type-II) yellow Chl catabolite
FCC fluorescent (type-I) Chl catabolite
HCAR 7^1-hydroxy Chl *a* reductase
***hm*FCC** hypermodified FCC
HOChl *a* 7^1-hydroxy Chl *a*
***i*DNCC** iso-DNCC (an NDCC)
***i*PBon** isophyllobilanone
LHC light-harvesting complex
MES methylesterase
***m*FCC** modified FCC
NCC nonfluorescent (type-I) Chl catabolite
NDCC nonfluorescent DCC
NMR nuclear magnetic resonance (spectroscopy)
NOL NYC1-like
NYC1 non-yellow coloring 1
PAO pheophorbide *a* oxygenase
PB phyllobilin
***p*FCC** primary FCC
Pheide pheophorbide
Phein pheophytin
PiCC pink (type-I) Chl catabolite
PPH pheophytinase
***py*NCC** pyro-NCC
***py*Pheide** pyro-pheophorbide

RCC	red Chl catabolite
RCCR	RCC-reductase
sFCC	secondary FCC
SGR	stay-green (protein)
TIC55	translocon at the inner chloroplast membrane 55
UV/Vis	ultraviolet/visible (absorbance spectroscopy)
YCC	yellow (type-I) Chl catabolite

1. Introduction

The appearance of the fall colors is a puzzling and fascinating phenomenon (Matile, 2000). The seasonally synchronized disappearance of chlorophyll (Chl) in autumn and the widespread re-greening of the vegetation in spring are probably the most visual signs of life, which are even observable on Earth from outer space (Kräutler & Matile, 1999). Indeed, the temporal apparent "recycling" of the green pigment Chl by seasonally alternating *de novo* Chl biosynthesis and Chl degradation, on land and in the oceans, involves an estimated 1000 million tons, worldwide (Hendry, Houghton, & Brown, 1987; Morel, 2006). Strikingly, until about 30 years ago, Chl seemed to disappear in higher plants without leaving a trace, and non-green products of Chl breakdown remained elusive (Hendry et al., 1987; Matile, Ginsburg, Schellenberg, & Thomas, 1987). As we know today, the bilin-type Chl breakdown products, now generally named phyllobilins (PBs) (Kräutler, 2014), accumulate in higher plants as easily overlooked, colorless linear tetrapyrroles, primarily (Hörtensteiner, 2013; Hörtensteiner & Kräutler, 2011; Kräutler, 2016; Kräutler, Jaun, Bortlik, Schellenberg, & Matile, 1991; Kräutler & Matile, 1999; Kuai, Chen, & Hörtensteiner, 2018), the structures of which are related in a remarkable way to those of bilins, the products from heme breakdown (Falk, 1989).

First traces of presumed non-green Chl breakdown products were detected by Matile, Thomas and coworkers, when these authors compared the pigment pattern of senescent leaves of wild type *Festuca pratensis* with those in "stay green" mutants of this grass (Matile, 1987; Thomas, Bortlik, Rentsch, Schellenberg, & Matile, 1989). Colorless compounds that were absent in the green mutant could be spotted in the wild type and were considered to represent some of the elusive Chl catabolites. They were called "rusty pigments," as they rapidly converted into rust-colored products (Bortlik, Peisker, & Matile, 1990). One of the "rusty pigments" from senescent leaves of barley (*Hordeum vulgare*) could be isolated without degradation and color formation, and its structure was elucidated in 1991 with the help of modern spectroscopic means (Kräutler et al., 1992, 1991). It was confirmed to be a Chl catabolite, as it was a deconjugated bilin-type linear tetrapyrrole that carried the characteristic ring C/E-portion and other structural hallmarks of the Chls (Kräutler et al., 1991). This compound from barley was named the "nonfluorescent" Chl catabolite (NCC) *Hv*-NCC-1 (**1**) (Scheme 1). The structure of NCC **1** indicated oxidative ring opening of the Chl macrocycle between C4 and C5 and **1** was first described as a 4,5-secophytoporphyrinoid (Kräutler et al., 1992, 1991). This "northern" cleavage site contrasted with all expectations from Chl chemistry (Brown, Houghton, & Hendry, 1991), but turned out to be strikingly reminiscent of the site of ring opening in heme breakdown (Wilks, 2008). However, the *meso*-carbon C5 of Chl *a* was retained in the formyl group of the NCC **1**, in contrast to the typical bilins from heme breakdown (Falk, 1989; Wilks, 2008), but similar to some heme-cleavage products observed in bacteria (Wilks & Ikeda-Saito, 2014). Additional polar groups at the molecular periphery of **1** pointed to puzzling catabolic processes in the formation of **1** from Chl, raising further questions with respect to the biochemistry of Chl breakdown in higher plants. As the NCC **1** was produced in barley leaves in experimental darkness, its key relevance for typical Chl breakdown in the fall leaves was not unambiguous.

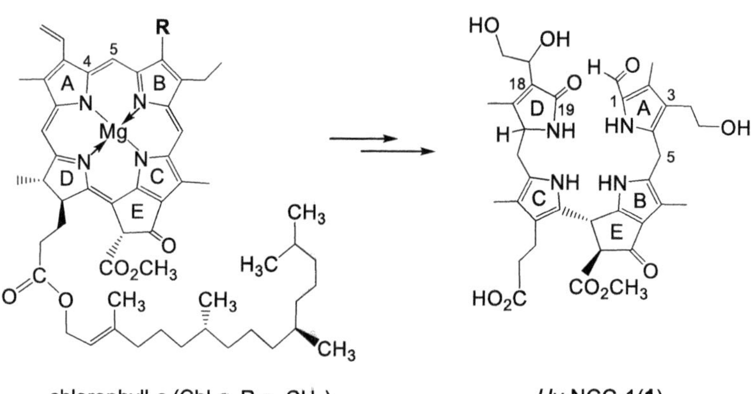

Scheme 1 Structural formulae of Chls *a* and *b*, and of the first identified bilin-type Chl catabolite, the phyllobilane *Hv*-NCC-1 (**1**). Atom numbering of the two tetrapyrroles and labels of their 5-membered rings (A-E) follow the conventions for natural porphyrinoids (left) and bilins (right), differing systematically in the two types of compounds.

In the event, similar NCCs were soon found in naturally senescent leaves of the dicot oilseed rape (*Brassica napus*) (Mühlecker & Kräutler, 1996; Mühlecker, Kräutler, Ginsburg, & Matile, 1993). All of these NCCs carried a methyl group at their ring A, indicating their more direct lineage with Chl *a* (Kräutler, 2016). Much of the puzzle is now deciphered, how the NCCs are generated in the higher plants (Hörtensteiner & Kräutler, 2011), what happens to Chl *b* (Hörtensteiner, 2013; Kuai et al., 2018; Scheumann, Schoch, & Rüdiger, 1999; Tanaka, Kobayashi, & Masuda, 2011; Tanaka & Tanaka, 2006), and how the outset of senescence-induced Chl breakdown is subjected to interactive regulatory processes (Hörtensteiner, 2009; Tanaka & Tanaka, 2011). Pheophorbide *a* (Pheide *a*) has been identified as the ultimate intermediate with an intact porphyrinoid macrocycle (see Scheme 2 and Section 3.1) (Hörtensteiner, 2013). The specific oxygen-dependent cleavage of the macrocycle of Pheide *a* by Pheide *a* oxygenase (PAO) generates the "red Chl catabolite" (RCC) (Kräutler, 2019; Kräutler, Mühlecker, Anderl, & Gerlach, 1997; Rodoni, Mühlecker, et al., 1997), a cryptic red formyloxo-bilin that represents the common precursor of the natural Chl-derived PBs in angiosperms (Hörtensteiner, Wüthrich, Matile, Ongania, & Kräutler, 1998; Pružinská, Anders, Tanner, Roca, & Hörtensteiner, 2003). Efficient enzymatic reduction (in a form of "metabolic channeling") of RCC by RCC reductases (RCCRs) (Rodoni, Vicentini, Schellenberg, Matile, & Hörtensteiner, 1997; Sugishima, Kitamori, Noguchi, Kohchi, & Fukuyama, 2009; Wüthrich, Bovet, Hunziker, Donnison, & Hörtensteiner, 2000) next produces a first type of a fluorescent formyloxobilin (fluorescent type-I PB), the "*primary*" fluorescent Chl catabolites (*p*FCCs). Depending on the plant species, the blue fluorescent *p*FCCs is formed stereospecifically in one of two C16-epimeric versions, **2** or ***epi*-2** (Kräutler, 2016; Mühlecker, Kräutler, Moser, Matile, & Hörtensteiner, 2000; Mühlecker, Ongania, Kräutler, Matile, & Hörtensteiner, 1997). Typical "*modified*" FCCs (*m*FCCs) are produced from the two epimeric *p*FCCs by the action of chloroplastic or cytosolic enzymes (Hörtensteiner & Kräutler, 2011; Kuai et al., 2018). Most FCCs exist only fleetingly and isomerize spontaneously in the weakly acidic media of the vacuoles to the corresponding NCCs (Oberhuber, Berghold, Breuker, Hörtensteiner, & Kräutler, 2003; Oberhuber, Berghold, & Kräutler, 2008). Alternatively, *p*FCCs and 3^2-hydroxy-*p*FCCs (**3** and ***epi*-3**) are deformylated effectively by a cytochrome P450 mono-oxygenase (CYP)—identified in *Arabidopsis thaliana* (*Arabidopsis*) as CYP89A9 (Christ et al., 2013)—that furnishes corresponding dioxobilin-type fluorescent Chl catabolites (DFCCs),

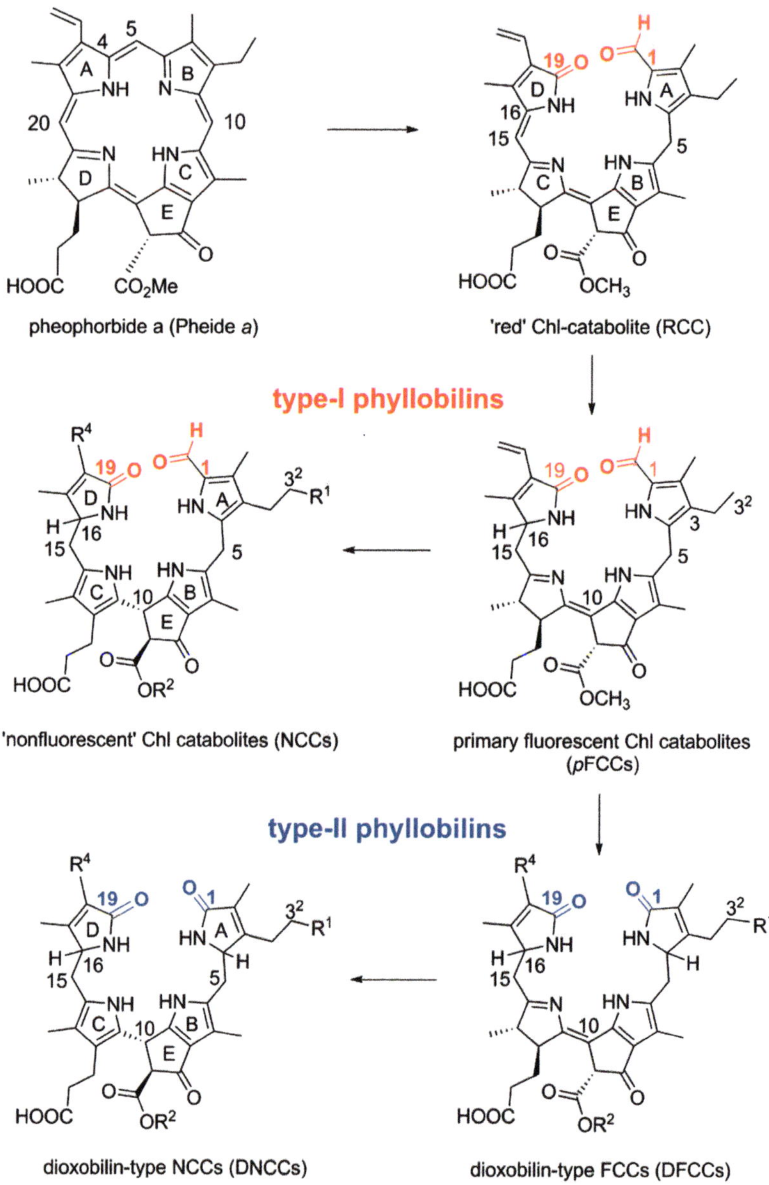

Scheme 2 Structural outline of central steps of Chl breakdown in higher plants. The "red" Chl catabolite (RCC) is the common 1-formyl-19-oxobilin type precursor of all phyllobilins (PBs) along the PAO/phyllobilin pathway. The latter branches into two major lines at the stage of the colorless FCCs (fluorescent 1-formyl-19-oxobilin type PBs), which can undergo deformylation to fluorescent 1,19-dioxobilin type PBs (DFCCs). Both types of fluorescent PBs isomerize rapidly to corresponding "nonfluorescent" 1-formyl-19-oxobilin type (NCCs) or 1,19-dioxobilin type PBs (DNCCs), representing type-I or type-II PBs, respectively.

which isomerize rapidly and stereospecifically to natural dioxobilin-type NCCs (DNCCs) (Süssenbacher, Hörtensteiner, & Kräutler, 2015). Indeed, two major lines of colorless, "nonfluorescent" Chl catabolites (or phyllobilanes) are now known to accumulate in senescent leaves of angiosperms, representing NCCs ("nonfluorescent" type-I PBs) (Kräutler, 2014) or dioxobilin-type NCCs (DNCCs, "nonfluorescent" type-II PBs) (Kräutler, 2016). As it was the case for the first NCC to be discovered, the first DNCCs, which were called "urobilinogenoidic" Chl catabolites, also happened to be found in barley leaves (Losey & Engel, 2001).

2. Structures of Chl catabolites (phyllobilins) from higher plants and their nomenclature

A mostly phenomenological nomenclature was used in the earlier investigations on Chl catabolites (phyllobilins), which were named by the specification of the plant of origin, the basic structure type and a measure of polarity, e.g., as *Bn*-NCC-1 to *Bn*-NCC-3 for the three NCCs from oilseed rape (Mühlecker & Kräutler, 1996). Since then, several lines of practical phenomenological nomenclature have been developed in different laboratories (Christ, Hauenstein, & Hörtensteiner, 2016; Moser, Müller, Holzinger, Lutz, & Kräutler, 2012; Roca, Rios, & Perez-Galvez, 2018). In addition, a first structure-based semi-systematic nomenclature of Chl catabolites described them as 4,5-seco-phytoporphyrins (Kräutler et al., 1991; Kräutler & Matile, 1999). Following the introduction of the general naming of tetrapyrrolic Chl catabolites as "phyllobilins" (PBs) (Kräutler, 2014), nomenclature rules from the bilin literature (Falk, 1989) are used, as also exemplified in this review. In line with this transition, atom renumbering was also adapted in 2014 (Kräutler, 2014). NCCs, such as **1** (Scheme 1), were more systematically classified on the basis of their structures as type-I phyllobilanes (Kräutler, 2014). Hence, **1** is named the 8^2S,$10R$,$16n$-3^2,18^1,18^2-trihydroxy-1-formyl-19-oxo-16,19-dihydrophyllobilane (Kräutler, 2014, 2016). This structure-based nomenclature clearly identifies PBs as specific compounds, independent of their origin from particular plants, and is used, nowadays, in parallel with the established phenomenological names, like *Hv*-NCC-1 (Christ et al., 2016; Kräutler, 2016; Moser et al., 2012; Roca et al., 2018). For practical reasons the chemical formulae of PBs are generally presented in a pseudocyclic form, in spite of the flexible nature of many of them and their presumed preference in more stable, extended structures (Kräutler, 2014). The crystal structure of an

RCC-derivative (Engel, Gossauer, Gruber, & Kratky, 1993) of two yellow- (Li et al., 2018, 2016) and a pink-colored PB (Li et al., 2014), as well as some structures calculated by quantum chemical methods (Erhart, Mittelberger, et al., 2018; Scherl et al., 2016), has helped to convey a more precise three-dimensional picture of selected PBs (Kräutler, 2016) (see below).

2.1 Colorless fluorescent phyllobilins from higher plants and their structures

2.1.1 Fluorescent (formyloxobilin-type) Chl catabolites

In the late 1980s colorless metabolites with a strong blue fluorescence, named FCs, were detected by Matile and coworkers in senescent leaves of barley, and were discussed in the context of Chl breakdown (Düggelin, Schellenberg, Bortlik, & Matile, 1988). In a targeted preparative assay, using an active extract of senescent leaves of oilseed rape, the first fluorescent Chl catabolite was isolated and identified as *p*FCC (**2**, 8^2S,10*Z*,12*S*,13*R*,16*n*-1-formyl-19-oxo-12,13,16,19-tetrahydrophyllobilene-*b*) by detailed structural characterization (Mühlecker et al., 1997). The phyllobilene-*b* **2** featured a structure that was key for assigning the hypothetical role of its direct precursor to the elusive RCC, a crucial cryptic intermediate of Chl breakdown (Mühlecker et al., 1997) along the common PAO/phyllobilin pathway (Hörtensteiner & Kräutler, 2011) (Scheme 2). In a related study with senescent leaves of sweet pepper (*Capsicum annum*), its main FCC was characterized as the C16-epimer of **2**, called *epi-p*FCC (***epi*-2**) (Mühlecker et al., 2000). Hence, the *p*FCCs are produced in one of two C16-epimeric forms by the stereospecific reduction of the C15=C16 double bond of RCC by one of two lines of cofactor-free RCCRs, establishing the deep rooted relevance of two stereo-types of RCCRs (Hörtensteiner et al., 2000). An electrochemical reduction process modeled a stereo-unspecific version of RCC reduction by the RCCRs (Kräutler, 2019; Oberhuber et al., 2008). Typical FCCs, now also classified as fluorescent type-I PBs (Kräutler, 2016), do not accumulate in plant cells and disappear quickly, either by rapid isomerization to corresponding NCCs (in the weakly acidic medium of the vacuoles, in particular) (Christ et al., 2012; Oberhuber et al., 2003, 2008) or by deformylation to corresponding DFCCs on the way to type-II PBs (Christ et al., 2013; Hörtensteiner & Kräutler, 2011). For such reasons, the number of FCCs detected and characterized in studies of senescent leaves and ripening fruit is still inferior to the ones of the generally accumulating NCCs and DNCCs (Christ et al., 2016; Kräutler, 2016; Roca et al., 2018). Only a handful of *m*FCCs have been isolated from plant extracts and structurally characterized, such as *Mc*-FCC-62 (***epi*-3**, 3^2-hydroxy-*epi*-*p*FCC) (Moser et al., 2012), *At*-FCC-1 (**4**, 3^2-hydroxy-8^4-desmethyl-*p*FCC) and *At*-FCC-2 (**5**, 8^4-desmethyl-*p*FCC) (Pružinská et al., 2005) (see Table 1). UV/Vis-spectra of FCCs display two closely spaced maxima, one around 360 nm (due to the conjugated system of rings B and C of FCCs), and a second absorption maximum around 320 nm, which is due to the characteristic α-formylpyrrole unit (ring A) of type-I PBs. The nearly colorless FCCs show a strong blue fluorescence with an emission maximum near 450 nm and photoexcited FCCs sensitize the formation of singlet oxygen very effectively (Jockusch, Turro, Banala, & Kräutler, 2014).

In ripe yellow bananas (*Musa acuminata* cv. Cavendish) first persistent FCCs were detected in 2008, as well as the strong blue fluorescence of ripe bananas was discovered, which is easily seen, when observed under "black (UV)-light" (Moser, Müller, et al., 2008). This new type of exceptionally persistent FCCs was revealed to carry an unprecedented sugar-containing ester function at their propionate side chain (Scheme 3 and Table 1). They were classified as "hypermodified" FCCs (*hm*FCCs), such as the banana FCCs *Mc*-FCC-56 (**6**), *Mc*-FCC-53 (***iso*-6**), *Mc*-FCC-49 (**7**) and M*c*-FCC-46 (***iso*-7**) (Moser, Müller, et al., 2008; Moser et al., 2009). Similar persistent *hm*FCCs were also detected as the natural FCCs *Ma*-FCC-61 (**8**) (Banala et al., 2010), *Ma*-FCC-64 (**9a**) (Vergeiner et al., 2013) and *Ma*-FCC-63 (**9b**) (Vergeiner et al., 2013), and *Sw*-FCC-62 (**10**) (Kräutler et al., 2010), respectively, in extracts of senescent leaves of banana plants (*Musa acuminate*) or of peace lily (*Spathiphyllum wallisii*). However, the likewise identified persistent methyl ester **11** of ***epi*-3** was classified as an isolation artifact in the methanolic banana peel extract (Moser et al., 2012). Interestingly, some of the *hm*FCCs from bananas carried an additional β-glucopyranosyl unit appended to their 3^2-hydroxygroup, such as *Mc*-FCC-49 (**7**) and *Mc*-FCC-46 (***iso*-7**) (Moser et al., 2009, 2012). Such sugar appendages have frequently been observed at $C3^2$ in the more abundant NCCs (Kräutler, 2016), and a corresponding 3^2-β-glucopyranosylated *m*FCC (3^2-β-glucopyranosyl-*p*FCC **12**) has also tentatively been identified in *Arabidopsis* (Christ et al., 2012). Remarkably, all of the known *hm*FCCs still carry at their ring D a vinyl group and a methyl ester group at $C8^2$ of ring E, as present in the Chls (see Table 1).

Table 1 Structures of natural fluorescent 1-formyl-19-oxobilin type (type-I) phyllobilins (FCCs): the labels R^1 to R^4 refer to the generalized formula of FCCs, as shown in Scheme 3.

#[a]	R^1	R^2	R^3	R^4	C16[b]	Provisional names[c]	References
2	H	CH_3	H	$CH{=}CH_2$	n	*Bn*-FCC-2 (*p*FCC)	Mühlecker et al. (1997)
***epi*-2**	H	CH_3	H	$CH{=}CH_2$	epi	*Ca*-FCC-2 (*epi-p*FCC)	Mühlecker et al. (2000)
5	H	H	H	$CH{=}CH_2$	n	*At*-FCC-2	Pružinská et al. (2005)
4	OH	H	H	$CH{=}CH_2$	n	*At*-FCC-1	Pružinská et al. (2005)
***epi*-3**	OH	CH_3	H	$CH{=}CH_2$	epi	*Mc*-FCC-62	Moser et al. (2012)
12	O-Glc[d]	CH_3	H	$CH{=}CH_2$	n	*At*-FCC-40	Christ et al. (2012)
11	OH	CH_3	CH_3[e]	$CH{=}CH_2$	epi	*Mc*-FCC-62	Vergeiner, Banala, and Kräutler (2013)
9a	OH	CH_3	6′-βGlc	$CH{=}CH_2$	epi	*Ma*-FCC-64	Vergeiner et al. (2013)
9b	OH	CH_3	6′-βGlc	$CH{=}CH_2$	epi	*Ma*-FCC-63	Vergeiner et al. (2013)
6	OH	CH_3	4′-Daucyl[f]	$CH{=}CH_2$	epi	*Mc*-FCC-56	Moser et al. (2008)
***iso*-6**	OH	CH_3	4′-Daucyl[g]	$CH{=}CH_2$	epi	*Mc*-FCC-53	Moser, Müller, et al. (2008)
7	O-Glc[d]	CH_3	5′-Daucyl[f]	$CH{=}CH_2$	epi	*Mc*-FCC-49	Moser et al. (2009)
***iso*-7**	O-Glc[d]	CH_3	4′-Daucyl[g]	$CH{=}CH_2$	epi	*Mc*-FCC-46	Moser et al. (2009)
8	OH	CH_3	–[h]	$CH{=}CH_2$	epi	*Ma*-FCC-61	Banala et al. (2010)
10	OH	CH_3	–[i]	$CH{=}CH_2$	n	*Sw*-FCC-62	Kräutler et al. (2010)
13	O-Glc*[j]	CH_3	6′-Glc*[j]	$CH{=}CH_2$	epi	*Vv*-FCC-55	Erhart, Mittelberger, et al. (2018)

[a]Compound-number.
[b]Relative configuration at C16: n = "normal," i.e., derived from *p*FCC (**2**); *epi* = "epimeric," i.e., derived from *epi-p*FCC (***epi*-2**).
[c]*Bn*-FCC-2 from oilseed rape (*Brassica napus*) (Mühlecker et al., 1997); *Ca*-FCC-2 from sweet pepper (*Capsicum annuum*) (Mühlecker et al., 2000); *At*-FCCs from *Arabidopsis thaliana* (Pružinská et al., 2005); *Mc*-FCCs from banana peels (*Musa acuminata*, Cavendish cultivar) (Moser, Müller, et al., 2008; Moser et al., 2009, 2012); *Ma*-FCCs from banana leaves (*Musa acuminata*, Cavendish cultivar) (Banala et al., 2010); *Sw*-FCC-62 from peace lily (*Spathiphyllum wallisii*) (Kräutler et al., 2010); *Vv*-FCC-55 from grapevine (*Vitis vinifera*) (Erhart, Mittelberger, et al., 2018).
[d]Glc = 1′β-D-glucopyranosyl.
[e]Likely to be an artifact, from methanolysis of "persistent" FCC-daucyl-esters, such as *Mc*-FCC-56 and *Mc*-FCC-53.
[f]Bound at daucyl 4′-OH.
[g]Bound at daucyl 5′-OH (Moser et al., 2012).
[h]R^4 = glyceryl-di-galactosyl (Banala et al., 2010).
[i]R^4 = dihydroxyphenylethyl-glucosyl (Kräutler et al., 2010).
[j]Glc* = twice attached as 1′β,6′-D-glycopyranosyl moiety (Erhart, Mittelberger, et al., 2018).

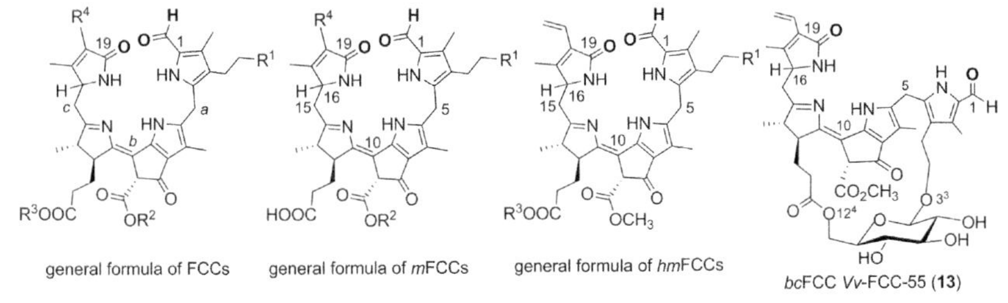

Scheme 3 General formula of FCCs, *m*FCCs, and *hm*FCCs, and formula of the *bc*FCC *Vv*-FCC-55 (**13**) (see Table 1). FCCs constitute the PB-class of the type-I phyllobilenes-*b*.

A unique combination of the substituent pattern of *hm*FCCs and of 3^2-glycopyranosyl FCCs was discovered in senescent leaves of grapevine (*Vitis vinifera* cv. Chardonnay), where the *Vv*-FCC-55 (**13**) was characterized, a so-called bicyclo[17.3.1]-1′,6′-glycopyranosyl-FCC ("bicyclo"-FCC, *bc*FCC), in which a β-glucopyranosyl unit is attached twice, once—glycosidic—at $O3^2$, and once—as an O6′-glucosyl-ester moiety at the propionate substituent (see Scheme 3) (Erhart, Mittelberger, et al., 2018). The bicyclo[17.3.1]-1′,6′-glycopyranosyl moiety imposes a fairly rigid three dimensional structure upon *Vv*-FCC-55 (**13**), a remarkably stable, novel type of a *hm*FCC (Erhart, Mittelberger, et al., 2018), conspicuously related to *Vv*-DFCC-53 (**14**) the fluorescent type-II analog (see below).

2.1.2 Dioxobilin-type fluorescent Chl catabolites

The first DFCCs were identified in (wild type) *Arabidopsis* (Christ et al., 2013; Süssenbacher, Hörtensteiner, et al., 2015). Typical DFCCs or fluorescent type-II PBs were assigned the role of the direct precursors of the more abundant DNCCs (Christ et al., 2013; Kräutler, 2016). The polar and short lived *At*-DFCC-33 (**15**), with "n"-configuration at C16, but unknown at C4, named 8^2S,10*Z*,12*S*,13*R*,16*n*-1,19-dioxo-1,4,12,13,16,19-hexahydro-phyllobilene-*b*, was observed to isomerize stereospecifically and rapidly to a corresponding DNCC, named *At*-DNCC-33 (**16**) (Süssenbacher, Hörtensteiner, et al., 2015). Likewise, in the *mes16* mutant of *Arabidopsis* (see below), the astoundingly hydroxymethylated *At*-2HM-iso-DFCC (**17**) was identified and characterized (Scheme 4) (Süssenbacher, Christ, Hörtensteiner, & Kräutler, 2014), which appears to be the direct metabolic precursor of the corresponding *At*-2HM-iso-DNCC (**18**) (see below) (Süssenbacher et al., 2014). UV/Vis-spectra of DFCCs display a maximum at around 360 nm, like FCCs, but lack the second absorption maximum of the type-I PBs, at around 320 nm, which is due to their α-formylpyrrole unit. However, photoexcited DFCC **15** shows a similar strong blue emission as FCCs (Süssenbacher, Hörtensteiner, et al., 2015).

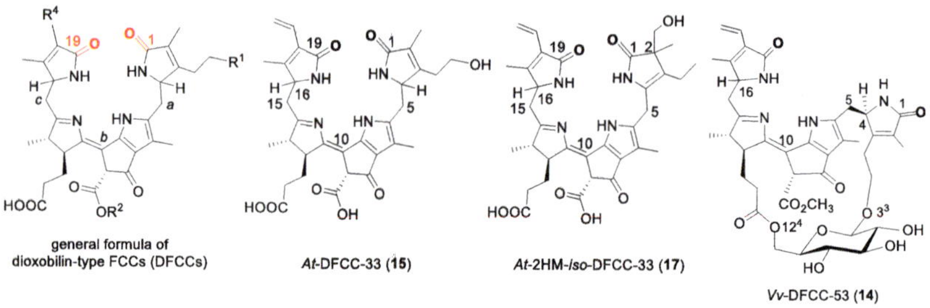

Scheme 4 General formula of DFCCs and formula (left to right) of *At*-DFCC-33 (**15**), of the iso-DFCC *At*-2HM-iso-DFCC (**17**), and of the *bc*DFCC *Vv*-DFCC-53 (**14**). DFCCs constitute the class of the type-II phyllobilenes-*b*.

In senescent grapevine leaves the unique DFCC **14** was detected, an analog of the bicycloglycosidic *Vv*-FCC-55 (**13**). It was structurally characterized as a bicyclo[17.3.1]-1′,6′-glycopyranosyl-DFCC (a *bc*DFCC) (Scheme 4) (Erhart, Mittelberger, et al., 2018), assigning **14** the role of the formal direct product of the oxidative deformylation of the *bc*FCC **13**. Interestingly, in the *bc*DFCC **14** the absolute configuration of the newly introduced stereocenter at C4 could indirectly be deduced as *R* from NMR-studies in combination with model structures from high level theoretical calculations (Erhart, Mittelberger, et al., 2018). However, the biochemistry of the metabolic formation of such extraordinary bicyclo [17.3.1]-1′,6′-glycopyranosyl-PBs still is obscure, as are the enzymes catalyzing attachment of the bicyclo[17.3.1]-1′,6′-glycopyranosyl moiety to the bilin-type tetrapyrrole core.

2.2 Colorless nonfluorescent phyllobilins from higher plants and their structures

NCCs, now alternatively classified as formyloxobilin-type phyllobilanes (or as type-I phyllobilanes) (Kräutler, 2016), are the most numerous group among the known natural Chl catabolites (see Table 2). The colorless NCCs accumulate in senescent leaves, and one of them also happens to represent the first bilin-type Chl catabolite to be characterized (Kräutler et al., 1991).

2.2.1 *Formyloxobilin-type nonfluorescent Chl catabolites*

Over 20 structurally different natural NCCs have been identified, so far, in senescent leaves (Christ et al., 2016; Curty & Engel, 1996; Kräutler, 2016; Kräutler et al., 1991; Mittelberger et al., 2017; Roca et al., 2018), vegetables

Table 2 Structures of natural nonfluorescent type-I Chl catabolites (NCCs): the labels **R¹** to **R⁴** refer to the generalized formula of type-I phyllobilanes, atom numbering shown in Scheme 6.

#[a]	R^{1b}	R^2	R^3	R^4	C16[c]	Provisional name[d]	References
27	H	H	H	$CH{=}CH_2$	n	*At*-NCC-5[e], *Bo*-NCC-2	Pružinská et al. (2005); Roiser, Müller, and Kräutler (2015)
***epi*-27**	H	H	H	$CH{=}CH_2$	epi	*Oe*-NCC-4[e]	Vergara-Dominguez, Rios, Gandul-Rojas, and Roca (2016)
28	H	H	H	$CH{=}CH_2$	n	*At*-NCC-3[f]	Müller et al. (2006)
19	H	CH_3	H	$CH{=}CH_2$	n	syn[g]	Oberhuber et al. (2008)
***epi*-19**	H	CH_3	H	$CH{=}CH_2$	*epi*	*Cj*-NCC-2	Oberhuber et al. (2003)
29	OH	H	H	$CH{=}CH_2$	n	*Bn*-NCC-3	Mühlecker and Kräutler (1996)
***epi*-29**	OH	H	H	$CH{=}CH_2$	epi	*So*-NCC-3	Berghold, Breuker, Oberhuber, Hörtensteiner, and Kräutler (2002)
30	OH	H	H	$CH(OH){-}CH_2OH$	epi	*So*-NCC-1	Berghold et al. (2002)
1	OH	CH_3	H	$CH(OH){-}CH_2OH$	n	*Hv*-NCC-1	Kräutler et al. (1991)
***epi*-1**	OH	CH_3	H	$CH(OH){-}CH_2OH$	epi	*So*-NCC-2	Oberhuber, Berghold, Mühlecker, Hörtensteiner, and Kräutler (2001)
26	OH	CH_3	H	$CH{=}CH_2$	n	*Sw*-NCC-58	Kräutler et al. (2010)
***epi*-26**	OH	CH_3	H	$CH{=}CH_2$	epi	*Cj*-NCC-1	Curty and Engel (1996)
20	O-Glc	H	H	$CH{=}CH_2$	n	*Bn*-NCC-2	Mühlecker and Kräutler (1996)
31	O-Glc	CH_3	H	$CH{=}CH_2$	n	*At*-NCC-4[e]	Pružinská et al. (2005)
32	O-Glc	H	H	$CH(OH){-}CH_2OH$	epi?	*Co*-NCC-2[e]	Rios, Perez-Galvez, et al. (2014)

Continued

Table 2 Structures of natural nonfluorescent type-I Chl catabolites (NCCs): the labels $\mathbf{R^1}$ to $\mathbf{R^4}$ refer to the generalized formula of type-I phyllobilanes, atom numbering shown in Scheme 6.—cont'd

#[a]	R^{1b}	R^2	R^3	R^4	C16[c]	Provisional name[d]	References
33	O-Glc	CH_3	H	$CH(OH)\text{-}CH_2OH$	epi	*Zm*-NCC-1	Berghold, Müller, Ulrich, Hörtensteiner, and Kräutler (2006)
22	O-Glc	CH_3	H	$CH(OH)CH_2$-O-Glc[h]	epi	*Pd*-NCC-32	Erhart et al. (2016)
34	O-(6′-OMal)Glc	CH_3	H	$CH{=}CH_2$	epi	*Nr*-NCC-1	Berghold, Eichmüller, Hörtensteiner, and Kräutler (2004)
35	O-Mal	H	H	$CH{=}CH_2$	n	*Bn*-NCC-1	Mühlecker et al. (1993)
36	O-Mal	CH_3	H	$CH{=}CH_2$	epi	*Ej*-NCC-2[e]	Rios, Roca, et al. (2014)
23	OH	CH_3	–[i]	$CH{=}CH_2$	epi	*Mc*-NCC-58	Moser et al. (2012)
24	OH	CH_3	–[j]	$CH{=}CH_2$	epi	*Mc*-NCC-55	Moser et al. (2012)
25	O-Glc[k]	CH_3	6′-βGlc[k]	$CH{=}CH_2$	epi	*Ug*-NCC-53	Scherl et al. (2016)
21	OH	–[l]	H	$CH{=}CH_2$	epi[h]	*py*NCC (syn[g])	Li et al. (2018)

[a]Compound number.
[b]Abbreviations: Mal = malonyl; Glc = 1′β-D-glucopyranosyl.
[c]Configuration at C16 from correlation with NCCs derived from *p*FCC (**2**) (n = "normal") or from *epi-p*FCC (***epi*-2**) (epi = "epimeric"), the absolute configuration at C16 is still unknown.
[d]*Hv*-NCC-1 from barley (*Hordeum vulgare*) (Kräutler et al., 1992, 1991); *So*-NCCs from spinach (*Spinacia oleracea*) (Berghold et al., 2002; Oberhuber et al., 2001); *Cj*-NCCs from Katsura tree (*Cercidiphyllum japonicum*) (Curty & Engel, 1996); *Sw*-NCC-58 from peace lily (*Spathiphyllum wallisii*) (Kräutler et al., 2010); *Pc*-NCCs from pear (*Pyrus communis*) (Müller, Ulrich, Ongania, & Kräutler, 2007); *Md*-NCCs from apple (*Malus domestica*) (Müller et al., 2007); *At*-NCCs (from *Arabidopsis thaliana* (Müller et al., 2006; Pružinská et al., 2005)); *Nr*-NCCs from tobacco (*Nicotiana rustica*) (Berghold et al., 2004); *Zm*-NCCs from maize (*Zea mays*) (Berghold et al., 2006); *Bn*-NCCs from oilseed rape (*Brassica napus*) (Mühlecker & Kräutler, 1996; Mühlecker et al., 1993); *Ej*-NCC from Loquat Fruits (*Eriobotrya japonica*) (Rios, Roca, & Perez-Galvez, 2014); *Mc*-NCCs from banana peels (*Musa acuminata*, Cavendish cultivar) (Moser et al., 2012); *Pd*-NCC from plum tree (*Prunus* × *domestica*) (Erhart et al., 2016); and *Ug*-NCC from wych elm (*Ulmus glabra*) (Scherl et al., 2016).
[e]Identified by UV/Vis and MS.
[f]At-NCC-3 carries a HO-CH_2-group at C2 (Müller et al., 2006).
[g]From partial synthesis (Li et al., 2018; Oberhuber et al., 2003).
[h]Glc = 1′β-D-glucopyranosyl.
[i]R^4 = daucyl unit and *R*-configuration at C10.
[j]R^4 = daucyl unit and *S*-configuration at C10 (as derived from CD-spectra (Moser et al., 2012)).
[k]1′,6′-Glycopyranosyl bridge attached at O^3 and $O12^4$ (Scherl et al., 2016).
[l]Unsubstituted H_2C8^2 (Li et al., 2018).

(Berghold et al., 2002; Roiser et al., 2015) and fruit (Moser et al., 2012; Müller et al., 2007; Rios, Perez-Galvez, & Roca, 2014; Rios, Roca, & Perez-Galvez, 2014; Roca, Rios, Chahuaris, & Perez-Galvez, 2017; Vergara-Dominguez et al., 2016) (see Table 2). Many of these NCCs could be isolated and their actual chemical structures were determined (Kräutler, 2016). They all show not only a common tetrapyrrolic core but feature, in addition, a common stereochemistry at their *meso* C10 position. However, there are two C16 *epi*meric lines of NCCs, just as with their FCC precursors. It has been proposed (Oberhuber et al., 2003) and is generally agreed today that typical natural NCCs are directly formed from their corresponding FCC precursors by a remarkable non-enzymatic isomerization that is acid-activated and presumed to take place predominantly in the vacuole (Kuai et al., 2018; Oberhuber et al., 2003). This stereospecific chemical process transforms the apparently less stable FCCs, or type-I phyllobilenes-*b*, into the more stable NCCs, where ring C is a third aromatic pyrrole ring (Kräutler, 2016, 2019). The isomerization deconjugates the eponymous fluorophore of the blue luminescent FCCs into a non-luminescent dipyrromethane unit (see Scheme 5). At the same time it establishes the configuration at the new asymmetric center C10, which has *R*-configuration in the known type-I phyllobilanes (Kräutler, 2016). The adjacent asymmetric $C8^2$ has *S*-configuration in the predominant forms of the NCCs. However, $C8^2$ is part of a β-ketoester grouping at ring E, which promotes slow equilibration of the main isomer with the minor, $C8^2$-*R*-epimer (Moser, Scherzer, & Kräutler, 2017).

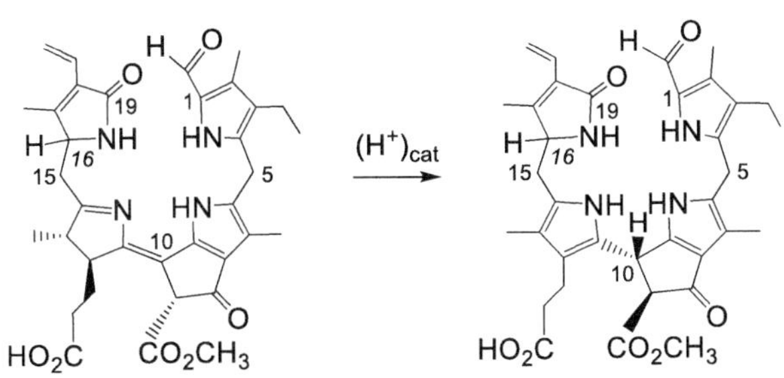

*p*FCC (**2**) → NCC **19**

epi-*p*FCC (***epi*-2**) → NCC ***epi*-19**

Scheme 5 The typical natural NCCs are generated by a non-enzymatic isomerization of FCCs that carry the free propionic acid substituent derived from Pheide *a* and RCC (Kräutler, 2016). In this process, *p*FCC (**2**) isomerizes to the NCC **19** (retaining C16-"n" configuration), and *epi*-*p*FCC (***epi*-2**) to the corresponding C16-epimer ***epi*-19** (retaining C16-"epi" configuration).

In most natural NCCs the propionic acid group is available as the free acid (R^3 = H; Scheme 6). In a few relatively apolar NCCs the 3-position is decorated with an ethyl group (R^1 = H). Much more common are more polar NCCs with R^1 at $C3^2$ represented by a hydroxygroup (such as in **1**) or with a β-glycopyranosyl or malonyl group attached to it (such as in *Bn*-NCC-2, **20** or *Bn*-NCC-1, **35**) (Mühlecker & Kräutler, 1996). At ring E the substituent may be a methyl ester function (as in **1**, directly derived from the Chls) or it is hydrolyzed to a free carboxylic acid group, such as in the NCCs from oilseed rape, such as **20** and **35**. The β-keto-carboxylate groups in the latter are surprisingly resistant against loss as carbon dioxide (Mühlecker & Kräutler, 1996), and the preparation of corresponding (decarboxylated) pyro-NCCs (*py*NCCs, such as **21**), by partial synthesis, required rather harsh synthetic conditions (Li et al., 2018). Surprisingly, genuine natural *py*NCCs are not known (Li et al., 2018). The substituent R^4 at ring D often is a vinyl group, a stereouniform 1,2-dihydroxy-ethyl group (as in **1**), or an 18^1-hydroxy,18^2-β-glycopyranosyl-ethyl group, as discovered in the 3^2,18^2-diglycosylated type-I phyllobilane *Pd*-NCC-32 (**22**) from plum (*Prunus × domestica*) (Erhart et al., 2016). *Mc*-NCC-58 (**23**) and *Mc*-NCC-55 (**24**), two minor NCCs in extracts of the outer peels of ripe bananas [*Mc*-NCC-58 (**23**) and *Mc*-NCC-55 (**24**)], were provisionally identified as isomers of the main *hm*FCC from this source, hence, tentatively characterized as NCCs carrying a daucic acid ester function at their propionic acid side chain (Moser et al., 2012). A first bicyclo[17.3.1]-1′,6′-glycopyranosyl-NCC [the *bc*NCC *Ug*-NCC-53 (**25**)] was identified in extracts of wych elm (*Ulmus glabra*) leaves, an exceptional type-I phyllobilane, in which a 1′,6′-glycopyranosyl-linker bridges the phyllobilane skeleton at $O3^3$ (at ring A) and $O12^4$ (at ring C) (Scherl, Müller, & Kräutler, 2012)

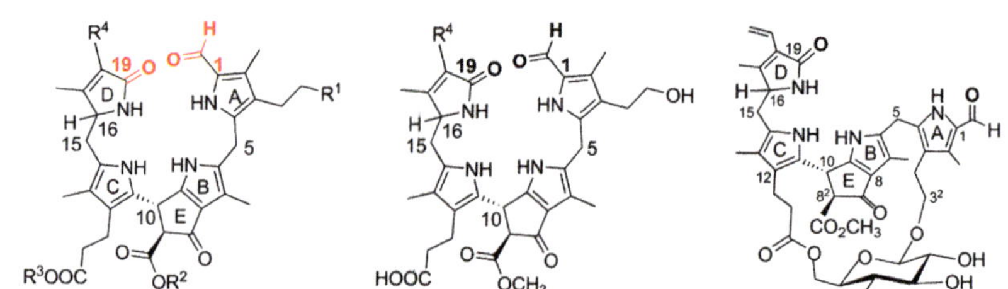

Scheme 6 General formula of NCCs and formulae of the typical natural NCCs **1**, ***epi*-26**, and of the *bc*NCC *Ug*-NCC-53 (**25**) (see Table 2). NCCs constitute the class of the type-I phyllobilanes.

(see Scheme 6). Stable NCCs from a variety of different plant species may have identical structures, such as ***epi*-26** (*Cj*-NCC-1, discovered in *Cercidiphyllum japonicum*; Curty & Engel, 1996), but also identified in other plants, as, e.g., *So*-NCC-4, *Pc*-NCC-2, *Md*-NCC-2, *Mc*-NCC-61, *Ej*-NCC-4 (Kräutler, 2016; Roca et al., 2018). In other cases, NCCs differ characteristically by their configuration at C16, i.e., they may be pairwise C16-epimers, such as the isomeric NCCs ***epi*-26** (*Cj*-NCC-1) and **26** (detected as *Sw*-NCC-58 in senescent leaves of the peace lily) (see Table 2).

2.2.2 Dioxobilin-type nonfluorescent Chl catabolites

In senescent leaves of barley two isomeric forms of DNCCs were first described by Losey and Engel, which classified the two type-II phyllobilanes **37** and **38** as "urogenobilinoidic" Chl catabolites. This finding was considered a first sign of a further oxidative deformylation of the NCC **1** (Scheme 7) (Losey & Engel, 2001). Strikingly, in senescent leaves of Norway maple no NCCs were detected, but the *Ap*-DNCC **39**, a further stereo-isomer of the two "urogenobilinoidic" *Hv*-DNCCs (Müller, Rafelsberger, Vergeiner, & Kräutler, 2011). The (pseudo-) mirror image relationship of **39** (from CD-spectra) with the two DNCCs from barley was considered to be remarkably incompatible with an NCC precursor, and suggested the involvement of unknown transformations of Chl catabolites that would occur earlier on the PAO/phyllobilin pathway, e.g., at the stage of a properly structured FCC (Müller et al., 2011). Indeed, as is described in more detail below, the massive accumulation of a range of DNCCs in senescent leaves of *Arabidopsis* could soon be rationalized by the activity of a CYP that deformylated some FCCs very effectively (Christ et al., 2013).

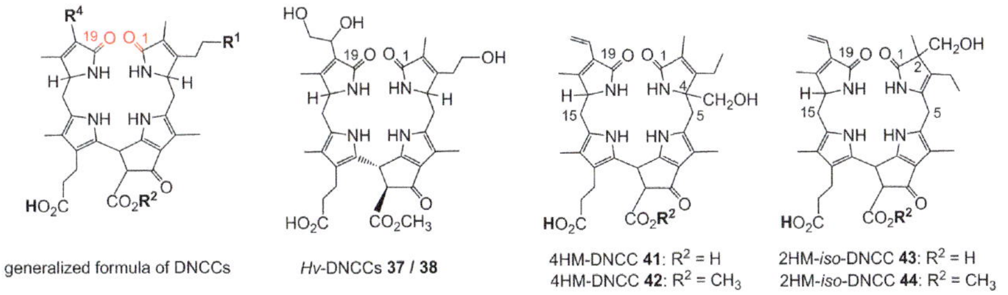

Scheme 7 General formula of dioxobilin-type NCCs (DNCCs), and formulae of (left to right) *Hv*-DNCCs **37** and **38**, the 4HM-DNCCs**41** and **42**, and the 2HM-isoDNCCs **43** and **44**. DNCCs constitute the class of the type-II phyllobilanes.

In the meantime, a variety of DNCCs have been described from various plant sources (see Table 3), including senescent leaves (Christ et al., 2016; Kräutler, 2016; Losey & Engel, 2001; Roca et al., 2018), vegetables (Roiser et al., 2015) and fruits (Rios, Perez-Galvez, et al., 2014; Roca et al., 2018), establishing, together with their presumed fluorescent precursors (DFCCs), the "second" line of the dioxobilin-type PBs, now all also classified as type-II PBs (Kräutler, 2014, 2016; Kuai et al., 2018). Interestingly, an intimate structural and chemical relationship between the DFCC *At*-DFCC-33 (**15**) and the main DNCC, *At*-DNCC-33 (**16**) of *Arabidopsis* could be established in the chemical isomerization of **15** to **16**. This indicated the same type of stereochemical correlation and non-enzymatic mechanism of DNCC formation (Süssenbacher, Hörtensteiner, et al., 2015), as was first deduced for the relationship between FCCs and NCCs (Oberhuber et al., 2003). Accordingly, in further agreement with the chiroptical properties of the presently known natural NCCs and DNCCs (with the exception of *Ap*-DNCC **39**), the stereochemistry at C10 is tentatively deduced to be the same, i.e., to display *R*-configuration (Kräutler, 2016; Süssenbacher, Hörtensteiner, et al., 2015). In the case of the *bc*DFCC *Vv*-DFCC-53 (**14**) from senescent leaves of grape vine, the new asymmetric C4 at ring A of this type-II PB was also deduced to have *R*-configuration (see Scheme 4). The relevance of the tentative stereochemical assignment *R* of C4 in *Vv*-DNCC-51 (**40**) from the same plant (Erhart, Mittelberger, et al., 2018) remains to be established, as is the configuration of C4 of other DFCCs and DNCCs. The stereochemistry of this methine carbon is an indirect result of the oxidative deformylation by a CYP enzyme (CYP89A9 in *Arabidopsis*) (Christ et al., 2013), which may exhibit different signs and degrees of stereoselectivity in different angiosperms.

In *Arabidopsis* type-II PBs with a strikingly altered ring A structure have been observed as results of (an apparently aberrant) metabolism of *p*FCC (**2**) molecules that escape hydroxylation by TIC55 (see below). This catabolic line produces dioxobilin-type phyllobilanes that carry an additional hydroxymethyl group at C4 of their ring A, such as *At*-4HM-DNCC-41 (**41**) and At_{mes16}-4HM-DNCC-44 (**42**), or in their regioisomers *At*-2HM-iso-DNCC-43 (**43**) and At_{mes16}-2HM-iso-DNCC-46 (**44**) (see Scheme 7 and Table 4). These nonfluorescent DCCs (NDCCs) are presumed to be isomers and descendants of the correspondingly altered fluorescent DCCs, such as *At*-2HM-iso-DFCC (**16**) (Süssenbacher, Kreutz, et al., 2015).

Table 3 Structures of natural dioxobilin-type nonfluorescent Chl catabolites (DNCCs); the labels R^1, R^2 and R^4 refer to the generalized constitutional formula of type-II phyllobilanes (see Scheme 7).

#[a]	R^1	R^2	R^4	C16[b]	Provisional names[c]	References
45	H	H	$CH{=}CH_2$	n	*At*-DNCC-45[d]	Süssenbacher, Kreutz, Christ, Hörtensteiner, and Kräutler (2015)
***iso*-45**	H	H	$CH{=}CH_2$	n	*At*-DNCC-48[d]	Süssenbacher, Kreutz, et al. (2015)
16	OH	H	$CH{=}CH_2$	n	*At*-NDCC-1, *At*-DNCC-33	Christ et al. (2013); Süssenbacher, Hörtensteiner, et al. (2015)
40	OH	H	$CH{=}CH_2$	*epi*	*Hvir*-UNCC[e], *Vv*-DNCC-51	Djapic, Pavlovic, Arsovski, and Vujic (2009); Erhart, Mittelberger, et al. (2018)
46	OH	CH_3	$CH{=}CH_2$	n	At_{mes16}-DNCC-38	Süssenbacher et al. (2014)
37	OH	CH_3	$CH(OH){-}CH_2OH$	n	*Hv*-DNCC-a[d,f]	Losey and Engel (2001)
38	OH	CH_3	$CH(OH){-}CH_2OH$	n	*Hv*-DNCC-b[d,f]	Losey and Engel (2001)
39	OH	CH_3	$CH(OH){-}CH_2OH$	epi	*Ap*-DNCC	Müller et al. (2011)

[a]Compound number.
[b]Configuration at C16: n = "normal," derived from *p*FCC; epi = "*epi*meric," derived from *epi-p*FCC.
[c]For specific provisional names see references: e.g. *Ap*-DNCC (from *Acer platanoides*); *At*-PBs (from *Arabidopsis thaliana*); At_{mes16}-PBs from *mes16* mutant of *Arabidopsis*; *Hv*-DNCCs from barley (*Hordeum vulgare*); *Vv*-DNCC from grapevine (*Vitis vinifera*).
[d]Presumed to be C4-stereoisomers pairwise.
[e]Abbreviation used in the original paper (Djapic et al., 2009).
[f]A name was not given in the original paper (Losey & Engel, 2001), adaptation to nomenclature used here.

Table 4 Nonfluorescent dioxobilin-type Chl catabolites that carry a new CH_2OH substituent.

#[a]	R^1	R^2	R^4	C16[b]	Provisional names[c]	References
41	H	H	$CH{=}CH_2$	n	*At*-4HM-DNCC-41[d]	Süssenbacher, Kreutz, et al. (2015)
43	H	H	$CH{=}CH_2$	n	*At*-2HM-*iso* DNCC-43[e]	Süssenbacher, Kreutz, et al. (2015)
42	H	CH_3	$CH{=}CH_2$	n	At_{mes16}-4HM-DNCC-44[d]	Süssenbacher et al. (2014)
44	H	CH_3	$CH{=}CH_2$	n	At_{mes16}-2HM-*iso*-DNCC-46[e]	Süssenbacher et al. (2014)

These type-II PBs were also classified provisionally as NDCCs (Christ et al., 2013; Süssenbacher et al., 2014), they are listed here as HM-DNCCs and HM-*iso*-DNCCs (for formulae see Scheme 7).

[a]Compound number.

[b]Configuration at C16: n = "normal," derived from *p*FCC; epi = "epimeric," derived from *epi*-*p*FCC.

[c]For specific provisional names see references: e.g. *At*-PBs (from *Arabidopsis thaliana*); At_{mes16}-PBs from *mes16* mutant of *Arabidopsis*.

[d]Carries an extra CH_2OH unit at C4.

[e]Is unsaturated between C3 and C4 and carries the additional CH_2OH unit at C2.

2.2.3 Isophyllobilanones

So far, studies of Chl breakdown in higher plants, as well as of their bilin-type Chl catabolites (Kräutler, 2016), have nearly exclusively dealt with angiosperms (Hörtensteiner & Kräutler, 2011; Kuai et al., 2018; Tanaka et al., 2011). Chl catabolites have been very recently investigated in the extracts of senescent leaves of bracken (*Pteridium aquilinum*), i.e., from a member of the large plant division of the ferns (Erhart, Vergeiner, Kreutz, Kräutler, & Müller, 2018). The structures of two of the tetrapyrrolic catabolites turned out to differ remarkably from the so far identified PBs. Their structures displayed an unprecedented (C6 to C7)-rearranged carbon skeleton. Furthermore, the new structures reflected the basic elements of a *py*NCC lacking a substituent at $C8^2$ of ring E, and unprecedented with natural PBs (Li et al., 2018). In addition, an extra oxygen atom decorated their ring B moiety in a novel vinylogous imide functionality (see Scheme 8). Therefore the two fern Chl catabolites were classified as isophyllobilanones (*i*PBons) (Erhart, Vergeiner, et al., 2018). However, the structures of both of the *i*PBons (**47**, **48**) relate to the formyloxobilin pattern known in type-I PBs from angiosperm. This matter of fact indicates the key activity of enzymes in the fern that are related to PAO and cleave the porphyrinoid macroring oxygenolytically at the northern *meso*-position (Hörtensteiner et al., 1998). Furthermore, a peripheral hydroxylation at $C3^2$ reminds of the activity of the hydroxylase TIC55 (see below), and a chromophore reduction analogous to the one of the known RCCRs (Rodoni, Vicentini, et al., 1997) appears to saturate C15 and C16 at the "western" *meso*-position. The lack of a substituent at $C8^2$ of the *i*PBons suggests a more direct relationship with pyro-Pheide (*py*Pheide) *a* (Scheer, 2006) than with Pheide *a* itself, a finding reminiscent of the structures of some red Chl catabolites isolated from the green alga *Auxenochlorella protothecoides* (Engel, Curty, & Gossauer, 1996).

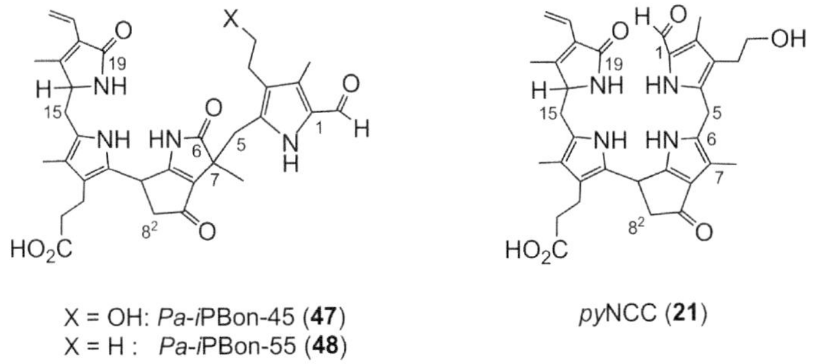

X = OH: *Pa-i*PBon-45 (**47**)
X = H : *Pa-i*PBon-55 (**48**)

*py*NCC (**21**)

Scheme 8 Formulae of isophyllobilanones (*i*PBons) **47** and **48** from bracken fern, and of the semisynthetic pyro-NCC (*py*NCC) **21**.

2.3 Phyllochromobilins—Colored phyllobilins in leaves and fruit

Consistent with the original name of the colorless NCCs as "rusty pigments," NCCs typically undergo a ready transformation upon storage with exposure to air, when also exposed to day light, in particular, furnishing yellow-colored Chl catabolites (YCCs), first, and subsequently pink-colored Chl catabolites (PiCCs) (Ulrich, Moser, Müller, & Kräutler, 2011). The "spontaneous" formation of such colored PBs (collectively named phyllochromobilins) is the visual sign of the extension of π-conjugated chromophores as a result of an oxidation, which has been related to the antioxidant properties of NCCs and YCCs. First, the *meso*-position C15 is unsaturated, giving YCCs (Moser, Ulrich, Müller, & Kräutler, 2008), which are subsequently oxidized further at C10, to furnish PiCCs (Ulrich et al., 2011). In these colored PBs new intensive absorption bands near 430 nm and near 520 nm, respectively, are responsible for the bright colors of YCCs and PiCCs (Kräutler, 2016; Li et al., 2014).

2.3.1 *Yellow phyllobilins*

In some senescent leaves, where PBs typically accumulate as colorless linear tetrapyrroles, yellow formyloxobilin-type Chl catabolites (YCCs), also classified as type-I phylloxanthobilins (PxBs), have been detected as a new type of yellow plant pigments (Moser, Ulrich, et al., 2008) (see Scheme 9). A first natural YCC was isolated from senescent leaves of *C. japonicum* and named *Cj*-YCC-2 (**50**). YCCs are induced by day light to interconvert photoreversibly with their slightly more polar isomer, such as **50** with **51**. The YCCs **50** and **51** were identified by NMR-spectroscopy as 15*Z*- and 15*E*-double bond isomers (Moser, Ulrich, et al., 2008). The structures of the methyl esters (**Me-50** and **Me-52**) of the natural YCC **50** (Li et al., 2016) and of the related semisynthetic *pyro*-YCC **52** (Li et al., 2018), respectively, have been analyzed by crystallography confirming the solution structures, derived earlier, and revealing the formation of unique homochiral π-staked and H-bonded dimers (Li et al., 2016). More polar PxBs related to **50**, such as the YCCs **53** (Scherl et al., 2012) and **54** (Wakana et al., 2014), have also been detected in extracts of some senescent leaves (Kräutler, 2016; Scherl et al., 2012; Wakana et al., 2014) and ripening fruit (Roca et al., 2017). The YCC **50** is the indirect product of an endogenous oxidation, either of the NCC ***epi*-26** or of the epimeric NCC **26** (Vergeiner et al., 2015). A still puzzling regiospecific reaction oxidizes NCCs at their

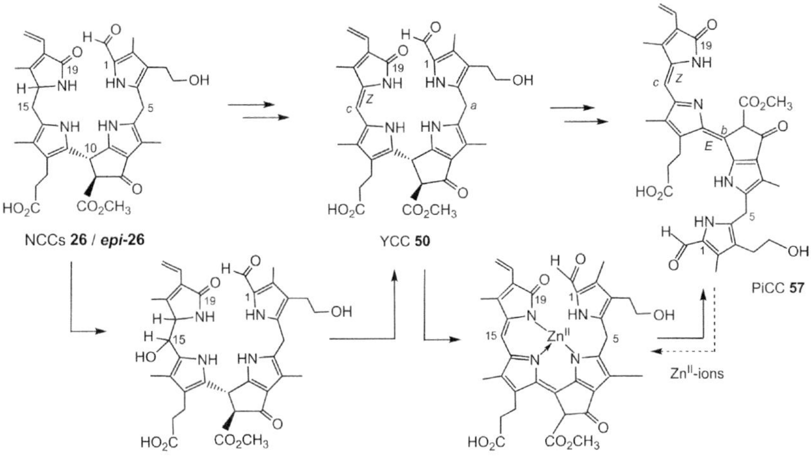

Scheme 9 Left: Formation of YCC **50** by oxidation of the C16-epimeric NCCs **26** and ***epi*-26** to corresponding intermediate 15-hydoxyl-NCCs. Right: Synthesis of PiCC **57** from YCC **50** via the Zn-complex Zn-PiCC **58**. YCCs are type-I PxBs or type-I phyllobilenes-*c*, PiCCs are type-I phyllobiladienes-*b,c*.

C15 to furnish polar analogs, which, by acid-induced elimination, convert easily and highly selectively into the 15*Z*-isomers of corresponding YCCs, such as **50** (Vergeiner et al., 2015) (Scheme 9).

More recently, yellow Chl catabolites that lack the absorption band near 320 nm (that is due to the α-formylpyrrole unit in type-I PBs) were detected in extracts of senescent leaves of grapevine (Erhart, Mittelberger, et al., 2018) and of *Arabidopsis* (I. Süssenbacher, D. Menghini, S. Hörtensteiner, & B. Kräutler, unpublished). They were provisionally described as dioxobilin-type YCCs (DYCCs or type-II PxBs). Similar to the mechanism of formation of YCCs from NCCs (Moser, Ulrich, et al., 2008), the pathway of the endogenous formation of DYCCs (yellow type-II PxBs) was proposed to involve oxidation of DNCCs. This was verified by the partial "green" synthesis of the DYCC **55** from the main grapevine leaf DNCC, *Vv*-DNCC-51 (**40**), using an extract of *S. wallisii* leaves in an aerated solution (Li, Erhart, Liu, & Kräutler, 2019). The synthetic DYCC was identified with the DYCC detected in extracts of senescent grapevine leaves (Li et al., 2019). Upon photoexcitation, the DYCC **55** (15*Z* form) underwent photoreversible 15*Z*/15*E* double bond isomerization to the less stable 15*E*-isomer **56**, as had been observed with YCCs (see Scheme 10).

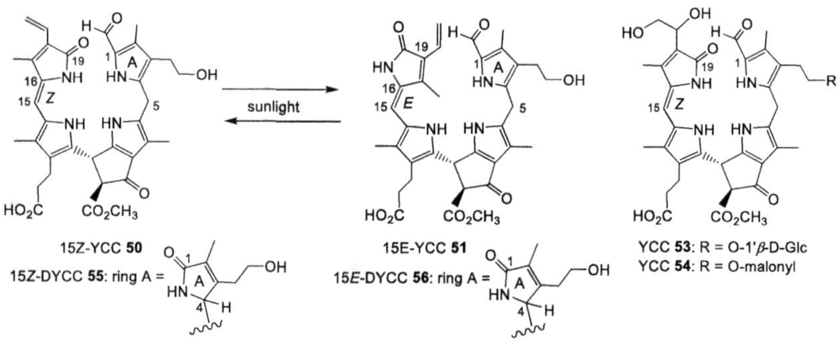

Scheme 10 Top: Formulae of the 15*Z*- and 15*E*-isomeric YCC-forms **50** and **51**, representing two natural YCCs, of the corresponding DYCCs **55** and **56**, and of the polar YCCs **53** and **54**. DYCCs constitute the class of the type-II phyllobilenes-*c*.

The yellow Chl catabolites either of type-I or type-II carry a chromophore, first seen in bilirubin (Falk, 1989), and they also exhibit related (photo) chemistry.

2.3.2 Pink phyllobilins

The chromophore of YCCs lends itself for oxidation, causing extension of the conjugated π-system to rings B and E in PiCCs, or type-I phylloroseobilins (PrBs; see Scheme 11). Oxidation of aerated solutions of type-I PxBs leads to type-I PrBs (Kräutler, 2016), first described as the formyloxobilin-type PiCC **57** (Ulrich et al., 2011). The structure of this pink-colored type-I 10*E*,15*Z*-phyllobiladiene-*b*,*c*, first deduced from NMR-data, was confirmed by X-ray crystallography (Li et al., 2014).

PiCC **57** binds transition metal ions remarkably well in a tridentate fashion, so that in the resulting blue metal complexes the ligand is present in a 10*Z*,15*Z*-phyllobiladien-*b*,*c* arrangement (Li et al., 2014). Transition metal-binding induces a bathochromic shift of the long wavelength absorption maximum by about 100 nm from about 520 nm, in the practically non-luminescent PiCC (**57**) to about 620 nm, in the strongly red-fluorescent Zn-PiCC complex (**58**). Binding of a variety of divalent first and second row transition metal ions occurs rather tightly down to sub-nmol concentrations, and the YCC converted into the Zn-complex **58** directly and efficiently in an aerated methanolic solution at room temperature (Li et al., 2014). The Zn^{2+}-ion can be removed by treatment with phosphate, so that the sequence YCC **50** to Zn-PiCC **58** to PiCC **57** is an efficient synthetic route to PiCCs (Li et al., 2014).

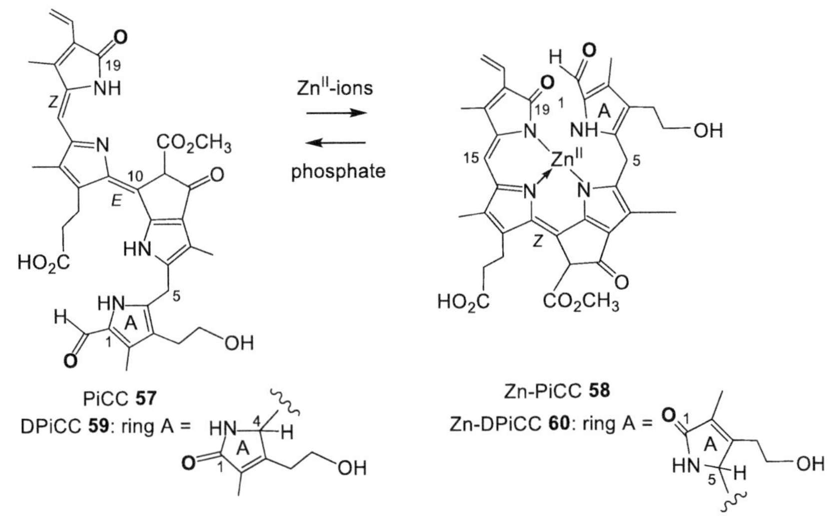

Scheme 11 The pink-colored PBs, PiCC **57** and DPiCC **59**, represent 10*E*,15*Z*-phyllobiladienes-*b,c*. Both PiCC **57** and DPiCC **59** bound Zn^{2+}-ions remarkably well in a tridentate fashion, giving the blue Zn-complexes **58** and **60**, in which a 10*Z*,15*Z*-phyllobiladien-*b,c* represents the isomerized ligand (Li et al., 2014).

Similar experiments as those with the PiCC **57** and its Zn-complex **58** have been performed lately with the semisynthetic DYCC **55** that lacked the absorption band near 320 nm (of the α-formylpyrrole unit of type-I PBs). As described for the preparation of PiCCs from YCCs (Li et al., 2014), the oxidation of aerated solutions of the type-II PxB **55** led to the first dioxobilin-type pink Chl catabolite (DPiCC), the type-II PrB **59**. The structure of the PrB **59** was again deduced from NMR-data to represent a 10*E*,15*Z*-phyllobiladiene-*b,c* (Li & Kräutler, 2019). The DPiCC bound Zn^{2+}-ions remarkably well in a tridentate fashion, similar as observed with the PiCC **57**, giving blue metal complexes, in which the ligand was present in a 10*Z*,15*Z*-phyllobiladiene-*b,c* arrangement (C. J. Li & B. Kräutler, unpublished). As seen with the PiCC **57**, Zn^{2+}-ion binding caused a large bathochromic shift of the long wavelength absorption maximum from 522 nm for the practically non-luminescent DPiCC **59**, to about 619 nm for the strongly red fluorescent Zn-DPiCC complex **60** (C. J. Li & B. Kräutler, unpublished). As originally found with the related type-I phyllochromobilins (Li et al., 2014), the DYCC **40** directly converts into the Zn-complex **60** in an aerated methanolic solution at room temperature, and the bound Zn-ion of the complex **60** could be removed by treatment with phosphate. Hence, the sequence DYCC **40** to Zn-DPiCC **60** to DPiCC **59** turned out to be an efficient semisynthetic route to the DPiCC **59** (Li & Kräutler, 2019) (see Scheme 11).

3. Biochemistry of Chl breakdown in higher plants

The pathway of Chl breakdown is nowadays called the "PAO/phyllobilin" pathway (Hörtensteiner & Kräutler, 2011; Kuai et al., 2018). The name is derived from the key enzyme of the pathway, PAO, which is responsible for the formation of the basic backbone structure of the degradation products of Chl, i.e., the phyllobilins (PBs) (Kräutler, 2014). The first part of the PAO/phyllobilin pathway describes the enzymatic steps converting colored and phototoxic Chl to *p*FCC—taking place within the senescent chloroplast. The second part includes FCC-modifying reactions mainly happening in the cytosol followed by the relocalization of the FCCs to the vacuole, where the isomerization to analogous NCCs takes place, thereby causing the loss of the fluorescence property of the FCCs.

3.1 Colorless, fluorescent Chl catabolites from Chl breakdown in chloroplasts

3.1.1 Pheophorbide and other green catabolites with intact Chl macrocycle

3.1.1.1 Chl *b* to Chl *a* conversion

Besides Chl *a*, green algae and plants also contain Chl *b* in their light-harvesting complexes (LHCs). However, prior to degradation, Chl *b* must be converted to Chl *a* as the enzymes downstream in the pathway are specific in accepting this form as substrate (Hörtensteiner, Vicentini, & Matile, 1995; Shimoda, Ito, & Tanaka, 2012). This is in line with the observation that, with the exception of one NCC from *Arabidopsis* (Müller et al., 2006), PBs derive from Chl *a* (Kräutler, 2016). The conversion of Chl *b* to Chl *a* is catalyzed by two enzymatic steps (Scheme 12). In the first step, Chl *b* reductase (CBR) reduces the C7-formyl group of Chl *b* to 7^1-hydroxy Chl *a* (HOChl *a*) (Scheumann et al., 1999; Tanaka et al., 2011). In the second step, 7^1-hydroxy Chl *a* reductase (HCAR) converts HOChl *a* to Chl *a* (Meguro, Ito, Takabayashi, Tanaka, & Tanaka, 2011).

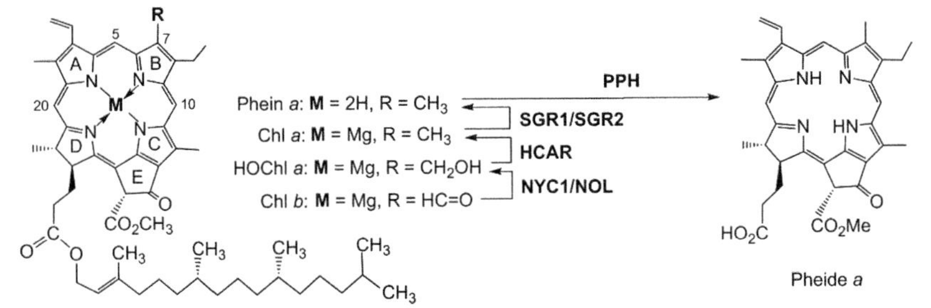

Scheme 12 Outline of the early steps of Chl breakdown in higher plants, depicting chemical structures and critical enzymes (see text and list for abbreviations used).

Arabidopsis has two CBR isoenzymes, i.e., non-yellow coloring 1 (NYC1) and NYC1-like (NOL) (Kusaba et al., 2007; Sato et al., 2009), and both are thought to catalyze the reduction of Chl *b*. However, only the activity of NOL was confirmed in in vitro assays. When analyzing the effect of *cbr* mutations in *Arabidopsis*, *nyc1* mutants show retention of Chl *b* during senescence, whereas Chl *b* levels of *nol* mutants only slightly differ from wild-type plants (Horie, Ito, Kusaba, Tanaka, & Tanaka, 2009). An *Arabidopsis nyc1 nol* double mutant shows almost complete suppression of Chl *b* breakdown and a stay-green phenotype during leaf senescence and seed maturation. Interestingly, the double mutant exhibited an unaltered Chl *a*/*b* ratio during vegetative growth, suggesting a crucial role in Chl breakdown but a negligible function in maintaining the Chl *a*/*b* ratio during the vegetative growth phase. This observation is in line with the fact that at least NYC1 is transcriptionally upregulated during senescence. The lack of Chl *b*-reducing activity in *nyc1* mutants results in the stabilization of the LHC proteins to which Chl is bound. This indicates that CBR, indirectly, also triggers subsequent LHC proteolysis (Horie et al., 2009).

The second step catalyzed by HCAR was initially characterized in *Arabidopsis*. In vitro assays showed the ability of the enzyme to convert HOChl *a* to Chl *a*, and *Arabidopsis HCAR* T-DNA insertion mutants accumulate HOChl *a*. Furthermore, the mutant line also accumulated Pheide *a*, a downstream Chl catabolite and substrate of PAO. Interestingly, PAO protein levels were not altered in these plants (Meguro et al., 2011).

3.1.1.2 Magnesium dechelation and dephytylation

The release of the central Mg^{2+}-ion from Chl *a* results in the formation of pheophytin *a* (Phein *a*) (Scheme 12). Non-enzymatic release of Mg^{2+} from Chl *a* occurs slowly under mild acidic conditions (Saga & Tamiaki, 2012).

However, until recently it was unknown which enzyme is responsible for catalyzing this reaction as part of regular Chl breakdown. The riddle was solved when Shimoda and coworkers identified stay-green (SGR) as the Mg-dechelatase involved in senescent leaves (Shimoda, Ito, & Tanaka, 2016).

Even though identification of the molecular function of SGR is only a recent discovery, the stay-green phenotype of *sgr* mutants [also described as *non-yellowing* (*nye*)] mutants has been described long time ago (Hörtensteiner, 2006). The most famous *sgr*-type mutant among them is Gregor Mendel's green cotyledon pea (*Pisum sativum*) mutant that was used as a marker for determining the law of genetics (Mendel, 1866). Defects in orthologous loci were further assigned to cause stay-green phenotypes in *Lolium/Festuca* (Armstead et al., 2006), rice (*Oryza sativa*) (Jiang et al., 2007; Park et al., 2007), *Arabidopsis* (Ren et al., 2007) and over a dozen of other plant species (Hörtensteiner, 2009).

The *Arabidopsis* genome encodes three SGR proteins, i.e., SGR1 and SGR2, which are highly homologous, as well as SGR-like (SGRL), which is more distantly related (Sakuraba, Kim, Kim, Hörtensteiner, & Paek, 2014) (also described as NYE1, NYE2 and NYE-like, respectively) (Ren et al., 2007; Wu et al., 2016). SGR1 and 2 share their substrate specificity for Chl, whereas SGRL has a much higher activity for Chlide. However, all three SGRs are specific for the "*a*" form. It has experimentally been shown that SGR acts on free Chl, but also on Chl bound to Chl-protein complexes, i.e., in the photosystem core complexes and LHCs. Therefore, the role of SGR is not only to catalyze the step in which Chl is committed to degradation, but also to initiate the destabilization of the photosystems similar to NYC1 (Shimoda et al., 2016). Previously, interaction of SGR with six Chl catabolic enzymes (CCEs), i.e., HCAR, NYC1, NOL, PPH, PAO, and RCCR, together with LHCII was observed, which led to the hypothesis that SGR recruits these CCEs to allow metabolic channeling at the thylakoid membrane during Chl breakdown (Sakuraba, Kim, Yoo, Hörtensteiner, & Paek, 2013; Sakuraba et al., 2012).

Among a collection of *Arabidopsis* mutants deficient in a gene for phytol-hydrolase candidates that were selected based on their predicted chloroplast localization and upregulation during senescence, pheophytinase (PPH) mutants exhibited a stay-green phenotype during induced leaf senescence. In following in vitro experiments, the enzyme was shown to specifically hydrolyze Mg^{2+}-free Phein leading to the formation of Pheide (Schelbert et al., 2009). At about the same time, PPH was independently described in *Arabidopsis* (Ren et al., 2010) and its ortholog in rice (Morita, Sato, Masuda, Nishimura, & Kusaba, 2009). Since then the indispensable function of PPH during Chl breakdown was shown in a range of different plant species, such as Yali pear (*Pyrus* × *bretschneideri*), tomato (*Solanum lycopersicum*), broccoli (*Brassica oleracea*) and ryegrass (*Lolium perenne*) (Büchert, Civello, & Martínez, 2011; Cheng & Guan, 2014; Guyer et al., 2014; Zhang et al., 2016). Catalytic and structural properties of PPH were recently analyzed in detail (Guyer, Salinger, Krügel, & Hörtensteiner, 2018). Notably, in fruits, redundant phytol-hydrolyzing activity seems to be present: while *pph* mutants of tomato (*S. lycopersicum*) showed an impairment in Chl breakdown and accumulated Phein in their senescent leaves, they were still able to degrade Chl in the fruits (Guyer et al., 2014). As a consequence, it was hypothesized that further chloroplastic hydrolases might be involved; however their nature remains elusive. In a genetic screen for heat sensitivity in *Arabidopsis*, Chl dephytylase 1 (CLD1) was identified and proposed to be involved in Chl turnover (Lin, Wu, & Charng, 2016). CLD1 is able to dephytylate both Chl and Phein, but it is unclear so far whether CLD1 might also be involved in leaf senescence- or fruit/seed ripening-related Chl breakdown.

3.1.2 Primary *fluorescent Chl catabolite formation from pheophorbide* a

For the process of photosynthesis the light absorption capacity of Chl is essential. This is due to the characteristic conjugated π-system of the macroring of Chls, which still persists in the early Chl catabolites Phein *a* and Pheide *a*. Their spectral properties cause phototoxicity during the degradation of the photosynthetic apparatus. Therefore, impairment of these steps of Chl breakdown is considered to be crucial for Chl-dependent photodynamic damage. Chl detoxification is accomplished by the opening of the macrocycle of Pheide *a,* a two-step path to the first colorless Chl degradation intermediate *p*FCC. This conversion requires the correlated action of two independent (Hörtensteiner et al., 1995; Rodoni, Mühlecker, et al., 1997), but physically interacting enzymes (Pružinská et al., 2007; Sakuraba et al., 2012). The first step is the opening of the porphyrin ring by PAO (Pružinská et al., 2003), thereby producing the hardly detectable RCC (Kräutler et al., 1997; Rodoni, Mühlecker, et al., 1997), and the second, the reduction of RCC by RCCR (Wüthrich et al., 2000) (see Scheme 2).

Mutants in either of these two enzymes show a light-dependent accelerated cell death (ACD) phenotype during senescence, and indeed, PAO and RCCR mutants were originally identified as *acd1* and *acd2*, respectively (Greenberg & Ausubel, 1993; Greenberg, Guo, Klessig, & Ausubel, 1994). The observed cell death was proposed to be due to highly photodynamic properties considered for both Pheide *a* and RCC (Mach, Castillo,

Hoogstraten, & Greenberg, 2001; Pružinská et al., 2003; Tanaka, Hirashima, Satoh, & Tanaka, 2003). Interestingly, in addition to light-dependent lesion formation, *pao/acd1*, but not *rccr/acd2* mutants also show light-independent cell death, when leaf senescence is induced in constant darkness (Aubry et al., 2018; Hirashima, Tanaka, & Tanaka, 2009). This led to the hypothesis (Kuai et al., 2018) that Pheide *a*—apart from being a strong phototoxin (Xodo, Rapozzi, Zacchigna, Drioli, & Zorzet, 2012)—could also act as a cell-death signaling molecule. Christ et al. (2012) showed that Pheide *a* is not able to leave the chloroplast and, therefore, it is considered unlikely that the molecule itself acts as a retrograde chloroplast-to-nucleus signal, but it could be a sensor of impairment of chloroplast integrity. Since no further signaling molecules are known in this context, it still remains unclear how light-independent cell death is triggered. Interestingly, in *Arabidopsis pao1*, hundreds of genes are already differentially expressed in green tissues prior to the onset of senescence (Aubry et al., 2018).

PAO is a Rieske-type mono-oxygenase containing an iron-sulfur cluster, a mononuclear iron center, and two C-terminally located transmembrane domains that anchor the enzyme in the thylakoid membrane with the active site facing to the chloroplast stroma (Pružinská et al., 2003; Sakuraba et al., 2012). PAO is specific for the "*a*" form of Pheide, which is opened at the C5/C6 position. In vitro enzyme assays demonstrated that this activity depends on electron supply by the cofactor ferredoxin (Fd) (Pružinská et al., 2003). Furthermore, in $^{18}O_2$-labeling experiments it was shown that the oxygen incorporated by PAO at the C1 formyl position of *p*FCC is derived from molecular oxygen (Hörtensteiner et al., 1998). The formation of RCC, which is the first linear tetrapyrrolic PB, marks the step when Chl catabolites can pass the chloroplast membrane. This was demonstrated by the finding of RCC-type PBs in the vacuole in *acd2* (Pružinská et al., 2007).

RCCR is a distantly related member of the Fd-dependent bilin reductase family that includes phytochromobilin synthase of higher plants and different phycobilin synthases of red algae and cyanobacteria (Frankenberg, Mukougawa, Kohchi, & Lagarias, 2001). RCCR is localized in the chloroplast stroma (Wüthrich et al., 2000). It reduces the C15/C16 double bond of RCC in a stereospecific fashion, whereby the introduced stereocenter at C16 is dependent on the RCCR ortholog. The stereospecificity of two different RCCR orthologs, termed type I or type II RCCRs, can be manipulated by a Phe-to-Val exchange at the residue 218, as shown for the *Arabidopsis* RCCR (Pružinská et al., 2007). Later it was shown that this manipulation indeed concerns the substrate binding pocket of the enzyme

(Sugishima et al., 2009, 2010). Whether one or the other stereospecific enzyme is present depends on the species. For example, *Arabidopsis* has a type I RCCR and, therefore, produces *p*FCC (Mühlecker et al., 1997), while the RCCR from sweet pepper is a type II and produces *epi-p*FCC (Mühlecker et al., 2000). The formation of the *p*FCCs is considered to be the last step of the early PAO/phyllobilin pathway. If not defined otherwise the term *p*FCCs also includes its epimer *epi-p*FCC, as further enzymes are able to use both epimers as substrates. The stereochemistry introduced at the level of the two *p*FCCs is also found in the downstream PBs.

3.2 Peripheral modifications of fluorescent Chl catabolites

The following enzymatic reactions of the PAO/phyllobilin pathway describe the steps leading to the structural diversity of the final PBs. The sources of these variations are modifications at peripheral side chains on derivatives of the *p*FCCs. One of these modifying reactions still occurs in the chloroplast and—after the release from the chloroplast stroma—the subsequent ones in the cytosol. From then onward the downstream fluorescent PBs are termed *m*FCCs or DFCCs. Last, *m*FCCs (and corresponding DFCCs) are transported into the vacuole, where they isomerize non-enzymatically to NCCs (and DNCCs, respectively) (see Schemes 2 and 5).

The first presumed catabolites derived from Chl were discovered in a senescing variety of *F. pratensis* back in 1987 (Matile et al., 1987). Soon after, pink pigments were detected in the vacuoles of barley (*H. vulgare*) leaves whose appearance correlated with rates of Chl degradation (Matile, Ginsburg, Schellenberg, & Thomas, 1988). This was followed by further reports on Chl catabolites in different species, but the breakthrough was marked by the structural characterization of a Chl catabolite found in senescent barley leaves, termed *Hv*-NCC-1 (**1**) (Kräutler et al., 1991; see above). This initial discovery allowed conclusions to be drawn concerning enzymatic activities along the breakdown pathway, but also, with subsequent structural elucidations, revealed the catabolite diversity among different species.

The previously described part of the PAO/phyllobilins pathway is widely conserved in plants and the involved enzymes are rather well described. By contrast, much less is known about the enzymes involved in the side chain-modifying reactions described in the following sections. So far, three of these enzymes have been characterized in *Arabidopsis*, i.e., CYP89A9 (Christ et al., 2013), methylesterase 16 (MES16) (Christ et al., 2012) and TIC55 (Hauenstein, Christ, Das, Aubry, & Hörtensteiner, 2016).

3.2.1 Terminal hydroxylation of the ethyl side chain at C3

Hydroxylation at the terminal carbon $C3^2$ is the first modification of *p*FCCs and furnishes 3^2-hydroxy-*p*FCCs (**3** and ***epi*-3**) within the senescing chloroplast (Hauenstein et al., 2016). A fluorescent compound that was more polar than the subsequently characterized *p*FCCs, presumably the hydroxy-*p*FCC **3**, was already observed in 1992 in isolated barley chloroplasts (Matile, Schellenberg, & Peisker, 1992). The 3^2-hydroxy-*p*FCCs have long evaded characterization (Christ et al., 2012; Moser et al., 2012), but their intermediary existence has been derived from the occurrence of the corresponding hydroxylated NCCs, for which consistent evidence has been found in all species investigated (Christ et al., 2016; Curty & Engel, 1996; Kräutler, 2016; Roca et al., 2018). The common occurrence of the hydroxylation and its localization within the plastid (Hauenstein et al., 2016) connects this reaction to the first part of the PAO/phyllobilin pathway and, thus, 3^2-hydroxy-*p*FCCs can be considered "secondary" FCCs (*s*FCCs). In senescent wild-type leaves of *Arabidopsis*, sFCC (=3^2-hydroxy-*p*FCC) is the precursor of all hydroxylated PBs which compose more than 50% of the catabolites. Only recently, the responsible enzyme catalyzing the $C3^2$ hydroxylation of *p*FCC in *Arabidopsis* was characterized and identified as TIC55 (Hauenstein et al., 2016). TIC55 belongs to the class of Rieske-type mono-oxygenases, like PAO. It was shown to catalyze the $C3^2$ hydroxylation of *p*FCC in vitro in a Fd-dependent way producing 3^2-hydroxy-*p*FCC (*s*FCC). The 3^2-hydroxylated PBs are completely absent in senescent leaves of *tic55* mutants, demonstrating the requirement of TIC55 for their production. Apart from the clear phenotype in the catabolite profile, *tic55* mutants show no other physiological differences to wild-type plants (Hauenstein et al., 2016). Previously, TIC55 (translocon at the inner chloroplast membrane 55) was assigned to be a potential component of the protein import machinery at the inner chloroplast membrane with a speculated role in redox sensing (Balsera, Soll, & Buchanan, 2010; Bartsch et al., 2008; Calibe et al., 1997). However, experimental evidence for this function is missing (Boij, Patel, Garcia, Jarvis, & Aronsson, 2009).

3.2.2 O-Glycosylation and malonylation reactions

After the release of *p*FCC or its hydroxylated form from the plastid to the cytosol, further species-specific side chain modifications occur effectively. The hydroxyethyl side chain at C3 of 3^2-hydroxy-*p*FCCs can subsequently be glucosylated and/or malonylated, as found in NCCs of some species such as *Arabidopsis*, tobacco (*Nicotiana rustica*) (Berghold et al., 2004) and oilseed rape (Mühlecker et al., 1993), as well as in ripe fruits, such as apples and pears (Müller et al., 2007), loquat (*Eriobotrya japonica*) and quince (*Cydonia oblonga*)

(Rios, Perez-Galvez, et al., 2014; Rios, Roca, et al., 2014). The relative amount of glucosylated catabolites varies greatly across species. In different *Arabidopsis* ecotypes only around 5% of the catabolites are glucosylated (Christ et al., 2016), whereas in tobacco all of the known Chl catabolites are glucosylated or additionally malonylated (Berghold et al., 2004). Although malonyltransferase activity has been characterized in oilseed rape (Hörtensteiner, 1998), the responsible enzyme for malonylation still remains unknown. For glucosylation, it has been speculated that UDP-dependent glycosyltransferases might play a role, since they are known to catalyze the addition of glucose to hydroxyl groups. However, so far none of the UDP-dependent glycosyltransferases could be assigned to this function. Based on the fact that isolated senescing chloroplasts do not accumulate glucosylated FCCs and that *tic55* mutants lack glucosylated NCCs and DNCCs (Hauenstein et al., 2016), the proposed glycosyltransferases are likely cytosolic enzymes and FCC hydroxylation is an obvious prerequisite for subsequent glucosylation.

3.2.3 Oxidative deformylation to dioxobilin-type Chl catabolites

Deformylation at C1 leads to the conversion of FCCs to their respective dioxobilin-type analogs, the DFCCs. Respective nonfluorescent DNCCs had originally been identified in barley (Losey & Engel, 2001), but were thought to be oxidation products of NCCs, probably for a long time, even though they are the major Chl catabolites accounting for over 90% of all PBs in this species (Christ et al., 2013; Süssenbacher et al., 2014). Only in 2013 DNCCs were described in the published literature as the main products of Chl breakdown in *Arabidopsis* and the responsible enzyme activity identified in the cytosol (Christ et al., 2013). DNCC formation in senescent leaves was inhibited in a carbon monoxide-enriched atmosphere, indicating that CYPs could be responsible for deformylation. Based on these results, *CYP* gene candidates were selected and respective mutants screened for altered catabolite profiles. It was found that *cyp89a9* mutants are devoid of DNCCs, while they accumulate accordingly more NCCs. Further in vitro enzyme assays confirmed this encoded candidate protein (Christ et al., 2013). CYP89A9 is localized in the ER and oxidatively deformylates FCCs, but not NCCs. The occurrence of DNCCs and NCCs varies among different plant species (Das, Christ, & Hörtensteiner, 2018). In contrast to *Arabidopsis*, which accumulates both types of PBs, in *C. japonicum* exclusively NCCs are found (Curty & Engel, 1996; Oberhuber et al., 2003), while in Norway maple one single DNCC accumulates, but no NCCs (Müller et al., 2011).

3.2.4 Hydrolysis of the methyl ester function

MES16 catalyzes the $C8^2$-carboxymethyl group demethylation of PBs in *Arabidopsis*. As revealed by the recently published crystal structure, it belongs to the α/β-hydrolase superfamily (Li & Pu, 2016). It was experimentally shown that the natural substrates of MES16 are cytosolic fluorescent PBs, even though mistargeted chloroplast-localized MES16 was able to convert Pheide *a* to *py*Pheide *a* in a *pao* mutant background (Christ et al., 2012). In senescent *mes16* mutant leaves, isomerization of FCCs (and, presumably, DFCCs) to corresponding phyllobilanes was slowed down. This led to the accumulation of fluorescent PBs in the vacuole. Under UV light it also caused senescent *mes16* leaves to fluoresce (Christ et al., 2012). The demethylation of PBs is, as the deformylation, highly species-specific. Interestingly, even within the species of *Arabidopsis*, the Landsberg *erecta* (L*er*) ecotype seems to be a natural *mes16* mutant. Out of 18 analyzed *Arabidopsis* ecotypes all except L*er* showed virtually 100% demethylated PBs (Christ et al., 2016, 2012). However, this may only be an artifact, since it was shown that many lab strains of L*er* are derived from an ethane-methylsulfonic acid-induced mutagenesis event that explains the occurrence of many single nucleotide polymorphisms in this accession (Zapata et al., 2016).

Besides its role in Chl breakdown, MES16 also accepts methyl-indole acetic acid as a substrate (Abbas et al., 2018; Christ et al., 2012; Yang et al., 2008). Together with the finding of its upregulation during the night this led to the hypothesis that MES16 could play a role in auxin activation during nocturnal growth (S. Hörtensteiner, unpublished).

3.2.5 Dihydroxylation of the vinyl group at C18

C18-dihydroxylated PBs have been identified in many plants; in particular, they have been found in all grass species analyzed to date (Berghold et al., 2006; Das et al., 2018; Kräutler et al., 1991), while they are absent from several families of angiosperms such as the Brassicaceae (Christ et al., 2016; Mühlecker & Kräutler, 1996; Pružinská et al., 2005). This indicates the presence of a respective dihydroxylating activity in only certain plant clades. The $^{18}O_2$-labeling experiments showed that both oxygen atoms incorporated into the C18-dihydroxyethyl group are derived from molecular oxygen, indicating that the responsible enzyme(s) could be a dioxygenase (Das et al., 2018). However, the molecular identity of such a dioxygenase remains elusive.

3.2.6 Esterification at the propionyl group

Typical FCCs usually only exist fleetingly before they are further converted to NCCs (Oberhuber et al., 2003). However, banana and peace lily accumulate hypermodified blue fluorescent catabolites, i.e., *hm*FCCs, in their senescent leaves (Banala et al., 2010) and also in the ripe banana fruit peel (Moser, Müller, et al., 2008; Moser et al., 2009). The accumulation of *hm*FCCs in ripe banana peels was shown to be the dominant cause of their fluorescent blue glow (Moser, Müller, et al., 2008; Moser et al., 2009). The persistency of *hm*FCCs is caused by the esterification of their propionic acid group with different functional groups, such as daucyl, glucosyl, digalactosyl-glyceryl, or glucosyl-(dihydroxyphenyl) ethyl, which greatly interferes with the FCC-to-NCC isomerization (Oberhuber et al., 2008). In the exceptional bicycloglycosidic PBs **13**, **14** and **25** (see above and Schemes 3, 4 and 6), both esterification and glycosidation are present. However, the mechanisms of glycosylation and propionate esterification remain unsolved and require the identification of the respective enzyme(s) catalyzing these modifications.

3.3 Core transformations of fluorescent Chl catabolites

3.3.1 Isomerization to nonfluorescent Chl catabolites

As described in Section 3.2, the typical FCCs are prone to spontaneous and stereoselective isomerization to the apparently more stable corresponding NCCs (see Scheme 5) (Oberhuber et al., 2003). Hence, this chemical reaction has been proposed to account for the formation of NCCs and, indeed, there is no evidence for an enzyme catalyzing the FCC-to-NCC isomerization (Hörtensteiner & Kräutler, 2011; Kräutler, 2016; Kuai et al., 2018). Consistent with the subcellular location of NCCs (Hinder et al., 1996), transport of FCCs into the acidic vacuole and their chemical isomerization to NCCs has been accepted as the biochemical model of the NCC formation step in senescent leaves (Kräutler, 2016). A related mechanism has been deduced to concern the formation of DNCCs from the corresponding DFCCs (Süssenbacher, Hörtensteiner, et al., 2015).

3.3.2 Carbon skeleton rearrangement and pyro formation

Recent analysis of the fern *P. aquilinum* revealed an unprecedented type of PBs, which were classified as *i*PBons (Erhart, Vergeiner, et al., 2018). These novel PBs are devoid of the typical carboxylic acid or ester function at $C8^2$ classifying them as *pyro*-PBs. So far, evidence for the natural occurrence of *py*NCCs was missing and explained by the surprising resistance of NCCs for the loss of this carboxyl unit (Li et al., 2018). Thus, pyro formation may occur on the level of Pheide *a*, hence inside plastids. However, this would require a MES16-like activity inside the plastid as well (see above) (Christ et al., 2012). Alternatively, *P. aquilinum* may possess a so-called demethoxycarbonylase, as described in the green alga *Chlamydomonas reinhardtii* (Suzuki, Doi, & Shioi, 2002) and suggested to be able to cleave the entire carboxymethyl group. Possibly, such an enzyme might act even at a later stage of Chl breakdown, i.e., on the level of a linear tetrapyrrole, but a corresponding activity has not been identified.

Furthermore, *i*PBons show a unique carbon skeleton rearrangement, where the connection of ring A to ring B, which is usually the C5/C6 carbon bond, is shifted to a C5/C7 bond (Erhart, Vergeiner, et al., 2018). Except this difference, the carbon backbone is identical to the known type-I Chl catabolites and it has been presumed that these *i*PBons derive from a precursor PB generated in the "early" part of the PAO/phyllobilin pathway (Erhart, Vergeiner, et al., 2018). Although both pyro formation and carbon skeleton shift were deduced from the structures of the *i*PBons, these reactions probably are not synchronized. However, hydroxylation at C6 is a prerequisite for the ring A migration and may occur on the stage of FCC (Erhart, Vergeiner, et al., 2018). The nature of such a novel enzyme remains elusive. From an evolutionary point of view carbon skeleton shift and pyro formation are interesting, as these enzyme activities occur in lower plants, such as ferns and some gymnosperms, including *Ginkgo biloba* (T. Erhart, T. Müller & B. Kräutler, unpublished; Erhart, Vergeiner, et al., 2018; M. Hauenstein & S. Hörtensteiner, unpublished), but have been lost during the evolution of angiosperms.

4. Biology of Chl breakdown in plants

4.1 Subcellular localization of enzymes involved in Chl breakdown and catabolite transport

Already in 1988, when Chl catabolites were detected in senescent barley leaves, it was shown that the majority of the catabolites are present in the vacuolar fraction (Matile et al., 1988), implying the vacuole to be the typical place for their end storage (Matile, Hörtensteiner, Thomas, & Kräutler, 1996). Later this was also confirmed in other plant species (Christ et al., 2012; Hinder et al., 1996; Pružinská et al., 2007). Thus, typical Chl breakdown starts in senescing chloroplasts and ends inside the vacuole. A combination of biochemical approaches, cell fractionation studies and, recently, molecular cloning and localization studies of CCEs uncovered a cellular topology of Chl breakdown as depicted in Fig. 1. The first part of the PAO/phyllobilin pathway is localized inside the plastid. Protein-protein interaction analysis and localization studies using green fluorescent protein fusions indicated that the enzymes of the first part of the pathway required to form *p*FCC, i.e., NYC1, NOL, HCAR, SGR, PPH, PAO

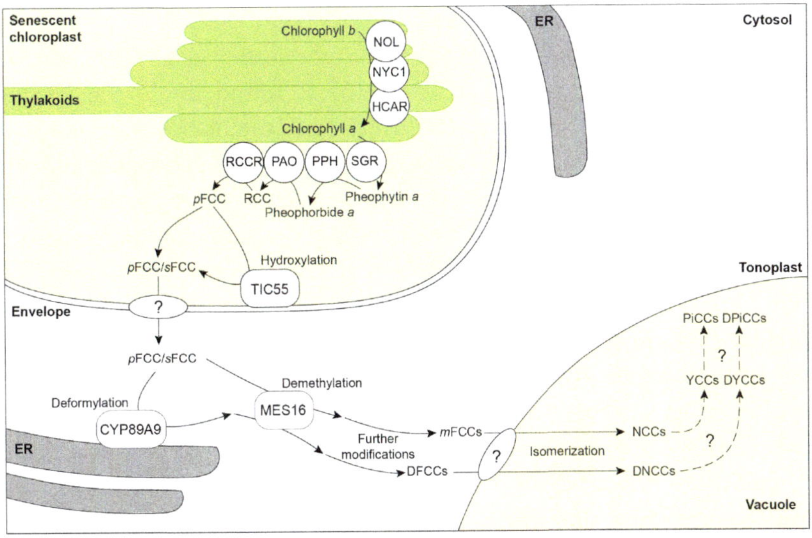

Fig. 1 Topology of the PAO/phyllobilin pathway of chlorophyll breakdown. Chl degradation starts at the thylakoid membrane of the senescing chloroplast. *p*FCC and/or 3^2-hydroxy-*p*FCCs ("secondary" FCCs; *s*FCCs) produced by TIC55 are exported from the plastid to the cytosol by unknown transporter(s). In the cytosol, side chain modifications may occur that lead to the formation of "modified" FCCs (*m*FCCs) or of DFCCs when involving the action of ER-localized CYP89A9. Modified fluorescent phyllobilins (*m*FCCs and DFCCs) are then imported into the vacuole by unknown transporter(s), where they are converted to respective nonfluorescent isomers (NCCs and DNCCs) due to the acidic milieu. NCCs (DNCCs) may further be converted to YCCs (DYCCs) and PiCCs (DPiCCs). For abbreviations, see text and list of abbreviations.

and RCCR, likely form a protein complex at the thylakoid membrane, which was considered to be important for metabolic channeling of otherwise potentially phototoxic Chl breakdown intermediates (Sakuraba et al., 2013, 2012). The localization of TIC55 is distinct from the other chloroplast-located CCEs, i.e., in the chloroplast envelope. Whether this links PB hydroxylation with FCC export from the plastid remains to be investigated. Based on the fact that Chl breakdown is not hampered in *tic55* mutants, it seems likely that catabolite export also occurs in the absence of TIC55 (Hauenstein et al., 2016). In *Arabidopsis*, additional *p*FCC-modifying CCEs are located outside the plastid, i.e., in the cytosol (MES16) or attached to the ER (CYP89A9), and it is likely to assume that also further activities, like $O3^3$-glucosylation, $O3^3$-malonylation and C18 vinyl group dihydroxylation, occur in the cytosol.

Given the facts that *p*FCCs/*s*FCCs are formed inside the plastid and that the isomerization process, which converts FCCs and DFCCs into their respective phyllobilane isomers (NCCs and DNCCs), happens non-enzymatically in an acidic environment (Oberhuber et al., 2003, 2008; Süssenbacher, Hörtensteiner, et al., 2015) as found inside the vacuole (Christ et al., 2013; Oberhuber et al., 2003), two membrane transport steps could be postulated, i.e., *p*FCC/sFCC export from the chloroplast and subsequent import of FCCs and DFCCs into the vacuole. Using isolated senescent chloroplasts of barley, Matile et al. (1992) demonstrated that export of the fluorescent presumed Chl catabolites requires supply and hydrolysis of ATP. However, the nature of transport remains unknown and up to now, envelope Chl catabolite transporters have not been identified at the molecular level. Nevertheless, based on the subcellular accumulation of intermediates in respective Chl breakdown mutants (Christ et al., 2012; Pružinská et al., 2007) it can be concluded that the envelope transporter(s) likely accept(s) linear tetrapyrroles, i.e., RCC- or FCC-type PBs as substrate, but not porphyrins, like Pheide *a*.

In contrast to chloroplast export, it was demonstrated that transport of Chl catabolites into isolated barley vacuoles is ATP-dependent. Inhibition studies of the transport indicated it to be a primary active process and, therefore, ATP-binding cassette (ABC) transporters are likely candidates for PB uptake into the vacuole (Hinder et al., 1996). Indeed, NCC transport could be demonstrated in yeast (*Saccharomyces cerevisiae*), which heterologously expressed ABCC2 and ABCC3 transporters of *Arabidopsis* (Lu et al., 1998; Tommasini et al., 1998). However, the characterization of ABC transporters is aggravated by their high substrate redundancy (Kang et al., 2011) and, therefore, the responsible transporter(s) of Chl catabolites in vivo has/have not been identified yet.

4.2 Regulatory aspects of Chl breakdown

Senescence in plants generally results in an increase of catabolic and a decrease of anabolic processes ultimately leading to cell death at a cellular, tissue, organ, or organismic level (Lim, Kim, & Nam, 2007). The complex molecular mechanisms involved in this process are orchestrated by internal or external factors. Important internal factors are stages of plant development and/or the action of phytohormones (Zhu, Chen, Qiu, & Kuai, 2017). External factors can be divided into biotic ones, such as pathogens, or abiotic ones, which include environmental stresses, like light, temperature, drought, ozone or nutrient deficiency (Lim et al., 2007). The transcriptional regulation of senescence and of Chl breakdown is often mediated by the same factors. The following sections outline the current knowledge of regulatory cues of Chl breakdown, i.e., the link and impact of transcriptional control of Chl catabolic genes (CCGs) encoding respective CCEs.

4.2.1 Ethylene

Ethylene is known for a long time as one of the major inducers of leaf senescence, fruit ripening and flower senescence (Abeles et al., 1988; Bleecker, Estelle, Somerville, & Kende, 1988; Chao et al., 1997). Likewise, ethylene-triggered Chl breakdown as the cause for the degreening process during leaf senescence and fruit maturation has been described (Azoulay Shemer et al., 2008; Cheng & Guan, 2014; Yin et al., 2016). Only recently, the underlying molecular genetic mechanisms, which link ethylene to Chl breakdown during senescence, have been elucidated. Generally, ethylene acts as a promoter of CCG expression through the transcription factors (TFs) EIN3 (ethylene insensitive 3) and ORE1 (oresara 1). EIN3 was shown to physically bind to the promoters of the genes encoding NYE1, NYC1 and PAO to activate their expression. ORE1, which is a direct target of EIN3, is also able to bind promoter regions of *NYE1*, *NYC1*, *NOL* and *PAO*, thereby activates their expression, and, in addition, increases ethylene production by inducing *ACS2* (encoding aminocyclopropane-1-carboxylic acid synthase 2), a rate-limiting gene of ethylene biosynthesis. Taken together EIN3, ORE1, and ACS2 create a feed-forward loop that regulates ethylene-mediated Chl breakdown during leaf senescence in *Arabidopsis* (Qiu et al., 2015). For Chl breakdown in ripening fruits, an ethylene-responsive factor was reported, probably acting via induction of *PPH* expression (Yin et al., 2016).

4.2.2 Abscisic acid

Involvement of abscisic acid (ABA) in stress-induced leaf senescence and seed maturation is deduced from a significant increase of ABA during these processes (Breeze et al., 2011; Gepstein & Thimann, 1980; Liu, Longhurst, Talavera-Rauh, Hokin, & Barton, 2016). The TFs ABF (ABA-responsive element-binding factor) and RAV1 (related to ABA-insensitive 3/viviparous 1) both upregulate expression of senescence-associated genes via ORE1, which connects the signaling pathway of ABA to ethylene signaling. It has been demonstrated that changes in ABA levels, i.e., either by modified endogenous levels or exogenous application, affect expression of CCGs (Raab et al., 2009; Takasaki et al., 2015; Ye et al., 2017). The molecular cause for these altered expression levels was recently described. The two TFs ABI5 (ABA insensitive

5) and EEL [enhanced EM (early methionine-labeled) level], both of which act in the ABA signaling pathway, directly upregulate *NYE1* and *NYC1* (Sakuraba, Jeong, et al., 2014). A further component of the ABA signaling pathway, NAC016 [NAM (no apical meristem), ATAF (*A. thaliana* activating factor) 1/2, CUC (cup-shaped cotyledon) 16], was shown to directly activate *NYE1* expression (Sakuraba, Han, Lee, Hörtensteiner, & Paek, 2016). Furthermore, *NYE1/2*, *NYC1* and *PAO* are activated during natural and ABA-triggered leaf senescence by ABF2/3/4 (Gao et al., 2016). During seed maturation the TF ABI3 seems to play an important role by regulating *NYE1/2* expression (Delmas et al., 2013).

4.2.3 Jasmonate

Already in 1980, methyl jasmonate (MeJA) and its precursor jasmonic acid (JA) were shown to be strong inducers of plant senescence (Ueda & Kato, 1980). He, Fukushige, Hildebrand, and Gan (2002) found that during leaf senescence in *Arabidopsis* measured JA levels were fourfold higher compared to non-senescing leaves, in line with the finding that during senescence JA biosynthesis-related genes are differentially activated. Furthermore, it was demonstrated that exogenous application of JA promotes senescence (He et al., 2002). Later, components of JA-induced senescence have been identified (Jiang, Liang, Yang, & Yu, 2014; Qi et al., 2015; Shan et al., 2011). How JA signaling impacts regulation of Chl degradation on the molecular genetic level was reported recently (Zhu et al., 2015). It was demonstrated that MYC (myelocytomatosis) proteins (MYC2/3/4)—key players in JA signaling—act as direct transcriptional regulators of *PAO* and, furthermore, are able to bind to the promoter regions of *NYE1* and *NYC1* and to regulate their expression. In the same study, it was shown that ANAC (*Arabidopsis* NAC) proteins (ANAC019/055/072), downstream targets of the MYCs in the JA signaling pathway, are able to directly activate *NYE1/2* and *NYC1*. Protein-protein interaction between ANAC019 and MYC2 causes upregulation of the expression of *NYE1* (Zhu et al., 2015).

4.2.4 Light

Already 35 years ago, light was shown to act as an environmental cue of senescence (Biswal & Biswal, 1984), i.e., red and far-red light pulses retarded and re-stimulated, respectively, leaf senescence. Also the fact that prolonged absence of light leads to senescence in plants is widely used to study senescence and Chl breakdown. This light-dependent signaling is mediated by phytochromes, which are important light receptors found in plants (Biswal & Biswal, 1984; Liebsch & Keech, 2016). Light impulses sensed by phytochromes are transmitted by PIFs (phytochrome-interacting factors), which are the major TFs controlling dark-induced senescence, with PIF3/4 and 5 acting as positive regulators of leaf senescence. PIF4 and 5 target *ABI5/EEL* and *EIN3*, products of which directly upregulate *NYE1* and *NYC1* (Sakuraba, Jeong, et al., 2014). Additionally, EIN3 also directly targets ORE1 and thereby activates expression of *NYE1*, *NYC1*, *NOL* and *PAO* as described above in the context of ethylene signaling.

4.3 Significance of chlorophyll breakdown in plants

While senescence processes in different plant organs such as leaves, fruits and seeds lead to distinct ultrastructural changes within the specific organ tissue, it has been shown that the expression pattern of a very wide range of genes is regulated in an analogous way (Krupinska, Biswal, & Biswal, 2013). These senescence-associated genes encode proteins involved in macromolecular degradation, removal of oxidative metabolites, induction of defense mechanisms, and signaling and regulatory events (Gepstein et al., 2003). Recycling and detoxification are seen as the main consequences of structural and biochemical changes occurring in leaves and fruits prior to their abscission. Visually they share the most obvious signs of the so-called senescence syndrome (Lim et al., 2007), i.e., the degradation of Chl.

4.3.1 Chl breakdown and leaf senescence

Senescence in leaves usually progresses in an age-dependent manner or is seasonally induced by short day length in deciduous plants. But senescence can also be a premature response to different biotic and abiotic stresses, such as lack of light, drought or pathogen attack (Guo & Gan, 2012; Lim et al., 2007). In leaves, signs of senescence usually start from the tip and progress to the base, and within the mesophyll from the inside toward the epidermis, excluding tissues along vascular bundles (Inada, Sakai, Kuroiwa, & Kuroiwa, 1998), or occur locally upon pathogen infection. In *Arabidopsis*, infection with *Pseudomonas syringae* was shown to lead to senescence-associated Chl breakdown (Mecey et al., 2011; Mur et al., 2010). In terms of Chl breakdown, the leaf is the best-characterized plant organ, since it served as a starting point for research and characterization of the PAO/phyllobilin pathway.

4.3.2 Chl breakdown and fruit ripening

The ripening process of fruits relates to senescence processes. Ripening of dry fruits, as found, for example, in the siliques of *Arabidopsis*, is much more similar to leaf senescence (Mizzotti et al., 2018), while in fleshy fruits, ripening is accompanied by many more physiological changes, such as carotenoid and anthocyanin synthesis, pericarp softening, cell wall degeneration and starch to sugar conversion (Gomez, Vera-Sirera, & Perez-Amador, 2014). In fleshy fruits, such as tomato, the color change is caused by the transition of chloroplasts into chromoplasts. This differentiation includes the lysis of grana stacks and thylakoid membranes as well as the synthesis of new membranes derived from the former inner plastid membranes, which become the specialized site for carotenoid synthesis (Barsan et al., 2012; Kahlau & Bock, 2008). Only when Chl is broken down, previously hidden and newly synthesized carotenoids and anthocyanins get unmasked, revealing the eye-catching colors of fruits. It was demonstrated that fruit color is dependent on abiotic factors, but also a result of evolutionary adaption to the visual detection by specific seed-dispersing animals (Valenta et al., 2018) or of breeding (Stommel, Abbott, Saftner, & Camp, 2005). Additionally, it was shown that colored Chl catabolites, i.e., YCCs and PiCCs, contribute to the color of senescent leaves or ripened fruits (Moser, Ulrich, et al., 2008; Ulrich et al., 2011). Moreover, NCCs in ripe fruits were demonstrated to be powerful antioxidants (Müller et al., 2007).

However, the underlying biochemical processes of Chl breakdown in fruits are not entirely resolved. CCEs known to be involved in Chl breakdown in leaves have also been identified in fruit proteomic studies (Barsan et al., 2010; Li et al., 2017; Wang et al., 2013). While in some fruits, such as apple (*Malus domestica*) and pear (*Pyrus communis*), Chl is broken down to PBs that are identical to those from respective senescent leaves (Müller et al., 2007), in other species Chl breakdown differs in leaves and fruits. As an example, different PBs are found in ripening banana fruit (Moser, Müller, et al., 2008; Moser et al., 2009, 2012) and their senescing leaves (Banala et al., 2010). Furthermore, *pph* mutants in *Arabidopsis* and tomato exhibit a clear stay-green phenotype during leaf senescence, in contrast to their fruits and seeds (Guyer et al., 2014; Zhang et al., 2014). However, at least in *Arabidopsis pph*, lower amounts of catabolites and higher amounts of retained Chl were observed in the siliques as compared to wild type (D. Menghini & S. Hörtensteiner, unpublished). This demonstrates the involvement of PPH in the breakdown of Chl in the generative tissue, but also suggests the involvement of further hydrolytic activities in fruit, seeds and flowers. For ripening of citrus fruits, chlorophyllases have been proposed to be involved in dephytylation (Azoulay-Shemer et al., 2011, 2008).

4.3.3 Chl breakdown and seed development

Development of seeds can be described in two phases, the morphogenesis phase, involving the development of the embryo and the endosperm; and the maturation phase, in which desiccation tolerance and seed dormancy are established. This latter phase involves senescence processes where, particularly toward the end of maturation, Chl is rapidly degraded (Li, Ban, Limwachiranon, & Luo, 2017). Plant varieties with a green seed phenotype have been known for a long time, most famously, Mendel's green cotyledon mutant in pea, but also in certain varieties of other species, such as soybean (*Glycine max*) and *Arabidopsis*, in which NYE genes were shown to be mutated (Fang et al., 2014; Sato, Morita, Nishimura, Yamaguchi, & Kusaba, 2007). NOL was also shown to be involved in Chl degradation in seeds, which is in fact crucial for seed storability (Nakajima, Ito, Tanaka, & Tanaka, 2012). These findings demonstrate the involvement of the enzymes of the PAO/phyllobilin pathway that have been characterized in leaves. While it is clear that Chl breakdown in leaves and seeds share biochemical characteristics similar to fruits, there are also some inconsistencies. Thus, leaves of the *Arabidopsis nye1* mutant shows a stay-green phenotype, while during seed maturation Chl is only retained in a *nye1 nye2* double mutant, possibly due to tissue-specific redundancy (Delmas et al., 2013; Wu et al., 2016). Furthermore, it was shown that distribution of differentially modified PB is not the same in leaves and seeds, presumably due to different activities of the PB-modifying enzymes (D. Menghini & S. Hörtensteiner, unpublished).

4.4 Evolutionary aspects of Chl breakdown

Angiosperms, as the most diverse division of the plant kingdom, have been predominantly the main focus of studies on Chl breakdown (Hörtensteiner & Kräutler, 2011; Kuai et al., 2018). The structure-based information concerning Chl catabolites is, therefore, almost exclusively restricted to the PBs identified in angiosperms (Kräutler, 2016). So far, there are no published records of PBs in gymnosperms and mosses. However, only very recently the knowledge gap of PBs in ferns was filled with the first report on *i*PBons in *P. aquilinum*, revealing a novel type of Chl catabolites with a rearranged carbon backbone (Erhart, Vergeiner, et al., 2018). Interestingly, similar catabolites are also found in the primitive gymnosperm *Ginkgo biloba* (T. Erhart, T. Müller & B. Kräutler, unpublished; Erhart, Vergeiner, et al., 2018; M. Hauenstein & S. Hörtensteiner, unpublished). As these Chl catabolites seem to stem from a precursor of the PAO/phyllobilin pathway, the earlier hypothesis is supported that this pathway is highly conserved among land plants (Hörtensteiner, 2013; Thomas, Huang, Young, & Ougham, 2009).

Phylogenetic analysis demonstrated that CCEs, like NYEs, PPH, PAO, RCCR and TIC55, are highly conserved among them, whereby gene homology decreases in more basal photosynthetic organisms (Hauenstein et al., 2016; Pružinská et al., 2007; Schelbert et al., 2009; Thomas et al., 2009). Functional parallels of RCCR have been demonstrated in the gymnosperm *Pinus taeda*, the liverwort *Marchantia polymorpha*, and the bryophyte *Physcomitrella patens* (Pružinská et al., 2007). Based on these findings and despite the incomplete evidence of PBs in land plants (Erhart, Vergeiner, et al., 2018), it is assumed that the biochemical processes of the PAO/phyllobilin pathway are ubiquitous (Kuai et al., 2018).

The early part of the PAO/phyllobilin pathway seems to have established even further back on an evolutionary time scale, i.e., already before the evolution of land plants. The green alga *A. protothecoides* is known to be able to degrade Chl under N-limited, heterotrophic conditions in the dark (Curty & Engel, 1996). However, in contrast to land plants, *A. protothecoides* degrades Chl only to the stage of red catabolites (related to RCC), seemingly requiring the same biochemical processes as described for angiosperms (Curty, Engel, & Gossauer, 1995). The red catabolites formed are then excreted into the surrounding medium. A different way to eliminate phototoxic Chl catabolites, i.e., disposal in the vacuole, was only required when plants invaded terrestrial habitats and were no longer surrounded by aqueous environments (Hörtensteiner, 2006). In line with this hypothesis is the finding that RCCR is apparently missing in algae based on in silico predictions (S. Aubry & S. Hörtensteiner, unpublished; Gao et al., 2014). Homologues of the upstream enzymes have not been characterized yet. The most similar PAO-like enzyme in green algae exhibits 31% sequence identity to the one of *Arabidopsis* and its functional validation is required.

In contrast to the first part of the PAO/phyllobilin pathway, which seems to be highly conserved, the second part—mainly describing the side chain modifications of PBs (Kräutler, 2016)—is highly diverse among species. Since some of these modifications of PBs do not seem to be essential for the plant, i.e., obvious phenotypes of respective *Arabidopsis* mutants are missing (Christ et al., 2012, 2013; Hauenstein et al., 2016), their physiological function and the reason for their diversity in different species are still an enigma. Using the catabolite profiles of all plants identified so far and putting them into a phylogenetic context will allow tracing back the arise of these side chain-modifying activities and gain information on the specific recruitment of respective CCEs for Chl breakdown. Whether the known PBs represent the end stage of Chl breakdown or if degradation occurs beyond them in viable leaf tissue is not clear yet. It has been reported that the amount of degraded Chl does not always correlate with the amount of accumulated PBs (Berghold et al., 2002; Christ et al., 2016; Das et al., 2018; Kräutler et al., 1992; Losey & Engel, 2001). Catabolite analysis of *Nicotiana benthamiana* leaves that were selected along the stem representing a senescence gradient uncovered a peak of PB in mid-senescent leaves, which decreased again in the oldest leaves (M. Hauenstein & S. Hörtensteiner, unpublished). Furthermore, in senescent barley leaves, monopyrrolic compounds were found, which were inferred to be Chl catabolites (Suzuki & Shioi, 1999). Clearly, further degradation of Chl beyond the stage of the known PBs is rather likely and its elucidation remains to be a challenge (Kräutler, 2016; Li et al., 2019; C. J. Li & B. Kräutler, unpublished).

5. Conclusions and outlook

The fate of Chl during breakdown, considered an "enigma" until about 30 years ago (Hendry et al., 1987), has been largely resolved based on the compiled research of the last almost 30 years, which contributed to the biochemical, molecular and structural elucidation of Chl breakdown, now termed the PAO/phyllobilin pathway (Kräutler & Hörtensteiner, 2014), and its degradation products. The core pathway, i.e., the consecutive conversion of Chl to non-phototoxic intermediates within the chloroplast, was finally resolved with the recent discovery of the Mg dechelatase activity of SGR (Shimoda et al., 2016). This part of the degradation pathway is thought to be widely conserved in photosynthetically active organisms, down to unicellular algae, such as *A. protothecoides* (Matile, Hörtensteiner, & Thomas, 1999; Thomas et al., 2009), implying the importance of Chl degradation as detoxification process, in order to protect a cell from photooxidative damage by Chl (Hörtensteiner, 2004; Hörtensteiner & Kräutler, 2011). Interestingly, with the colonialization of terrestrial habitats, plants acquired further processing steps in order to store the accumulating Chl catabolites within the cell (Hinder et al., 1996; Oberhuber et al., 2003) versus excretion of RCC-like compounds in the surrounding aqueous environment as in *A. protothecoides* (Engel et al., 1996). These processing steps in higher land plants are further modifications of the first colorless degradation intermediate *p*FCC. These "late" Chl degradation processes create the large number of different PBs known to date (Kräutler, 2016). This variation is in clear contrast to the high conservation of the core first part of Chl degradation and is much less well understood. The only molecularly identified enzymes of this part of Chl breakdown so far are CYP89A9, MES16 and TIC55 from *Arabidopsis*. Their discovery will allow creating a *cyp89a9 mes16 tic55* triple mutant in the future to allow investigating potential biological roles of the modified downstream PBs.

One of the future challenges of research on Chl breakdown is to bring the occurrence of the PAO/phyllobilin pathway in an evolutionary context, i.e., to answer the standing question, when during evolution the pathway emerged. Data about the conservation and diversity of parts of the pathway (Gray et al., 2004; Thomas et al., 2009) let us draw the picture of a core first part that takes place within the chloroplast and, thus, likely evolved from genes that were present in the cyanobacterial endosymbiont; and of a second, much more diverse part, i.e., reactions of PB side chain modifications. Particularly, the evolutionary emergence of the latter—combined with a possible physiological relevance of PBs and their modifications (Kräutler, 2016; Moser et al., 2012; Vergeiner et al., 2013)—will be most challenging.

The observation is intriguing that abundance of PBs is low in senescent leaf tissue in many plant species despite the obvious visible signs of Chl breakdown (Das et al., 2018). This raises interesting questions regarding the final fate of Chl during breakdown, which 30 years ago was still enigmatic in many respects (Matile et al., 1987). Although our understanding of Chl breakdown has substantially increased since then (Kräutler, 2016; Kuai et al., 2018), we still may be "naïve" about nature's ability of diversification. This is seen by the fact that successively new types of colorless breakdown products were identified since the first structural description of an NCC almost 30 years ago (Kräutler, 2016; Kräutler et al., 1991), in particular, the *hm*FCCs (Moser, Müller, et al., 2008), the DNCCs (Christ et al., 2013; Losey & Engel, 2001) and the recently identified *i*PBons (Erhart, Vergeiner, et al., 2018), but also minor forms, such as the brightly colored YCCs and PiCCs (Moser, Ulrich, et al., 2008; Ulrich et al., 2011).

Nevertheless, the key step to all these (downstream) Chl catabolites is PAO (Pružinská et al., 2003), making it unlikely that, in higher plants, other pathways of Chl degradation that have been considered in the past (Hynninen, Kaartinen, & Kolehmainen, 2010; Johnson-Flanagan & Spencer, 1996; Matile, 1980) indeed exist besides the PAO/phyllobilin pathway. A notable diversification is the detection in heterotrophic protists of Chl breakdown (and detoxification) through the formation of an enol between the C17 propionic acid and ring E of Pheide *a* (Kashiyama et al., 2012, 2013). These so-called $13^2,17^3$-cyclopheophorbide enols are, in contrast to Chl, Phein and Pheide, nonfluorescent and do not produce singlet oxygen upon light excitation (Falk, Hoornaert, Isenring, & Eschenmoser, 1975). The biochemistry of the formation of these compounds is entirely unknown (Kashiyama & Tamiaki, 2014).

PBs display a remarkable diversity of structural properties, contrasting with the lack of knowledge of the phyllobilin-modifying enzymes, including those that achieve C18 dihydroxylation, $O3^3$ glucosylation and/or malonylation, as well as the remarkable $C12^3$ hypermodifications. PBs are also ubiquitous in nature, where annual Chl degradation amounts to an estimated 1000 million tons on Earth (Hendry et al., 1987). The question has, therefore, been asked repeatedly: "Are PBs not more than only degradation products?" and "Do they have physiological effects in the plants, or else, in the living nature?" (Kräutler, 2016). To date, biological activities of the abundant PBs are hardly explored. However, a striking case for the specific use of Chl-derived linear tetrapyrroles in marine organisms was discovered in the late 1980s (Nakamura, Kishi, Shimomura, Morse, & Hastings, 1989; Nakamura, Musicki, Kishi, & Shimomura, 1988), when it was shown that such bilin-type compounds are produced in different bioluminescing dinoflagellate species, and used as luciferins. Structurally these luciferins were considered to be formed through a remarkable ring opening reaction in the "Western" position of a Chl macrocycle (Topalov & Kishi, 2001), but their biosynthesis still remains to be clarified.

Several fields of research are attractive to address the quest for biological roles of the PBs. Specifically, plant and microbiological investigations concerned with potential functions of PBs in higher plants and other photosynthetic organisms would be of interest, possibly generating a range of agrotechnological and horticultural applications. Furthermore, PBs are part of our nutrition and may have metabolic roles in humans and higher animals. This recently acquired knowledge also calls for interest in physiological, pharmacological and biomedical investigations.

In summary, the biology of Chl catabolism and the potential physiological roles of Chl degradation products still harbor many topical questions. It will be interesting to learn more about enzymes that catalyze the respective PB-modifying reactions, about intracellular and other transport systems for PBs in higher plants, as well as the regulatory networks that control local and general onset and degree of Chl breakdown in response to external and internal signals. The further fate of the "late" PBs and their still unknown physiological roles in higher plants are, likewise, topics that invite further studies.

Acknowledgments

The group in Innsbruck would like to thank Thomas Müller, Simone Moser, Michael Oberhuber, Iris Süssenbacher, Chengjie Li and Clemens Vergeiner for their leading recent contributions to our work on chlorophyll breakdown. We are also grateful for the long lasting generous support by the Austrian Science Foundation (FWF), in particular from the current FWF project P-28522-N28 (to B.K.). The group in Zurich acknowledges the financial support by the Swiss National Science Foundation, currently with project 31003A_172977 (to S.H.).

References

Abbas, M., Hernandez-Garcia, J., Pollmann, S., Samodelov, S. L., Kolb, M., Friml, J., et al. (2018). Auxin methylation is required for differential growth in Arabidopsis. *Proceedings of the National Academy of Sciences of the United States of America, 115*, 6864–6869.

Abeles, F. B., Dunn, L. J., Morgens, P., Callahan, A., Dinterman, R. E., & Schmidt, J. (1988). Induction of 33-Kd and 60-Kd peroxidases during ethylene-induced senescence of cucumber cotyledons. *Plant Physiology, 87*, 609–615.

Armstead, I., Donnison, I., Aubry, S., Harper, J., Hörtensteiner, S., James, C., et al. (2006). From crop to model to crop: Identifying the genetic basis of the staygreen mutation in the *Lolium/Festuca* forage and amenity grasses. *New Phytologist, 172*, 592–597.

Aubry, S., Fankhauser, N., Ovinnikov, S., Zienkiewicz, K., Feussner, I., & Hörtensteiner, S. (2018). Pheophorbide *a*, a chlorophyll catabolite may regulate jasmonate signalling during dark-induced senescence in Arabidopsis. *BioRxiv*, 486886.

Azoulay Shemer, T., Harpaz-Saad, S., Belausov, E., Lovat, N., Krokhin, O., Spicer, V., et al. (2008). *Citrus* chlorophyllase dynamics at ethylene-induced fruit color-break: A study of chlorophyllase expression, post-translational processing kinetics and *in-situ* intracellular localization. *Plant Physiology, 148*, 108–118.

Azoulay-Shemer, T., Harpaz-Saad, S., Cohen-Peer, R., Mett, A., Spicer, V., Lovat, N., et al. (2011). Dual N- and C-terminal processing of citrus chlorophyllase precursor within the plastid membranes leads to the mature enzyme. *Plant and Cell Physiology, 52*, 70–83.

Balsera, M., Soll, J., & Buchanan, B. B. (2010). Redox extends its regulatory reach to chloroplast protein import. *Trends in Plant Science, 15*, 515–521.

Banala, S., Moser, S., Müller, T., Kreutz, C. R., Holzinger, A., Lütz, C., et al. (2010). Hypermodified fluorescent chlorophyll catabolites: Source of blue luminescence in senescent leaves. *Angewandte Chemie International Edition, 49*, 5174–5177.

Barsan, C., Sanchez-Bel, P., Rombaldi, C., Egea, I., Rossignol, M., Kuntz, M., et al. (2010). Characteristics of the tomato chromoplast revealed by proteomic analysis. *Journal of Experimental Botany, 61*, 2413–2431.

Barsan, C., Zouine, M., Maza, E., Bian, W., Egea, I., Rossignol, M., et al. (2012). Proteomic analysis of chloroplast-to-chromoplast transition in tomato reveals metabolic shifts coupled with disrupted thylakoid biogenesis machinery and elevated energy-production components. *Plant Physiology, 160*, 708–725.

Bartsch, S., Monnet, J., Selbach, K., Quigley, F., Gray, J., von Wettstein, D., et al. (2008). Three thioredoxin targets in the inner envelope membrane of chloroplasts function in protein import and chlorophyll metabolism. *Proceedings of the National Academy of Sciences of the United States of America, 105*, 4933–4938.

Berghold, J., Breuker, K., Oberhuber, M., Hörtensteiner, S., & Kräutler, B. (2002). Chlorophyll breakdown in spinach: On the structure of five nonfluorescent chlorophyll catabolites. *Photosynthesis Research, 74*, 109–119.

Berghold, J., Eichmüller, C., Hörtensteiner, S., & Kräutler, B. (2004). Chlorophyll breakdown in tobacco: On the structure of two nonfluorescent chlorophyll catabolites. *Chemistry & Biodiversity*, *1*, 657–668.

Berghold, J., Müller, T., Ulrich, M., Hörtensteiner, S., & Kräutler, B. (2006). Chlorophyll breakdown in maize: On the structure of two nonfluorescent chlorophyll catabolites. *Monatshefte für Chemie*, *137*, 751–763.

Biswal, U. C., & Biswal, B. (1984). Photocontrol of leaf senescence. *Photochemistry and Photobiology*, *39*, 875–879.

Bleecker, A. B., Estelle, M. A., Somerville, C., & Kende, H. (1988). Insensitivity to ethylene conferred by a dominant mutation in *Arabidopsis thaliana*. *Science*, *241*, 1086–1089.

Boij, P., Patel, R., Garcia, C., Jarvis, P., & Aronsson, H. (2009). In vivo studies on the roles of Tic55-related proteins in chloroplast protein import in *Arabidopsis thaliana*. *Molecular Plant*, *2*, 1397–1409.

Bortlik, K.-H., Peisker, C., & Matile, P. (1990). A novel type of chlorophyll catabolite in senescent barley leaves. *Journal of Plant Physiology*, *136*, 161–165.

Breeze, E., Harrison, E., McHattie, S., Hughes, L., Hickman, R., Hill, C., et al. (2011). High-resolution temporal profiling of transcripts during *Arabidopsis* leaf senescence reveals a distinct chronology of processes and regulation. *The Plant Cell*, *23*, 873–894.

Brown, S. B., Houghton, J. D., & Hendry, G. A. F. (1991). Chlorophyll breakdown. In H. Scheer (Ed.), *Chlorophylls* (pp. 465–489). Boca Raton: CRC Press.

Büchert, A. M., Civello, P. M., & Martínez, G. A. (2011). Chlorophyllase versus pheophytinase as candidates for chlorophyll dephytilation during senescence of broccoli. *Journal of Plant Physiology*, *168*, 337–343.

Calibe, A., Grimm, R., Kaiser, G., Lübeck, J., Soll, J., & Heins, L. (1997). The chloroplastic protein import machinery contains a Rieske-type iron-sulfur cluster and a mononuclear iron-binding protein. *The EMBO Journal*, *16*, 7342–7350.

Chao, Q. M., Rothenberg, M., Solano, R., Roman, G., Terzaghi, W., & Ecker, J. R. (1997). Activation of the ethylene gas response pathway in Arabidopsis by the nuclear protein ETHYLENE-INSENSITIVE3 and related proteins. *Cell*, *89*, 1133–1144.

Cheng, Y. D., & Guan, J. F. (2014). Involvement of pheophytinase in ethylene-mediated chlorophyll degradation in the peel of harvested 'Yali' pear. *Journal of Plant Growth Regulation*, *33*, 364–372.

Christ, B., Hauenstein, M., & Hörtensteiner, S. (2016). A liquid chromatography-mass spectrometry platform for the analysis of phyllobilins, the major degradation products of chlorophyll in *Arabidopsis thaliana*. *The Plant Journal*, *88*, 505–518.

Christ, B., Schelbert, S., Aubry, S., Süssenbacher, I., Müller, T., Kräutler, B., et al. (2012). MES16, a member of the methylesterase protein family, specifically demethylates fluorescent chlorophyll catabolites during chlorophyll breakdown in Arabidopsis. *Plant Physiology*, *158*, 628–641.

Christ, B., Süssenbacher, I., Moser, S., Bichsel, N., Egert, A., Müller, T., et al. (2013). Cytochrome P450 CYP89A9 is involved in the formation of major chlorophyll catabolites during leaf senescence in *Arabidopsis*. *The Plant Cell*, *25*, 1868–1880.

Curty, C., & Engel, N. (1996). Detection, isolation and structure elucidation of a chlorophyll *a* catabolite from autumnal senescent leaves of *Cercidiphyllum japonicum*. *Phytochemistry*, *42*, 1531–1536.

Curty, C., Engel, N., & Gossauer, A. (1995). Evidence for a monooxygenase-catalyzed primary process in the catabolism of chlorophyll. *FEBS Letters*, *364*, 41–44.

Das, A., Christ, B., & Hörtensteiner, S. (2018). Characterization of the pheophorbide *a* oxygenase/phyllobilin pathway of chlorophyll breakdown in grasses. *Planta*, *248*, 875–892.

Delmas, F., Sankaranarayanan, S., Deb, S., Widdup, E., Bournonville, C., Bollier, N., et al. (2013). ABI3 controls embryo degreening through Mendel's I locus. *Proceedings of the National Academy of Sciences of the United States of America*, *110*, E3888–E3894.

Djapic, N., Pavlovic, M., Arsovski, S., & Vujic, G. (2009). *Chlorophyll biodegradation product from Hamamelis virginiana* autumnal leaves. *Vol. 60* (pp. 398–402). Revista De Chimie.

Düggelin, T., Schellenberg, M., Bortlik, K.-H., & Matile, P. (1988). Vacuolar location of lipofuscin- and proline-like compounds in senescent barley leaves. *Journal of Plant Physiology, 133*, 492–497.

Engel, N., Curty, C., & Gossauer, A. (1996). Chlorophyll catabolism in *Chlorella protothecoides*. 8. Facts and artefacts. *Plant Physiology and Biochemistry, 34*, 77–83.

Engel, N., Gossauer, A., Gruber, K., & Kratky, C. (1993). X-ray molecular structure of a red bilin derivative from *Chlorella protothecoides*. *Helvetica Chimica Acta, 76*, 1–3.

Erhart, T., Mittelberger, C., Liu, X. J., Podewitz, M., Li, C. J., Scherzer, G., et al. (2018). Novel types of hypermodified fluorescent phyllobilins from breakdown of chlorophyll in senescent leaves of grapevine (*Vitis vinifera*). *Chemistry—A European Journal, 24*, 17268–17279.

Erhart, T., Mittelberger, C., Vergeiner, C., Scherzer, G., Holzner, B., Robatscher, P., et al. (2016). Chlorophyll catabolites in senescent leaves of the plum tree (*Prunus domestica*). *Chemistry & Biodiversity, 13*, 1441–1453.

Erhart, T., Vergeiner, S., Kreutz, C., Kräutler, B., & Müller, T. (2018). Chlorophyll breakdown in a fern—Discovery of phyllobilin isomers with a rearranged carbon skeleton. *Angewandte Chemie International Edition, 57*, 14937–14941.

Falk, H. (1989). *Chemistry of linear oligopyrroles and bile pigments*. Wien: Springer.

Falk, H., Hoornaert, G., Isenring, H. P., & Eschenmoser, A. (1975). Enol derivatives in chlorophyll series—Preparation of 13^2,17^3-cyclopheophorbide enols. *Helvetica Chimica Acta, 58*, 2347–2357.

Fang, C., Li, C. C., Li, W. Y., Wang, Z., Zhou, Z. K., Shen, Y. T., et al. (2014). Concerted evolution of D1 and D2 to regulate chlorophyll degradation in soybean. *The Plant Journal*, 77, 700–712.

Frankenberg, N., Mukougawa, K., Kohchi, T., & Lagarias, J. C. (2001). Functional genomic analysis of the HY2 family of ferredoxin-dependent bilin reductases from oxygenic photosynthetic organisms. *The Plant Cell, 13*, 965–978.

Gao, S., Gao, J., Zhu, X., Song, Y., Li, Z., Ren, G., et al. (2016). ABF2, ABF3, and ABF4 promote ABA-mediated chlorophyll degradation and leaf senescence by transcriptional activation of chlorophyll catabolic genes and senescence-associated genes in Arabidopsis. *Molecular Plant, 9*, 1272–1285.

Gao, C. F., Wang, Y., Shen, Y., Yan, D., He, X., Dai, J. B., et al. (2014). Oil accumulation mechanisms of the oleaginous microalga *Chlorella protothecoides* revealed through its genome, transcriptomes, and proteomes. *BMC Genomics, 15*, 582.

Gepstein, S., Sabehi, G., Carp, M. J., Hajouj, T., Nesher, M. F. O., Yariv, I., et al. (2003). Large-scale identification of leaf senescence-associated genes. *The Plant Journal, 36*, 629–642.

Gepstein, S., & Thimann, K. V. (1980). Changes in the abscisic-acid content of oat leaves during senescence. *Proceedings of the National Academy of Sciences of the United States of America*, 77, 2050–2053.

Gomez, M. D., Vera-Sirera, F., & Perez-Amador, M. A. (2014). Molecular programme of senescence in dry and fleshy fruits. *Journal of Experimental Botany, 65*, 4515–4526.

Gray, J., Wardzala, E., Yang, M., Reinbothe, S., Haller, S., & Pauli, F. (2004). A small family of LLS1-related non-heme oxygenases in plants with an origin amongst oxygenic photosynthesizers. *Plant Molecular Biology, 54*, 39–54.

Greenberg, J. T., & Ausubel, F. M. (1993). *Arabidopsis* mutants compromised for the control of cellular damage during pathogenesis and aging. *The Plant Journal, 4*, 327–341.

Greenberg, J. T., Guo, A., Klessig, D. F., & Ausubel, F. M. (1994). Programmed cell death in plants: A pathogen-triggered response activated coordinately with multiple defense functions. *Cell*, 77, 551–563.

Guo, Y. F., & Gan, S. S. (2012). Convergence and divergence in gene expression profiles induced by leaf senescence and 27 senescence-promoting hormonal, pathological and environmental stress treatments. *Plant, Cell & Environment, 35*, 644–655.

Guyer, L., Salinger, K., Krügel, U., & Hörtensteiner, S. (2018). Catalytic and structural properties of pheophytinase, the phytol esterase involved in chlorophyll breakdown. *Journal of Experimental Botany, 69*, 879–889.

Guyer, L., Schelbert Hofstetter, S., Christ, B., Silverstre Lira, B., Rossi, M., & Hörtensteiner, S. (2014). Different mechanisms are responsible for chlorophyll dephytylation during fruit ripening and leaf senescence in tomato. *Plant Physiology, 166*, 44–56.

Hauenstein, M., Christ, B., Das, A., Aubry, S., & Hörtensteiner, S. (2016). A role for TIC55 as a hydroxylase of phyllobilins, the products of chlorophyll breakdown during plant senescence. *The Plant Cell, 28*, 2510–2527.

He, Y. H., Fukushige, H., Hildebrand, D. F., & Gan, S. S. (2002). Evidence supporting a role of jasmonic acid in Arabidopsis leaf senescence. *Plant Physiology, 128*, 876–884.

Hendry, G. A. F., Houghton, J. D., & Brown, S. B. (1987). The degradation of chlorophyll—A biological enigma. *New Phytologist, 107*, 255–302.

Hinder, B., Schellenberg, M., Rodoni, S., Ginsburg, S., Vogt, E., Martinoia, E., et al. (1996). How plants dispose of chlorophyll catabolites. Directly energized uptake of tetrapyrrolic breakdown products into isolated vacuoles. *The Journal of Biological Chemistry, 271*, 27233–27236.

Hirashima, M., Tanaka, R., & Tanaka, A. (2009). Light-independent cell death induced by accumulation of pheophorbide *a* in *Arabidopsis thaliana*. *Plant and Cell Physiology, 50*, 719–729.

Horie, Y., Ito, H., Kusaba, M., Tanaka, R., & Tanaka, A. (2009). Participation of chlorophyll *b* reductase in the initial step of the degradation of light-harvesting chlorophyll *a*/*b*-protein complexes in *Arabidopsis*. *The Journal of Biological Chemistry, 284*, 17449–17456.

Hörtensteiner, S. (1998). NCC malonyltransferase catalyses the final step of chlorophyll breakdown in rape (*Brassica napus*). *Phytochemistry, 49*, 953–956.

Hörtensteiner, S. (2004). The loss of green color during chlorophyll degradation—A prerequisite to prevent cell death? *Planta, 219*, 191–194.

Hörtensteiner, S. (2006). Chlorophyll degradation during senescence. *Annual Review of Plant Biology, 57*, 55–77.

Hörtensteiner, S. (2009). Stay-green regulates chlorophyll and chlorophyll-binding protein degradation during senescence. *Trends in Plant Science, 14*, 155–162.

Hörtensteiner, S. (2013). Update on the biochemistry of chlorophyll breakdown. *Plant Molecular Biology, 82*, 505–517.

Hörtensteiner, S., & Kräutler, B. (2011). Chlorophyll breakdown in higher plants. *Biochimica et Biophysica Acta, 1807*, 977–988.

Hörtensteiner, S., Rodoni, S., Schellenberg, M., Vicentini, F., Nandi, O. I., Qiu, Y.-L., et al. (2000). Evolution of chlorophyll degradation: The significance of RCC reductase. *Plant Biology, 2*, 63–67.

Hörtensteiner, S., Vicentini, F., & Matile, P. (1995). Chlorophyll breakdown in senescent cotyledons of rape, *Brassica napus* L.: Enzymatic cleavage of phaeophorbide *a in vitro*. *New Phytologist, 129*, 237–246.

Hörtensteiner, S., Wüthrich, K. L., Matile, P., Ongania, K.-H., & Kräutler, B. (1998). The key step in chlorophyll breakdown in higher plants. Cleavage of pheophorbide *a* macrocycle by a monooxygenase. *The Journal of Biological Chemistry, 273*, 15335–15339.

Hynninen, P. H., Kaartinen, V., & Kolehmainen, E. (2010). Horseradish peroxidase-catalyzed oxidation of chlorophyll *a* with hydrogen peroxide. Characterization of the products and mechanism of the reaction. *Biochimica et Biophysica Acta, 1797*, 531–542.

Inada, N., Sakai, A., Kuroiwa, H., & Kuroiwa, T. (1998). Three-dimensional analysis of the senescence program in rice (Oryza sativa L.) coleoptiles. Investigations of tissues and cells by fluorescence microscopy. *Planta, 205*, 153–164.

Jiang, H., Li, M., Liang, N., Yan, H., Wei, Y., Xu, X., et al. (2007). Molecular cloning and function analysis of the *stay green* gene in rice. *The Plant Journal, 52*, 197–209.

Jiang, Y. J., Liang, G., Yang, S. Z., & Yu, D. Q. (2014). Arabidopsis WRKY57 functions as a node of convergence for jasmonic acid- and auxin-mediated signaling in jasmonic acid-induced leaf senescence. *The Plant Cell, 26*, 230–245.

Jockusch, S., Turro, N. J., Banala, S., & Kräutler, B. (2014). Photochemical studies of a fluorescent chlorophyll catabolite—Source of bright blue fluorescence in plant tissue and efficient sensitizer of singlet oxygen. *Photochemical & Photobiological Sciences, 13*, 407–411.

Johnson-Flanagan, A. M., & Spencer, M. S. (1996). Chlorophyllase and peroxidase activity during degreening of maturing canola (*Brassica napus*) and mustard (*Brassica juncea*) seed. *Physiologia Plantarum, 97*, 353–359.

Kahlau, S., & Bock, R. (2008). Plastid transcriptomics and translatomics of tomato fruit development and chloroplast-to-chromoplast differentiation: Chromoplast gene expression largely serves the production of a single protein. *The Plant Cell, 20*, 856–874.

Kang, J., Park, J., Choi, H., Burla, B., Kretzschmar, T., Lee, Y., et al. (2011). Plant ABC transporters. *The Arabidopsis Book, 9*, e0153.

Kashiyama, Y., & Tamiaki, H. (2014). Risk management by organisms of the phototoxicity of chlorophylls. *Chemistry Letters, 43*, 148–156.

Kashiyama, Y., Yokoyama, A., Kinoshita, Y., Shoji, S., Miyashiya, H., Shiratori, T., et al. (2012). Ubiquity and quantitative significance of detoxification catabolism of chlorophyll associated with protistan herbivory. *Proceedings of the National Academy of Sciences of the United States of America, 109*, 17328–17335.

Kashiyama, Y., Yokoyama, A., Shiratori, T., Inouye, I., Kinoshita, Y., Mizoguchi, T., et al. (2013). 13^2,17^3-Cyclopheophorbide *b* enol as a catabolite of chlorophyll *b* in phycophagy by protists. *FEBS Letters, 587*, 2578–2583.

Kräutler, B. (2014). Phyllobilins—The abundant bilin-type tetrapyrrolic catabolites of the green plant pigment chlorophyll. *Chemical Society Reviews, 43*, 6227–6238.

Kräutler, B. (2016). Breakdown of chlorophyll in higher plants—Phyllobilins as abundant, yet hardly visible signs of ripening, senescence, and cell death. *Angewandte Chemie International Edition, 55*, 4882–4907.

Kräutler, B. (2019). How chemistry has helped to decipher a striking biological enigma. *Synlett, 30*, 263–274.

Kräutler, B., Banala, S., Moser, S., Vergeiner, C., Müller, T., Lütz, C., et al. (2010). A novel blue fluorescent chlorophyll catabolite accumulates in senescent leaves of the peace lily and indicates a divergent path of chlorophyll breakdown. *FEBS Letters, 584*, 4215–4221.

Kräutler, B., & Hörtensteiner, S. (2014). Chlorophyll breakdown: Chemistry, biochemistry and biology. In G. C. Ferreira, K. M. Kadish, K. M. Smith, & R. Guilard (Eds.), *Vol. 28. Handbook of porphyrin science—Chlorophyll, photosynthesis and bio-inspired energy* (pp. 117–185). Singapore: World Scientific Publishing.

Kräutler, B., Jaun, B., Amrein, W., Bortlik, K., Schellenberg, M., & Matile, P. (1992). Breakdown of chlorophyll: Constitution of a secoporphinoid chlorophyll catabolite isolated from senescent barley leaves. *Plant Physiology and Biochemistry, 30*, 333–346.

Kräutler, B., Jaun, B., Bortlik, K.-H., Schellenberg, M., & Matile, P. (1991). On the enigma of chlorophyll degradation: The constitution of a secoporphinoid catabolite. *Angewandte Chemie International Edition, 30*, 1315–1318.

Kräutler, B., & Matile, P. (1999). Solving the riddle of chlorophyll breakdown. *Accounts of Chemical Research, 32*, 35–43.

Kräutler, B., Mühlecker, W., Anderl, M., & Gerlach, B. (1997). Breakdown of chlorophyll: Partial synthesis of a putative intermediary catabolite. *Helvetica Chimica Acta, 80*, 1355–1362.

Krupinska, K., Biswal, U. C., & Biswal, B. (2013). The dynamic role of chloroplasts in integrating plant growth and development. In B. Biswal, K. Krupinska, & U. C. Biswal (Eds.), *Vol. 36. Plastid development in leaves during growth and senescence* (pp. 3–16). Dordrecht, The Netherlands: Springer.

Kuai, B., Chen, J., & Hörtensteiner, S. (2018). The biochemistry and molecular biology of chlorophyll breakdown. *Journal of Experimental Botany*, *69*, 751–767.

Kusaba, M., Ito, H., Morita, R., Iida, S., Sato, Y., Fujimoto, M., et al. (2007). Rice NON-YELLOW COLORING1 is involved in light-harvesting complex II and grana degradation during leaf senescence. *The Plant Cell*, *19*, 1362–1375.

Li, L., Ban, Z., Limwachiranon, J., & Luo, Z. (2017). Proteomic studies on fruit ripening and senescence. *Critical Reviews in Plant Sciences*, *36*, 116–127.

Li, C. J., Erhart, T., Liu, X. J., & Kräutler, B. (2019). Yellow dioxobilin-type tetrapyrroles from chlorophyll breakdown in higher plants—A new class of colored phyllobilins. *Chemistry—A European Journal*, *25*, 4052–4057.

Li, C. J., & Kräutler, B. (2019). A pink colored dioxobilin-type phyllobilin from breakdown of chlorophyll. *Chemical Monthly*. https://doi.org/10.1007/s00706-019-02396-5. in press.

Li, H. M., & Pu, H. (2016). Crystal structure of methylesterase family member 16 (MES16) from *Arabidopsis thaliana*. *Biochemical and Biophysical Research Communications*, *474*, 226–231.

Li, C. J., Ulrich, M., Liu, X. J., Wurst, K., Müller, T., & Kräutler, B. (2014). Blue transition metal complexes of a natural bilin-type chlorophyll catabolite. *Chemical Science*, *5*, 3388–3395.

Li, Z. P., Wu, S. X., Chen, J. Y., Wang, X. Y., Gao, J., Ren, G. D., et al. (2017). NYEs/SGRs-mediated chlorophyll degradation is critical for detoxification during seed maturation in Arabidopsis. *The Plant Journal*, *92*, 650–661.

Li, C. J., Wurst, K., Berghold, J., Podewitz, M., Liedl, K. R., & Kräutler, B. (2018). Pyro-phyllobilins: Elusive chlorophyll catabolites lacking a critical carboxylate function of the natural chlorophylls. *Chemistry—A European Journal*, *24*, 2987–2998.

Li, C. J., Wurst, K., Jockusch, S., Gruber, K., Podewitz, M., Liedl, K. R., et al. (2016). Chlorophyll-derived yellow phyllobilins of higher plants as medium-responsive chiral photoswitches. *Angewandte Chemie International Edition*, *55*, 15760–15765.

Liebsch, D., & Keech, O. (2016). Dark-induced leaf senescence: New insights into a complex light-dependent regulatory pathway. *New Phytologist*, *212*, 563–570.

Lim, P. O., Kim, H. J., & Nam, H. G. (2007). Leaf senescence. *Annual Review of Plant Biology*, *58*, 115–136.

Lin, Y. P., Wu, M. C., & Charng, Y. Y. (2016). Identification of a chlorophyll dephytylase involved in chlorophyll turnover in Arabidopsis. *The Plant Cell*, *28*, 2974–2990.

Liu, T., Longhurst, A. D., Talavera-Rauh, F., Hokin, S. A., & Barton, M. K. (2016). The Arabidopsis transcription factor ABIG1 relays ABA signaled growth inhibition and drought induced senescence. *eLife*, *5*, e13768.

Losey, F. G., & Engel, N. (2001). Isolation and characterization of a urobilinogenoidic chlorophyll catabolite from *Hordeum vulgare* L. *The Journal of Biological Chemistry*, *276*, 27233–27236.

Lu, Y.-P., Li, Z.-S., Drozdowicz, Y.-M., Hörtensteiner, S., Martinoia, E., & Rea, P. A. (1998). AtMRP2, an Arabidopsis ATP binding cassette transporter able to transport glutathione *S*-conjugates and chlorophyll catabolites: Functional comparisons with AtMRP1. *The Plant Cell*, *10*, 267–282.

Mach, J. M., Castillo, A. R., Hoogstraten, R., & Greenberg, J. T. (2001). The *Arabidopsis*-accelerated cell death gene *ACD2* encodes red chlorophyll catabolite reductase and suppresses the spread of disease symptoms. *Proceedings of the National Academy of Sciences of the United States of America*, *98*, 771–776.

Matile, P. (1980). Catabolism of chlorophyll: Involvement of peroxidase? *Zeitschrift für Pflanzenphysiologie*, *99*, 457–478.

Matile, P. (1987). Seneszenz bei pflanzen und ihre bedeutung für den stickstoffhaushalt. *Chimia*, *41*, 376–381.

Matile, P. (2000). Biochemistry of Indian summer: Physiology of autumnal leaf coloration. *Experimental Gerontology*, *35*, 145–158.

Matile, P., Ginsburg, S., Schellenberg, M., & Thomas, H. (1987). Catabolites of chlorophyll in senescent leaves. *Journal of Plant Physiology, 129*, 219–228.

Matile, P., Ginsburg, S., Schellenberg, M., & Thomas, H. (1988). Catabolites of chlorophyll in senescing barley leaves are localized in the vacuoles of mesophyll cells. *Proceedings of the National Academy of Sciences of the United States of America, 85*, 9529–9532.

Matile, P., Hörtensteiner, S., & Thomas, H. (1999). Chlorophyll degradation. *Annual Review of Plant Physiology and Plant Molecular Biology, 50*, 67–95.

Matile, P., Hörtensteiner, S., Thomas, H., & Kräutler, B. (1996). Chlorophyll breakdown in senescent leaves. *Plant Physiology, 112*, 1403–1409.

Matile, P., Schellenberg, M., & Peisker, C. (1992). Production and release of a chlorophyll catabolite in isolated senescent chloroplasts. *Planta, 187*, 230–235.

Mecey, C., Hauck, P., Trapp, M., Pumplin, N., Plovanich, A., Yao, J., et al. (2011). A critical role of *STAYGREEN*/Mendel's *I* locus in controlling disease symptom development during *Pseudomonas syringae* pv *tomato* infection of Arabidopsis. *Plant Physiology, 157*, 1965–1974.

Meguro, M., Ito, H., Takabayashi, A., Tanaka, R., & Tanaka, A. (2011). Identification of the 7-hydroxymethyl chlorophyll *a* reductase of the chlorophyll cycle in *Arabidopsis*. *The Plant Cell, 23*, 3442–3453.

Mendel, G. (1866). Versuche über pflanzenhybriden. *Vol. 4* (pp. 3–47). Brünn: Verhandlungen des naturforschenden Vereines.

Mittelberger, C., Yalcinkaya, H., Pichler, C., Gasser, J., Scherzer, G., Erhart, T., et al. (2017). Pathogen-induced leaf chlorosis: Products of chlorophyll breakdown found in degreened leaves of phytoplasma-infected apple (*Malus x domestica* Borkh.) and apricot (*Prunus armeniaca* L.) trees relate to the pheophorbide *a* oxygenase/phyllobilin pathway. *Journal of Agricultural and Food Chemistry, 65*, 2651–2660.

Mizzotti, C., Rotasperti, L., Moretto, M., Tadini, L., Resentini, F., Galliani, B. M., et al. (2018). Time-course transcriptome analysis of Arabidopsis siliques discloses genes essential for fruit development and maturation. *Plant Physiology, 178*, 1249–1268.

Morel, A. (2006). Meeting the challenge of monitoring chlorophyll in the ocean from outer space. In B. Grimm, R. Porra, W. Rüdiger, & H. Scheer (Eds.), *Vol. 25. Chlorophylls and bacteriochlorophylls: Biochemistry, biophysics, functions and applications* (pp. 521–534). Dordrecht, The Netherlands: Springer.

Morita, R., Sato, Y., Masuda, Y., Nishimura, M., & Kusaba, M. (2009). Defect in non-yellow coloring 3, an a/b hydrolase-fold family protein, causes a stay-green phenotype during leaf senescence in rice. *The Plant Journal, 59*, 940–952.

Moser, S., Müller, T., Ebert, M. O., Jockusch, S., Turro, N. J., & Kräutler, B. (2008). Blue luminescence of ripening bananas. *Angewandte Chemie International Edition, 47*, 8954–8957.

Moser, S., Müller, T., Holzinger, A., Lutz, C., Jockusch, S., Turro, N. J., et al. (2009). Fluorescent chlorophyll catabolites in bananas light up blue halos of cell death. *Proceedings of the National Academy of Sciences of the United States of America, 106*, 15538–15543.

Moser, S., Müller, T., Holzinger, A., Lutz, C., & Kräutler, B. (2012). Structures of chlorophyll catabolites in bananas (*Musa acuminata*) reveal a split path of chlorophyll breakdown in a ripening fruit. *Chemistry—A European Journal, 18*, 10873–10885.

Moser, S., Scherzer, G., & Kräutler, B. (2017). On the nature of isomeric nonfluorescent chlorophyll catabolites in leaves and fruit—A study with a ubiquitous phylloleucobilin and its main isomerization product. *Chemistry & Biodiversity, 14*, e1700368.

Moser, S., Ulrich, M., Müller, T., & Kräutler, B. (2008). A yellow chlorophyll catabolite is a pigment of the fall colours. *Phytochemical & Phytobiological Sciences, 7*, 1577–1581.

Mühlecker, W., & Kräutler, B. (1996). Breakdown of chlorophyll: Constitution of non-fluorescing chlorophyll-catabolites from senescent cotyledons of the dicot rape. *Plant Physiology and Biochemistry, 34*, 61–75.

Mühlecker, W., Kräutler, B., Ginsburg, S., & Matile, P. (1993). Breakdown of chlorophyll: A tetrapyrrolic chlorophyll catabolite from senescent rape leaves. *Helvetica Chimica Acta, 76*, 2976–2980.

Mühlecker, W., Kräutler, B., Moser, D., Matile, P., & Hörtensteiner, S. (2000). Breakdown of chlorophyll: A fluorescent chlorophyll catabolite from sweet pepper (*Capsicum annuum*). *Helvetica Chimica Acta, 83*, 278–286.

Mühlecker, W., Ongania, K.-H., Kräutler, B., Matile, P., & Hörtensteiner, S. (1997). Tracking down chlorophyll breakdown in plants: Elucidation of the constitution of a 'fluorescent' chlorophyll catabolite. *Angewandte Chemie International Edition, 36*, 401–404.

Müller, T., Moser, S., Ongania, K.-H., Pružinská, A., Hörtensteiner, S., & Kräutler, B. (2006). A divergent path of chlorophyll breakdown in the model plant *Arabidopsis thaliana*. *Chembiochem, 7*, 40–42.

Müller, T., Rafelsberger, M., Vergeiner, C., & Kräutler, B. (2011). A dioxobilane as product of a divergent path of chlorophyll breakdown in Norway maple. *Angewandte Chemie International Edition, 50*, 10724–10727.

Müller, T., Ulrich, M., Ongania, K. H., & Kräutler, B. (2007). Colorless tetrapyrrolic chlorophyll catabolites found in ripening fruit are effective antioxidants. *Angewandte Chemie International Edition, 46*, 8699–8702.

Mur, L. A. J., Aubry, S., Mondhe, M., Kingston-Smith, A., Gallagher, J., Timms-Taravella, E., et al. (2010). Accumulation of chlorophyll catabolites photosensitizes the hypersensitive response elicited by *Pseudomonas syringae* in Arabidopsis. *New Phytologist, 188*, 161–174.

Nakajima, S., Ito, H., Tanaka, R., & Tanaka, A. (2012). Chlorophyll *b* reductase plays an essential role in maturation and storability of Arabidopsis seeds. *Plant Physiology, 160*, 261–273.

Nakamura, H., Kishi, Y., Shimomura, O., Morse, D., & Hastings, J. W. (1989). Structure of dinoflagellate luciferin and its enzymatic and nonenzymatic air-oxidation products. *Journal of the American Chemical Society, 111*, 7607–7611.

Nakamura, H., Musicki, B., Kishi, Y., & Shimomura, O. (1988). Structure of the light emitter in krill (*Euphausia pacifica*) bioluminescence. *Journal of the American Chemical Society, 110*, 2683–2685.

Oberhuber, M., Berghold, J., Breuker, K., Hörtensteiner, S., & Kräutler, B. (2003). Breakdown of chlorophyll: A nonenzymatic reaction accounts for the formation of the colorless "nonfluorescent" chlorophyll catabolites. *Proceedings of the National Academy of Sciences of the United States of America, 100*, 6910–6915.

Oberhuber, M., Berghold, J., & Kräutler, B. (2008). Chlorophyll breakdown by a biomimetic route. *Angewandte Chemie International Edition, 47*, 3057–3061.

Oberhuber, M., Berghold, J., Mühlecker, W., Hörtensteiner, S., & Kräutler, B. (2001). Chlorophyll breakdown—On a nonfluorescent chlorophyll catabolite from spinach. *Helvetica Chimica Acta, 84*, 2615–2627.

Park, S.-Y., Yu, J.-W., Park, J.-S., Li, J., Yoo, S.-C., Lee, N.-Y., et al. (2007). The senescence-induced staygreen protein regulates chlorophyll degradation. *The Plant Cell, 19*, 1649–1664.

Pružinská, A., Anders, I., Aubry, S., Schenk, N., Tapernoux-Lüthi, E., Müller, T., et al. (2007). In vivo participation of red chlorophyll catabolite reductase in chlorophyll breakdown. *The Plant Cell, 19*, 369–387.

Pružinská, A., Anders, I., Tanner, G., Roca, M., & Hörtensteiner, S. (2003). Chlorophyll breakdown: Pheophorbide *a* oxygenase is a Rieske-type iron-sulfur protein, encoded by the *accelerated cell death 1* gene. *Proceedings of the National Academy of Sciences of the United States of America, 100*, 15259–15264.

Pružinská, A., Tanner, G., Aubry, S., Anders, I., Moser, S., Müller, T., et al. (2005). Chlorophyll breakdown in senescent Arabidopsis leaves: Characterization of chlorophyll catabolites and of chlorophyll catabolic enzymes involved in the degreening reaction. *Plant Physiology, 139*, 52–63.

Qi, T. C., Wang, J. J., Huang, H., Liu, B., Gao, H., Liu, Y. L., et al. (2015). Regulation of jasmonate-induced leaf senescence by antagonism between bHLH subgroup IIIe and IIId factors in Arabidopsis. *The Plant Cell, 27*, 1634–1649.

Qiu, K., Li, Z., Yang, Z., Chen, J., Wu, S., Zhu, X., et al. (2015). EIN3 and ORE1 accelerate degreening during ethylene-mediated leaf senescence by directly activating chlorophyll catabolic genes in Arabidopsis. *PLoS Genetics, 11*, e1005399.

Raab, S., Drechsel, G., Zarepour, M., Hartung, W., Koshiba, T., Bittner, F., et al. (2009). Identification of a novel E3 ubiquitin ligase that is required for suppression of premature senescence in Arabidopsis. *The Plant Journal, 59*, 39–51.

Ren, G., An, K., Liao, Y., Zhou, X., Cao, Y., Zhao, H., et al. (2007). Identification of a novel chloroplast protein AtNYE1 regulating chlorophyll degradation during leaf senescence in Arabidopsis. *Plant Physiology, 144*, 1429–1441.

Ren, G. D., Zhou, Q., Wu, S. X., Zhang, Y. F., Zhang, L. G., Huang, J. R., et al. (2010). Reverse genetic identification of CRN1 and its distinctive role in chlorophyll degradation in *Arabidopsis*. *Journal of Integrative Plant Biology, 52*, 496–504.

Rios, J. J., Perez-Galvez, A., & Roca, M. (2014). Non-fluorescent chlorophyll catabolites in quince fruits. *Food Research International, 65*, 255–262.

Rios, J. J., Roca, M., & Perez-Galvez, A. (2014). Nonfluorescent chlorophyll catabolites in loquat fruits (*Eriobotrya japonica* Lindl.). *Journal of Agricultural and Food Chemistry, 62*, 10576–10584.

Roca, M., Rios, J. J., Chahuaris, A., & Perez-Galvez, A. (2017). Non-fluorescent and yellow chlorophyll catabolites in Japanese plum fruits (*Prunus salicina*, Lindl.). *Food Research International, 100*, 332–338.

Roca, M., Rios, J. J., & Perez-Galvez, A. (2018). Mass spectrometry: The indispensable tool for plant metabolomics of colourless chlorophyll catabolites. *Phytochemistry Reviews, 17*, 453–468.

Rodoni, S., Mühlecker, W., Anderl, M., Kräutler, B., Moser, D., Thomas, H., et al. (1997). Chlorophyll breakdown in senescent chloroplasts. Cleavage of pheophorbide *a* in two enzymic steps. *Plant Physiology, 115*, 669–676.

Rodoni, S., Vicentini, F., Schellenberg, M., Matile, P., & Hörtensteiner, S. (1997). Partial purification and characterization of red chlorophyll catabolite reductase, a stroma protein involved in chlorophyll breakdown. *Plant Physiology, 115*, 677–682.

Roiser, M. H., Müller, T., & Kräutler, B. (2015). Colorless chlorophyll catabolites in senescent florets of broccoli (*Brassica oleracea* var. italica). *Journal of Agricultural and Food Chemistry, 63*, 1385–1392.

Saga, Y., & Tamiaki, H. (2012). Demetalation of chlorophyll pigments. *Chemistry & Biodiversity, 9*, 1659–1683.

Sakuraba, Y., Han, S. H., Lee, S. H., Hörtensteiner, S., & Paek, N. C. (2016). *Arabidopsis* NAC016 promotes chlorophyll breakdown by directly upregulating *STAYGREEN1* transcription. *Plant Cell Reports, 35*, 155–166.

Sakuraba, Y., Jeong, J., Kang, M. Y., Kim, J., Paek, N. C., & Choi, G. (2014). Phytochrome-interacting transcription factors PIF4 and PIF5 induce leaf senescence in Arabidopsis. *Nature Communications, 5*, 4636.

Sakuraba, Y., Kim, D., Kim, Y. S., Hörtensteiner, S., & Paek, N. C. (2014). Arabidopsis STAYGREEN-LIKE (SGRL) promotes abiotic stress-induced leaf yellowing during vegetative growth. *FEBS Letters, 588*, 3830–3837.

Sakuraba, Y., Kim, Y. S., Yoo, S. C., Hörtensteiner, S., & Paek, N. C. (2013). 7-Hydroxymethyl chlorophyll *a* reductase functions in metabolic channeling of chlorophyll breakdown intermediates during leaf senescence. *Biochemical and Biophysical Research Communications, 430*, 32–37.

Sakuraba, Y., Schelbert, S., Park, S.-Y., Han, S.-H., Lee, B.-D., Besagni Andrès, C., et al. (2012). STAY-GREEN and chlorophyll catabolic enzymes interact at light-harvesting complex II for chlorophyll detoxification during leaf senescence in *Arabidopsis*. *The Plant Cell, 24*, 507–518.

Sato, Y., Moria, R., Katsuma, S., Nishimura, M., Tanaka, A., & Kusaba, M. (2009). Two short-chain dehydrogenase/reductases, NON-YELLOW COLORING 1 and NYC1-LIKE, are required for chlorophyll *b* and light-harvesting complex II degradation during senescence in rice. *The Plant Journal, 57*, 120–131.

Sato, Y., Morita, R., Nishimura, M., Yamaguchi, H., & Kusaba, M. (2007). Mendel's green cotyledon gene encodes a positive regulator of the chlorophyll-degrading pathway. *Proceedings of the National Academy of Sciences of the United States of America, 104*, 14169–14174.

Scheer, H. (2006). An overview of chlorophylls and bacteriochlorophylls: Biochemistry, biophysics, functions and applications. In B. Grimm, R. Porra, W. Rüdiger, & H. Scheer (Eds.), *Vol. 25. Chlorophylls and bacteriochlorophylls: Biochemistry, biophysics, functions and applications* (pp. 1–26). Dordrecht, The Netherlands: Springer.

Schelbert, S., Aubry, S., Burla, B., Agne, B., Kessler, F., Krupinska, K., et al. (2009). Pheophytin pheophorbide hydrolase (pheophytinase) is involved in chlorophyll breakdown during leaf senescence in Arabidopsis. *The Plant Cell, 21*, 767–785.

Scherl, M., Müller, T., & Kräutler, B. (2012). Chlorophyll catabolites in senescent leaves of the lime tree (*Tilia cordata*). *Chemistry & Biodiversity, 9*, 2605–2617.

Scherl, M., Müller, T., Kreutz, C. R., Huber, R. G., Zass, E., Liedl, K. R., et al. (2016). Chlorophyll catabolites in fall leaves of the wych elm tree present a novel glyosylation motif. *Chemistry—A European Journal, 22*, 9498–9503.

Scheumann, V., Schoch, S., & Rüdiger, W. (1999). Chlorophyll *b* reduction during senescence of barley seedlings. *Planta, 209*, 364–370.

Shan, X. Y., Wang, J. X., Chua, L. L., Jiang, D. A., Peng, W., & Xie, D. X. (2011). The role of Arabidopsis Rubisco activase in jasmonate-induced leaf senescence. *Plant Physiology, 155*, 751–764.

Shimoda, Y., Ito, H., & Tanaka, A. (2012). Conversion of chlorophyll *b* to chlorophyll *a* precedes magnesium dechelation for protection against necrosis in Arabidopsis. *The Plant Journal, 72*, 501–511.

Shimoda, Y., Ito, H., & Tanaka, A. (2016). Arabidopsis STAY-GREEN, Mendel's green cotyledon gene, encodes magnesium-dechelatase. *The Plant Cell, 28*, 2147–2160.

Stommel, J., Abbott, J. A., Saftner, R. A., & Camp, M. J. (2005). Sensory and objective quality attributes of beta-carotene and lycopene-rich tomato fruit. *Journal of the American Society for Horticultural Science, 130*, 244–251.

Sugishima, M., Kitamori, Y., Noguchi, M., Kohchi, T., & Fukuyama, K. (2009). Crystal structure of red chlorophyll catabolite reductase: Enlargement of the ferredoxin-dependent bilin reductase family. *Journal of Molecular Biology, 389*, 376–387.

Sugishima, M., Okamoto, Y., Noguchi, M., Kohchi, T., Tamiaki, H., & Fukuyama, K. (2010). Crystal structures of the substrate-bound forms of red chlorophyll catabolite reductase: Implications for site-specific and stereospecific reaction. *Journal of Molecular Biology, 402*, 879–891.

Süssenbacher, I., Christ, B., Hörtensteiner, S., & Kräutler, B. (2014). Hydroxymethylated phyllobilins: A puzzling new feature of the dioxobilin branch of chlorophyll breakdown. *Chemistry—A European Journal, 20*, 87–92.

Süssenbacher, I., Hörtensteiner, S., & Kräutler, B. (2015). A dioxobilin-type fluorescent chlorophyll catabolite as a transient early intermediate of the dioxobilin-branch of chlorophyll breakdown in *Arabidopsis thaliana*. *Angewandte Chemie International Edition, 54*, 13777–13781.

Süssenbacher, I., Kreutz, C., Christ, B., Hörtensteiner, S., & Kräutler, B. (2015). Hydroxymethylated dioxobilins in senescent *Arabidopsis thaliana* leaves—Sign of a puzzling biosynthetic intermezzo of chlorophyll breakdown. *Chemistry—A European Journal, 21*, 11664–11670.

Suzuki, Y., Doi, M., & Shioi, Y. (2002). Two enzymatic reaction pathways in the formation of pyropheophorbide *a*. *Photosynthesis Research, 74*, 225–233.

Suzuki, Y., & Shioi, Y. (1999). Detection of chlorophyll breakdown products in the senescent leaves of higher plants. *Plant and Cell Physiology, 40*, 909–915.

Takasaki, H., Maruyama, K., Takahashi, F., Fujita, M., Yoshida, T., Nakashima, K., et al. (2015). SNAC-As, stress-responsive NAC transcription factors, mediate ABA-inducible leaf senescence. *The Plant Journal, 84*, 1114–1123.

Tanaka, R., Hirashima, M., Satoh, S., & Tanaka, A. (2003). The *Arabidopsis-accelerated cell death* gene *ACD1* is involved in oxygenation of pheophorbide *a*: Inhibition of pheophorbide *a* oxygenase activity does not lead to the "stay-green" phenotype in *Arabidopsis*. *Plant and Cell Physiology, 44*, 1266–1274.

Tanaka, R., Kobayashi, K., & Masuda, T. (2011). Tetrapyrrole metabolism in *Arabidopsis thaliana*. *The Arabidopsis Book, 9*, e0145.

Tanaka, A., & Tanaka, R. (2006). Chlorophyll metabolism. *Current Opinion in Plant Biology, 9*, 248–255.

Tanaka, R., & Tanaka, A. (2011). Chlorophyll cycle regulates the construction and destruction of the light-harvesting complexes. *Biochimica et Biophysica Acta, 1807*, 968–976.

Thomas, H., Bortlik, K.-H., Rentsch, D., Schellenberg, M., & Matile, P. (1989). Catabolism of chlorophyll *in vivo*: Significance of polar chlorophyll catabolites in a non-yellowing senescence mutant of *Festuca pratensis* Huds. *New Phytologist, 111*, 3–8.

Thomas, H., Huang, L., Young, M., & Ougham, H. (2009). Evolution of plant senescence. *BMC Evolutionary Biology, 9*, 163.

Tommasini, R., Vogt, E., Fromenteau, M., Hörtensteiner, S., Matile, P., Amrhein, N., et al. (1998). An ABC transporter of *Arabidopsis thaliana* has both glutathione-conjugate and chlorophyll catabolite transport activity. *The Plant Journal, 13*, 773–780.

Topalov, G., & Kishi, Y. (2001). Chlorophyll catabolism leading to the skeleton of dinoflagellate and krill luciferins: Hypothesis and model studies. *Angewandte Chemie International Edition, 40*, 3892–3894.

Ueda, J., & Kato, J. (1980). Isolation and identification of a senescence-promoting substance from wormwood (*Artemisia absinthium* L.). *Plant Physiology, 66*, 246–249.

Ulrich, M., Moser, S., Müller, T., & Kräutler, B. (2011). How the colourless 'non-fluorescent' chlorophyll catabolites rust. *Chemistry—A European Journal, 17*, 2330–2334.

Valenta, K., Kalbitzer, U., Razafimandimby, D., Omeja, P., Ayasse, M., Chapman, C. A., et al. (2018). The evolution of fruit colour: Phylogeny, abiotic factors and the role of mutualists. *Scientific Reports, 8*, 14302.

Vergara-Dominguez, H., Rios, J. J., Gandul-Rojas, B., & Roca, M. (2016). Chlorophyll catabolism in olive fruits (var. Arbequina and Hojiblanca) during maturation. *Food Chemistry, 212*, 604–611.

Vergeiner, C., Banala, S., & Kräutler, B. (2013). Chlorophyll breakdown in senescent banana leaves: Catabolism reprogrammed for biosynthesis of persistent blue fluorescent tetrapyrroles. *Chemistry—A European Journal, 19*, 12294–12305.

Vergeiner, C., Ulrich, M., Li, C. J., Liu, X. J., Müller, T., & Kräutler, B. (2015). Stereo- and regioselective phyllobilane oxidation in leaf homogenates of the peace lily (*Spathiphyllum wallisii*): Hypothetical endogenous path to yellow chlorophyll catabolites. *Chemistry—A European Journal, 21*, 136–149.

Wakana, D., Kato, H., Momose, T., Sasaki, N., Ozeki, Y., & Goda, Y. (2014). NMR-based characterization of a novel yellow chlorophyll catabolite, Ed-YCC, isolated from *Egeria densa*. *Tetrahedron Letters, 55*, 2982–2985.

Wang, Y. Q., Yang, Y., Fei, Z. J., Yuan, H., Fish, T., Thannhauser, T. W., et al. (2013). Proteomic analysis of chromoplasts from six crop species reveals insights into chromoplast function and development. *Journal of Experimental Botany*, *64*, 949–961.
Wilks, A. (2008). Heme degradation: Mechanistic and physiological implications. In M. J. Warren & A. G. Smith (Eds.), *Tetrapyrroles: Birth, life and death* (pp. 101–115). Austin, Texas: Landes Bioscience.
Wilks, A., & Ikeda-Saito, M. (2014). Heme utilization by pathogenic bacteria: Not all pathways lead to biliverdin. *Accounts of Chemical Research*, *47*, 2291–2298.
Wu, S., Li, Z., Yang, L., Xie, Z., Chen, J., Zhang, W., et al. (2016). NON-YELLOWING2 (NYE2), a close paralog of NYE1, plays a positive role in chlorophyll degradation in Arabidopsis. *Molecular Plant*, *9*, 624–627.
Wüthrich, K. L., Bovet, L., Hunziker, P. E., Donnison, I. S., & Hörtensteiner, S. (2000). Molecular cloning, functional expression and characterisation of RCC reductase involved in chlorophyll catabolism. *The Plant Journal*, *21*, 189–198.
Xodo, L. E., Rapozzi, V., Zacchigna, M., Drioli, S., & Zorzet, S. (2012). The chlorophyll catabolite pheophorbide *a* as a photosensitizer for the photodynamic therapy. *Current Medicinal Chemistry*, *19*, 799–807.
Yang, Y., Xu, R., Ma, C. J., Vlot, A. C., Klessig, D. F., & Pichersky, E. (2008). Inactive methyl indole-3-acetic acid ester can be hydrolyzed and activated by several esterases belonging to the *At*MES esterase family of Arabidopsis. *Plant Physiology*, *147*, 1034–1045.
Ye, Y., Zhou, L., Liu, X., Liu, H., Li, D., Cao, M., et al. (2017). A novel chemical inhibitor of ABA signaling targets all ABA receptors. *Plant Physiology*, *173*, 2356–2369.
Yin, X. R., Xie, X. L., Xia, X. J., Yu, J. Q., Ferguson, I. B., Giovannoni, J. J., et al. (2016). Involvement of an ethylene response factor in chlorophyll degradation during citrus fruit degreening. *The Plant Journal*, *86*, 403–412.
Zapata, L., Ding, J., Willing, E. M., Hartwig, B., Bezdan, D., Jiao, W. B., et al. (2016). Chromosome-level assembly of *Arabidopsis thaliana* L*er* reveals the extent of translocation and inversion polymorphisms. *Proceedings of the National Academy of Sciences of the United States of America*, *113*, E4052–E4060.
Zhang, W., Liu, T. Q., Ren, G. D., Hörtensteiner, S., Zhou, Y. M., Cahoon, E. B., et al. (2014). Chlorophyll degradation: The tocopherol biosynthesis related phytol hydrolase in Arabidopsis seeds is still missing. *Plant Physiology*, *166*, 70–79.
Zhang, J., Yu, G., Wen, W., Ma, X., Xu, B., & Huang, B. (2016). Functional characterization and hormonal regulation of the *PHEOPHYTINASE* gene *LpPPH* controlling leaf senescence in perennial ryegrass. *Journal of Experimental Botany*, *67*, 935–945.
Zhu, X. Y., Chen, J. Y., Qiu, K., & Kuai, B. K. (2017). Phytohormone and light regulation of chlorophyll degradation. *Frontiers in Plant Science*, *8*, 1911.
Zhu, X., Chen, J., Xie, Z., Gao, J., Ren, G., Gao, S., et al. (2015). Jasmonic acid promotes degreening via MYC2/3/4- and ANAC019/055/072-mediated regulation of major chlorophyll catabolic genes. *The Plant Journal*, *84*, 597–610.

CHAPTER EIGHT

Evolution of tetrapyrrole pathway in eukaryotic phototrophs

Jaromír Cihlář[a,b,†], Zoltán Füssy[a,b,†], Miroslav Oborník[a,b,*]

[a]Faculty of Science, University of South Bohemia, České Budějovice, Czech Republic
[b]Institute of Parasitology, Biology Centre CAS, České Budějovice, Czech Republic
*Corresponding author: e-mail address: obornik@paru.cas.cz

Contents

Abstract

Tetrapyrroles such as chlorophyll and heme are involved in conversion of the light energy to chemical bonds by photosynthetic electron transfer as well as phosphorylation and participation as cofactor in various enzymatic activities and in cellular signaling, respectively. Therefore, they are crucial constituents of the cell metabolism in all three domains of cellular life. In eukaryotes, the pathway was shaped by past endosymbioses and consecutive endosymbiotic and non-endosymbiotic gene transfers. These processes resulted in a true mosaic of enzymes with diverse origins and reshuffled subcellular localizations. This chapter summarizes the current knowledge on the arrangement of tetrapyrrole biosynthesis in eukaryotic phototrophs, and discusses how endosymbiotic events modulate this pathway arrangement.

† Authors contributed equally.

Advances in Botanical Research, Volume 90
ISSN 0065-2296
https://doi.org/10.1016/bs.abr.2018.12.003

Abbreviations

ALA	δ-aminolevulinic acid
ALAD	ALA dehydratase
ALAS	ALA synthase
Chl	chlorophyll
CPOX	coproporphyrinogen oxidase
CSYN	Chl synthase
EGT	endosymbiotic gene transfer
FeCH	ferrochelatase
GSA-AT	glutamate-1-semialdehyde aminotransferase
GTR	glutamyl-tRNA reductase
MgCH	magnesium chelatase
PBGD	porphobilinogen deaminase
POR	protochlorophyllide oxidoreductase
PPIX	protoporphyrin IX
PPOX	protoporphyrinogen oxidase
UROD	uroporphyrinogen decarboxylase
UROS	uroporphyrinogen synthase

1. Introduction

Tetrapyrroles (also termed porphyrins) consists of four pyrrole rings attached to one another in a cyclic manner, often coordinated with a metal ion (Frankenberg, Moser, & Jahn, 2003) (Fig. 1). The most recognized tetrapyrroles are hemes and chlorophylls (Chls), metalated with iron and magnesium cations, respectively. Hemes as cofactors are directly involved in redox reactions such as the respiratory and photosynthetic electron transport chain and can possess regulatory roles. In phototrophs, a substantial part of tetrapyrroles is represented by Chls, pigments essential for light harvesting (Frankenberg et al., 2003; Mochizuki et al., 2010). In Eubacteria and Archaea, the pathway branches to produce cobalamin (vitamin B_{12}), an essential cofactor for enzymes involved in DNA synthesis and energy metabolism required by all organisms (Roth, Lawrence, & Bobik, 1996; Scott & Roessner, 2002), and siroheme used for nitrogen and sulfur fixation in plants and some Eubacteria (Frankenberg et al., 2003).

Tetrapyrroles are nearly ubiquitous and the ability to synthesize (or acquire) tetrapyrroles is therefore directly linked to life. Most organisms are capable of de novo tetrapyrrole synthesis (Yin & Bauer, 2013), while some parasites and predators are tetrapyrrole auxotrophs and must seek these compounds (primarily heme) in their hosts and prey (Kořený, Lukeš, & Oborník, 2010; Kořený, Oborník, & Lukeš, 2013; Sinclair & Hamza, 2015).

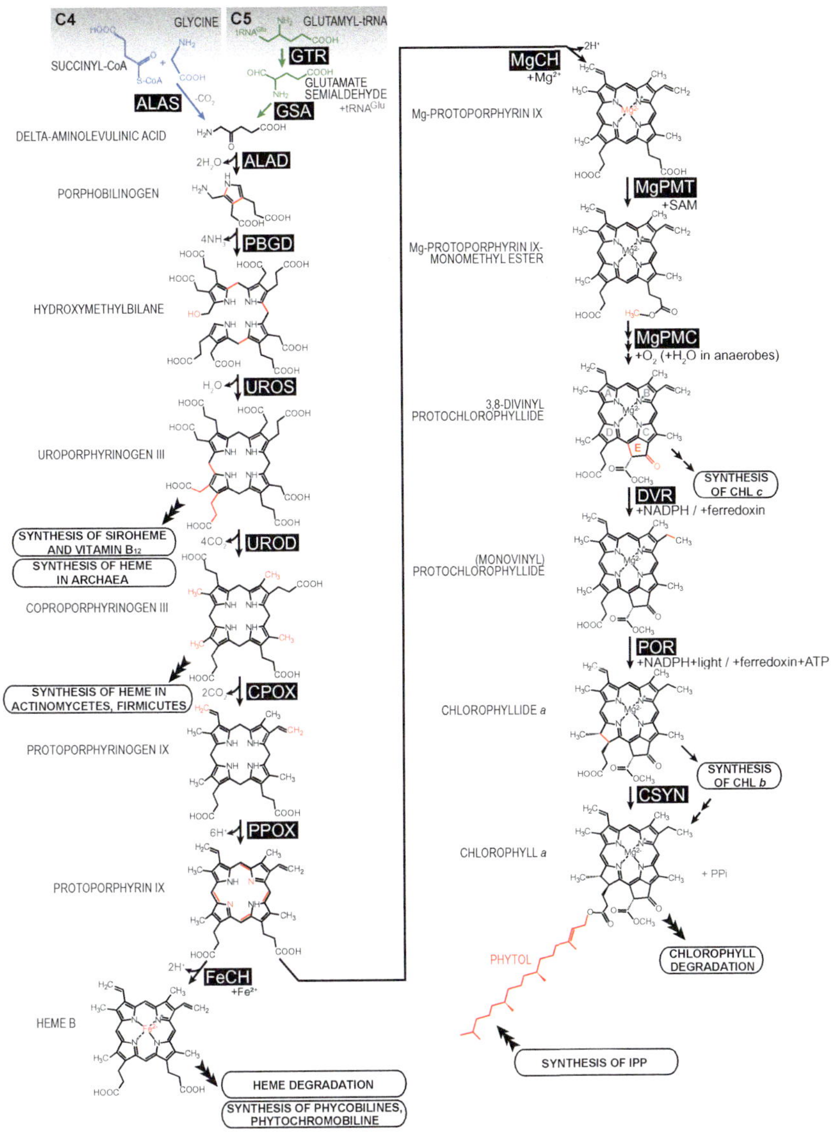

Fig. 1 Overview of tetrapyrrole synthesis. The pathway can be divided into three phases, the synthesis of ALA via either the C5 pathway or the C4 (Shemin) pathway, the synthesis of protoporphyrin IX, and the branches to heme B and chlorophyll *a* syntheses. The C5 pathway is, among other groups, present in cyanobacteria and most phototrophic eukaryotes, while the α-proteobacterial C4 pathway is adopted in aplastidic heterotrophic eukaryotes (but see text for exceptions). As noted, some Archaea (Storbeck et al., 2010), Actinobacteria and firmicutes (Dailey, Gerdes, Dailey, Burch, & Phillips, 2015) employ alternative routes starting from uroporphyrinogen III *(Continued)*

Similarly, some algal species depend on the supply of cobalamin from the environment (Helliwell, Wheeler, Leptos, Goldstein, & Smith, 2011). There are only rare cases of organisms that thrive entirely without tetrapyrroles, represented by some of pathogenic and anaerobic bacteria (Decker, Jungermann, & Thauer, 1970) and an exceptional aerobic eukaryote *Phytomonas serpens* (Kořený et al., 2012).

The arrangement of the tetrapyrrole pathway reveals many aspects of how an organism manages energy flows and synthesizes essential compounds. Therefore, tetrapyrrole synthesis proved to be a great tool to investigate metabolic reorganizations following endosymbiotic events (Cihlář, Füssy, Horák, & Oborník, 2016; Kořený & Oborník, 2011; Matsuo & Inagaki, 2018; Oborník & Green, 2005). This review concerns an overview of heme and Chl synthesis across eukaryotes and discusses how the evolution of tetrapyrrole biosynthesis supports our perception of processes that accompany plastid endosymbiosis.

2. Overview of tetrapyrrole synthesis

The biosynthesis of tetrapyrroles, here exemplified by heme B and Chl *a*, is a rather complex process. The intermediates are often unstable or cause oxidative stress, hence the pathway requires coordination of numerous enzymatic activities (Fig. 1). The pathway is divided into three main stages: the synthesis of δ-aminolevulinic acid (ALA) as the initial common precursor, the formation of protoporphyrin IX (PPIX) as the last common unmetalated intermediate, and the branches leading from PPIX to heme and Chl. The Chl branch is indeed unique to phototrophic lineages and in algae and plants (i.e., eukaryotic phototrophs) exclusively localizes to the plastid compartment (reviewed in Masuda, 2008). The enzymatic steps are remarkably conserved in all three domains of cellular life, and they are

Fig. 1—Cont'd and coproporphyrinogen III. The reaction sites are highlighted in red. Enzymes abbreviations: ALAD, aminolevulinate dehydratase; ALAS, aminulevulinic acid synthase; CPOX(i), coproporphyrinogen oxidase (oxygen-independent); CSYN, chlorophyll synthase; DVR, divinyl protochlorophyllide reductase; FeCH, ferrochelatase; GSA, glutamate semialdehyde-aminomutase; GTR, glutamyl-tRNA reductase; MgCH, magnesium chelatase; MgPMC, Mg-protoporphyrin monomethyl ester cyclase; MgPMT, Mg-protoporphyrin methyltransferase; PBGD, porphobilinogen deaminase; POR, protochlorophyllide oxidoreductase; PPOX, protoporphyrinogen oxidase; UROD, uroporphyrinogen decarboxylase; UROS, uroporphyrinogen synthase.

described in more detail in other chapters. From the evolutionary view, however, there are aspects to discuss here.

ALA can be formed by two principal ways (Fig. 1). Alpha-proteobacteria use the so-called C4 (or Shemin) pathway (Duncan, Faggart, Roger, & Cornell, 1999; Ferreira & Gong, 1995; Jordan & Shemin, 1972), whereby ALA is synthesized through the condensation of glycine and succinyl-CoA (succinate is a four-carbon compound, hence the pathway name) catalyzed by a single enzyme, ALA synthase (ALAS). Most heterotrophic eukaryotes use a mitochondrion-localized ALAS adopted from the α-proteobacterial predecessors of mitochondria. Archaea and other Eubacteria, including cyanobacteria (Beale, 1999; Panek & O'Brian, 2002), use the so-called C5 pathway for ALA synthesis, where tRNA-bound glutamate (a five-carbon compound) is used as the precursor. Consequently, most algae employ the C5 pathway for tetrapyrrole synthesis, as inherited from the cyanobacterial predecessors of plastids. Pre-mitochondrial proto-eukaryotes supposedly used the C5 pathway as well, since they share evolutionary origin with Archaea (Koonin, 2015).

The following six steps leading to the formation of PPIX are common to most organisms that are able to synthesize heme (Fig. 1) (Dailey et al., 2015; Heinemann, Jahn, & Jahn, 2008; Layer, Reichelt, Jahn, & Heinz, 2010; Storbeck et al., 2010). The canonical eukaryotic pathway employs ALA dehydratase (ALAD), porphobilinogen deaminase (PBGD), uroporphyrinogen synthase (UROS), uroporphyrinogen decarboxylase (UROD), coproporphyrinogen oxidase (CPOX), and protoporphyrinogen oxidase (PPOX). The insertion of a ferrous ion into PPIX by ferrochelatase (FeCH) is the next step of the heme branch and forms heme B (or protoheme), the most abundant type of heme (Beale & Weinstein, 1991).

Most organisms synthesize tetrapyrroles exclusively under aerobic conditions, mainly because two oxidative steps involved in the pathway, catalyzed by CPOX and PPOX (Fig. 1), require oxygen for their function (Heinemann et al., 2008). However, the oxygen-independent versions of both enzymes were identified in some bacteria (Frankenberg et al., 2003; Panek & O'Brian, 2002; Pereira, Teixeira, & Xavier, 1998). Notably, oxygen-independent CPOX was found also in algae and plants (Oborník & Green, 2005) and has an essential role in floral development, when energy demands on mitochondria are high (Pratibha, Singh, Srinivasan, Bhat, & Sreenivasulu, 2017).

Chls are magnesium-chelated cyclic tetrapyrroles (Sousa, Shavit-Grievink, Allen, & Martin, 2013). The Chl branch is initiated by magnesium chelatase

(MgCH) and continues with several modifications around the tetrapyrrole ring, including the formation of the fifth isocyclic ring E (see Fig. 1). Chl *a* synthesis is completed by the esterification of chlorophyllide with phytol catalyzed by Chl synthase (CSYN). This alcohol moiety is provided by the isoprenoid pathway, which is also frequently localized to plastids (Wang et al., 2014). Most oxygenic phototrophs employ also other chemically distinct Chls such as Chl *b*, *c*, *d*, or *f* (Chen & Blankenship, 2011). Chl *b*, *d*, and *f* differ from Chl *a* by formyl substitution at carbons C7, C3 or C2, respectively, and Chl *c* differs most obviously by lacking the phytyl appendix. These Chls have been developed by different lineages via appending several modifying enzymes to the late parts of the Chl *a* pathway (Loughlin, Lin, & Chen, 2013).

3. Endosymbiotic gene transfer and the evolution of algae

Mitochondria and plastids descend from free-living organisms through a transformation process called endosymbiosis (Vothknecht & Soll, 2007). Endosymbiotic partnerships of eukaryotic hosts and bacterial/eukaryotic endosymbionts are thought to boost evolutionary radiations by enhancing metabolic capacities of the host (Wolf & Koonin, 2013). For instance, endosymbioses gave eukaryotes the advantage of aerobic and phototrophic metabolism with effective carbohydrate decomposition and a virtually unlimited source of electrons and energy from water and the sun (see Box 1). While all extant mitochondria are believed to originate in a single endosymbiotic event and appear to have been vertically passed on from the last eukaryotic common ancestor, the history of plastids is much more colorful, documenting both vertical and horizontal transfers in quite disparate lineages of eukaryotes. Interested readers should refer to many excellent reviews on plastid endosymbioses in eukaryotes in an earlier issue of this series (Hirakawa, 2017).

Tetrapyrrole biosynthesis displays a somewhat higher complexity in eukaryotes than in Eubacteria and Archaea. This partly follows from the evolutionary history of eukaryotes vastly influenced by the endosymbiotic events that constituted mitochondria and plastids as new cellular compartments. It is believed that in early stages of endosymbiosis, both partner cells express independent sets of proteins needed for their existence and encoded by their respective genomes. During later stages, many genes from the endosymbiont are lost or transferred to the host nucleus in a process termed the endosymbiotic gene transfer (EGT) (Delwiche, 1999; Jiroutová, Kořený,

BOX 1 Endosymbiosis

Aplastidic eukaryotes, also referred to as primary heterotrophs, are present in most currently recognized eukaryotic kingdoms (Fig. 3). In fact, two of the currently recognized kingdoms (supergroups), Amoebozoa and Opisthokonta, are entirely heterotrophic. An endosymbiotic association of a heterotrophic host with a cyanobacterium gave rise to primary algae, represented by Archaeplastida and *Paulinella*. Plastids of primary phototrophs are typically bound by two-membrane envelopes. Through complex, or eukaryote-to-eukaryote endosymbioses, plastids spread to the majority of eukaryotic kingdoms. While in the two green-derived lineages complex endosymbioses occurred more recently and the plastids are clearly related to extant groups of chlorophyte algae (Jackson et al., 2018), the order of endosymbioses in the rhodophyte-derived lineages is currently unresolved. Hence, for clarity, we refer to higher-order endosymbiotic events and the resulting organisms as complex endosymbioses, complex algae. Complex plastids are typically bound by three, four, or five membranes, which is their most obvious difference from primary plastids. Some algal lineages lost photosynthesis and became secondarily heterotrophic, although they mostly retain their plastids for biochemical reasons. To avoid confusion, both primary and complex algae that lost photosynthesis are referred to as non-photosynthetic algae, and there are indeed non-photosynthetic plants as well.

Bowler, & Oborník, 2010; Martin et al., 1998, 2002; Oborník & Green, 2005; Timmis, Ayliffe, Huang, & Martin, 2004). EGT supposedly allows for enhanced host control over the organelle and reduces functional redundancy of the chimeric organism. As a result, a majority of the organellar proteomes is nuclear-encoded, translated in the cytosol and post-translationally imported into mitochondria and plastids. Some host nucleus-encoded proteins targeted to mitochondria and plastids originate from the genome of the endosymbiont, i.e., plastids are populated with cyanobacterial proteins, mitochondria with α-proteobacterial proteins, and others originated from the host nucleus. Though endosymbiont-derived proteins do not represent the majority of organellar proteomes (Gray, 2015; Qiu et al., 2013), the adhesion of proteins to a specific compartment possibly reflects a unique function in this cell part, or a need to interact in functional units (Suzuki & Miyagishima, 2010). Surprisingly, some biological processes recruit functionally more versatile proteins than others and show a higher degree of evolutionary mosaicism (Suzuki & Miyagishima, 2010). The origin of proteins thus not always corresponds to their distribution among cell compartments.

This can be documented by phylogenetic analyses of proteomes determined experimentally or predicted based on physicochemical properties of the N-termini of protein sequences, which corresponds to a mitochondrial or plastid targeting presequence (cytosolic proteins lack presequences).

A similar mosaicism principle applies to the tetrapyrrole pathway (Cihlář et al., 2016; Kořený, Sobotka, Janouškovec, Keeling, & Oborník, 2011; Oborník, & Green, 2005). The integration of mitochondria and plastids led to a redistribution of enzymatic steps among cellular compartments that supposedly facilitated responses to environmental changes (reviewed in, e.g., Brzezowski, Richter, & Grimm, 2015; Sun, Cheng, & Chen, 2015; Zhang, Zhang, Zhu, Zhang, & Yuan, 2015). Since these endosymbioses brought together organisms with somewhat overlapping protein collections, loss of redundant enzymes turned the tetrapyrrole pathway of eukaryotes into a mosaic of enzymes of various evolutionary origins (Oborník & Green, 2005). The discordances between the origin and localization of tetrapyrrole enzymes are discussed in the following sections.

4. Evolution of tetrapyrrole synthesis an eukaryotes

Rather than following horizontal evolution, the arrangement of tetrapyrrole synthesis of an organism reflects its trophic strategy. In this chapter we therefore distinguish eukaryotes with this respect. Heterotrophs (i.e., aplastidic primary heterotrophs) are mainly represented by animals and fungi (collectively termed Opisthokonta) and Amoebozoa. Aplastidic heterotrophs also include lineages that are closely or distantly related to phototrophic "crown" groups, such as ciliates, kinetoplastids, heterotrophic rhizarians, heterotrophic stramenopiles, and a collection of small groups that have been discovered over the past years (Archibald, 2015; Burki, 2017). Algae and plants are traditionally taken equivalent to primary and complex phototrophic organisms (elaborated in more detail in the following sections). We also consider eukaryotes that are no longer phototrophic to be non-photosynthetic algae (or plants), as they possess (relic) plastids in most cases.

4.1 Biosynthesis of heme in heterotrophic eukaryotes

Heme biosynthesis localizes in two different cellular compartments of aplastidic heterotrophs (see Box 1) (Dailey et al., 2015) (Fig. 2), reflecting the endosymbiotic origin of extant eukaryotes. The biosynthesis starts in mitochondria where ALA is synthesized via the C4 pathway (Jordan & Shemin, 1972).

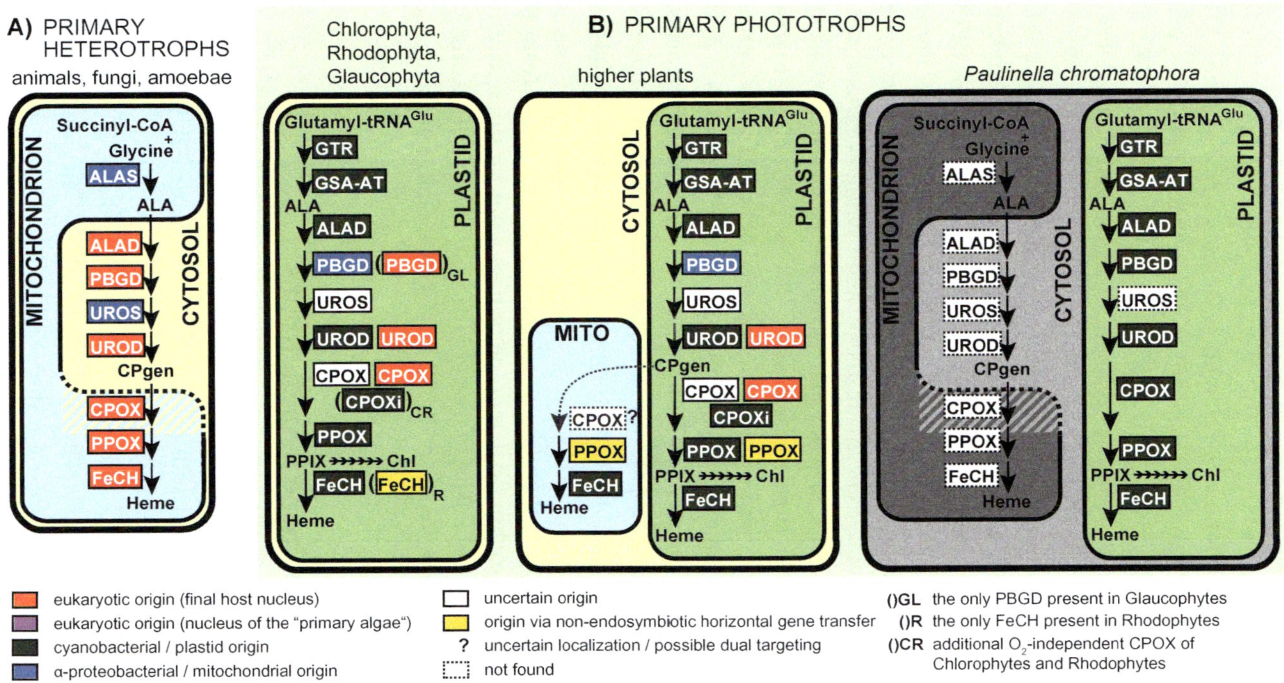

Fig. 2 Origins and subcellular locations of heme biosynthesis enzymes in aplastidic heterotrophs and primary phototrophs. Enzyme abbreviations are listed in Fig. 1. Colored rectangles indicate the origins of genes inferred from phylogenetic analyses, according to the color code below the schemes. Schemes of heme pathways in heterotrophs and phototrophs are based on Kořený and Oborník (2011). The origin of glaucophyte PBGD and the arrangement of heme pathway in *Paulinella chromatophora* were determined by Cihlář (2018).

ALAS is strictly bound to the mitochondria because one of its substrates, succinyl-CoA, is a product of the mitochondrial TCA cycle (Kořený et al., 2011). ALA is exported to the cytosol in exchange for glycine, which is a co-substrate for ALA synthesis (Chiabrando, Mercurio, & Tolosano, 2014; Guernsey et al., 2009; Hamza & Dailey, 2012). The next four steps of the pathway take place in the cytosol (ALAD, PBGD, UROS, and UROD). The intermediates of heme biosynthesis are highly reactive and this re-location of tetrapyrrole synthesis to the cytosol was hypothesized to reduce oxidative stress in the mitochondrion (Vavilin & Vermaas, 2002), which is still considerably high due to the reactive oxygen species production by the respiratory chain (Chen, Silva, & Klessig, 1993; Murphy, 2009; Turrens, 2003). CPOX is cytosolic in yeasts, rhizarians, and ciliates, but mitochondrial in animals, oomycetes, and some fungi, implying independent mitochondrial retargeting in these lineages. The last two enzymatic steps of heme synthesis are localized in mitochondria (Kořený & Oborník, 2011; Oborník & Green, 2005). This arrangement reflects the close interaction of PPOX and FeCH (Ferreira, Andrew, Karr, & Dailey, 1988), and the highest need of heme for the synthesis of cytochromes of the mitochondrial respiratory chain. The presence of the first and the last step of the pathway in the same compartment (the mitochondrion) is also important for the pathway regulation since both ALAS and FeCH are responsive to iron levels, and surplus heme mediates the feedback inhibition of ALA synthesis (Sun et al., 2015).

ALAS and UROS are the only two enzymes that originate from the α-proteobacterial/mitochondrial genome, while genes for the remaining enzymes are apparently of eukaryotic origin (Kořený et al., 2010; Kořený & Oborník, 2011; Oborník & Green, 2005). UROS was retargeted to function in the cytosol.

4.2 Biosynthesis of heme in phototrophs

4.2.1 Primary phototrophs

The evolution of tetrapyrrole biosynthesis in algae was largely influenced by the emergence of primary plastids (Box 1) in Archaeplastida, the eukaryotic supergroup currently encompassing glaucophytes, rhodophytes (red algae), chlorophytes (green algae), and higher plants (Archibald, 2009; Ponce-Toledo et al., 2017; Rodríguez-Ezpeleta et al., 2005). The entire pathway is plastid-localized, except for cases discussed below. During the primary plastid establishment, cyanobacterial genes of the tetrapyrrole pathway moved to the host nucleus by EGT. Being translated in the cytosol, all the enzymes obtained plastid-targeting presequences (i.e., transit peptides) and so can be delivered back to plastids (Fig. 2B).

Most tetrapyrrole-synthesizing enzymes in primary phototrophs are of cyanobacterial origin and, consistently, Archaeplastida synthesize ALA via the C5 pathway just like cyanobacteria. Genes coding for glutamyl-tRNA reductase (GTR), glutamate-1-semialdehyde aminotransferase (GSA-AT), ALAD, PPOX and FeCH show strong affiliation to their cyanobacterial counterparts (Cihlář et al., 2016; Kořený & Oborník, 2011; Kořený et al., 2011; Oborník & Green, 2005). Although the cyanobacterial origin of UROS is not well supported, UROS is very likely derived from the endosymbiont as well (Cihlář et al., 2016; Kořený & Oborník, 2011). Two copies encode some of the tetrapyrrole enzymes of primary algae. One copy of each UROD and CPOX is likely of cyanobacterial origin, the other copies apparently represent the original eukaryotic genes because they group with heterotrophs. These latter copies were therefore retargeted from the cytosol and contribute to the plastid pathway. Genes acquired from sources outside the endosymbiotic couple include the FeCH in rhodophytes (not directly related to chlorophyte and plant FeCH sequences, see below) and the PPOX in higher plants (PPII) that both apparently arose via non-endosymbiotic horizontal gene transfer from bacteria (Kořený & Oborník, 2011).

The phylogeny of PBGD in Archaeplastida illustrates how randomness affects gene replacement. PBGD of rhodophytes, chlorophytes and plants is clearly related to α-proteobacterial homologs, possibly derived from mitochondria, compared to PBGD in glaucophytes and heterotrophic lineages that are of host origin (Cihlář et al., 2016; Kořený et al., 2012; Kořený & Oborník, 2011; Oborník & Green, 2005). When plastid was established as an organelle, three copies of PBGD were apparently present: of eukaryotic (host), α-proteobacterial and cyanobacterial origin. If the eukaryotic and α-proteobacterial copies were encoded in the nucleus and expressed to localize in the cytosol, and the cyanobacterial copy encoded in the plastid genome and expressed in the stroma, none of these copies would possess a plastid-targeting presequence. In principle, if none of these copies were favored to gain this presequence (i.e., they were functionally equivalent), one of the nuclear copies may have been retargeted to the plastid before the cyanobacterial PBGD was transferred by EGT. In this situation, glaucophytes "chose" the eukaryotic gene, while the branch leading to other Archaeplastida maintained the α-proteobacterial homolog (Cihlář and Oborník, unpublished data; Oborník & Green, 2005).

Higher plants have developed a novel branch of heme synthesis localized to the mitochondrion (Fig. 2B). Experimental works demonstrated PPOX activity in mitochondrial fractions of pea leaves (Smith, Marsh, & Elder, 1993) and localization of one of the PPOX proteins of tobacco and spinach to the mitochondrion (Lermontova, Kruse, Mock, & Grimm, 1997; Watanabe et al., 2001). Similarly, transgenic oxygen-dependent CPOX was shown to localize to maize mitochondria (Williams, Hardeman, Fowler, & Rivin, 2006), and oxygen-independent CPOX might have the same role in other higher plants (Pratibha et al., 2017). Therefore, it is plausible that the last two (or three) enzymatic steps of the heme biosynthesis localize in parallel in both the chloroplast and the mitochondrion and that a portion of pathway intermediates can be exported from plastids to mitochondria (Chow, Singh, Walker, & Smith, 1998; Lermontova et al., 1997) (Fig. 2B). This necessitated the recruitment of additional copies of PPOX and FeCH to mitochondria and, consistently, there are two genes coding for PPOX (PPOX1 and PPOX2) and FeCH (FC1 and FC2) in higher plants (Hey et al., 2016; Lermontova et al., 1997; Narita et al., 1996). As for their phylogenetic origin, PPOX1 shows affinity with cyanobacteria and it is shared with other phototrophic eukaryotes, while PPOX2 seems to have been acquired horizontally from bacteria (Kořený & Oborník, 2011; Kořený et al., 2011). Similarly, the two copies of higher plant FeCH show distinct tissue-specific and development-dependent expression pattern (Chow et al., 1998; Hey et al., 2016; Nagai et al., 2007; Singh, Cornah, Hadingham, & Smith, 2002), though these copies arose by duplication of the cyanobacterial homologue. FC2 is bound to the plastid compartment, but FC1 is apparently dual-targeted (Chow, Singh, Roper, & Smith, 1997; Hey et al., 2016; Suzuki et al., 2002) and contributes to the mitochondrial FeCH activity (Cornah, Roper, Pal Singh, & Smith, 2002; Hey et al., 2016; Masuda, Suzuki, Shimada, Ohta, & Takamiya, 2003). This dual targeting may not be complicated to achieve in plants, as proteins targeted to mitochondria, plastids or both compartments largely differ only by the overall charge of their presequence (Garg & Gould, 2016). A systematic search for mitochondrial CPOX enzymes needs to be employed to find the factors that determined the emergence of this pathway branch in higher plants.

The plastids in Archaeplastida have long been considered the only case of primary endosymbiosis with cyanobacteria. This event has occurred between 900 and 1600 MYA (Hedges, Blair, Venturi, & Shoe, 2004; Parfrey, Lahr, Knoll, & Katz, 2011; Shih and Matzke, 2013; Yoon, Hackett, Ciniglia, Pinto, & Bhattacharya, 2004). Nonetheless, a more recent primary endosymbiotic event has been reported in *Paulinella chromatophora*, a freshwater filose thecamoeba of cercozoan (rhizarian) taxonomic affiliation (Martin, Nowack, & Melkonian, 2005; Nowack, Melkonian, & Glöckner, 2008; Yoon et al., 2009). *Paulinella* possess photosynthetically active organelles referred to as cyanelles or chromatophores (Kim & Park, 2016; Lhee et al., 2017; Nowack et al., 2008). The biosynthesis of tetrapyrroles in *Paulinella* spp. has not been addressed but our analyses of the cyanelle genome identified short sequences of the C5 pathway enzymes (GTR and GSA-AT) and almost all downstream enzymes of the pathway, except UROS (Cihlář, 2018). Indeed, all these enzymes show strong affiliation to cyanobacterial genera *Synechococcus* and *Prochlorococcus*, relatives to the ancestor of the chromatophore, consistent with a rather recent endosymbiosis in *P. chromatophora* dated between 60 and 200 MYA (Martin et al., 2005; Nowack et al., 2016; Yoon et al., 2009). The data suggest that the parallel host pathway in *Paulinella* supplies tetrapyrroles to remaining cellular compartments (Fig. 2B), though, in an absence of comprehensive genomic data (none of the heme pathway enzymes have been identified in the nuclear genomic data), one can only assume that the host pathway is not significantly different from mitochondrial-cytosolic pathways of other heterotrophic eukaryotes.

4.2.2 Complex algae with rhodophyte-derived plastids

Many other eukaryotic lineages were endowed with plastids through complex endosymbioses. These secondary or higher-order events mediated endosymbiotic partnerships of algae with another eukaryotic hosts (e.g., Archibald, 2009; Bodył, Stiller, & Mackiewicz, 2009; Keeling, 2013; Oborník, Janouškovec, Chrudimský, & Lukeš, 2009) (Fig. 3). Complex algae come in two "colors," green and red. Algae that belong to the two green lineages arose through engulfment of chlorophytes by rhizarian and excavate hosts to constitute Chlorarachniophyta and Euglenophyta, respectively (Rogers, Gilson, Su, McFadden, & Keeling, 2007). The relationships among the red lineages appear more ancient and less clear (Parfrey et al., 2011; Yoon et al., 2004). Cryptophytes, myzozoans, stramenopiles, and haptophytes (Fig. 3) house rhodophyte-derived plastids that share a common origin, at least based on the plastid genome phylogenies and some molecular features (Gould, Maier, & Martin, 2015). Nevertheless, plastid morphologies differ substantially among these groups, and several attempts failed to recover close evolutionary relationships of their hosts (reviewed in Burki, 2017). Most probably, multiple or serial complex endosymbioses led

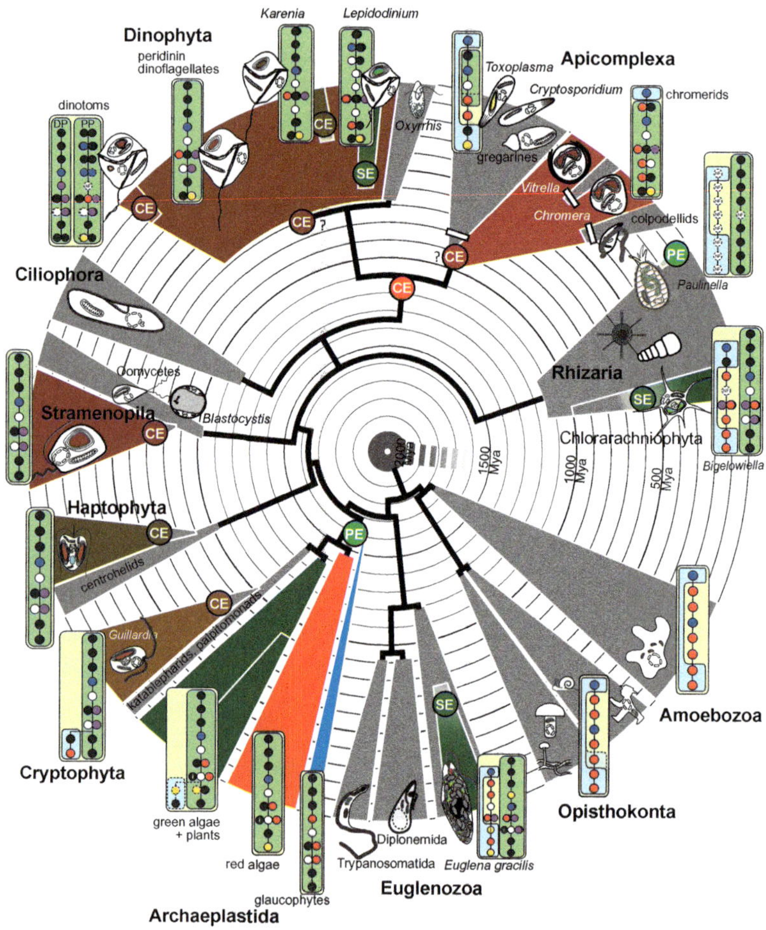

Fig. 3 Arrangements of heme biosynthesis pathway across eukaryotes. The schematic tree shows the relationships among major eukaryotic clades, with divergence times approximately inferred from several works (Jackson, Knoll, Chan, & Verbruggen, 2018; Parfrey et al., 2011; Yoon et al., 2004). Colored circles denote endosymbiotic events; PE, primary endosymbiosis; SE, secondary endosymbiosis; CE, complex endosymbiosis, i.e., the order of events resulting in the establishment of the rhodophyte-derived lineages (circles in shades of red/brown) are currently not resolved. This overview of heme biosynthetic pathways of particular groups is based on simplified Figs. 2, 4, and 6.

to the emergence of these lineages of complex algae, in an order of events still unresolved. It has been suggested that the plastid was horizontally transferred (stolen) between at least some of these lineages (Bodył, 2017; Oborník, 2018; Petersen et al., 2014).

Clear examples of horizontal plastid transfer (replacement) are found among dinoflagellates (Archibald, 2015; Delwiche, 1999; Keeling, 2010) (Fig. 3). The ancestral plastid of dinoflagellates is pigmented by a group-specific carotenoid, peridinin. Several dinoflagellate taxa, however, gained their present plastid from unrelated algal lineages and thus replaced the ancestral peridinin-pigmented organelle (Inagaki, Dacks, Doolittle, Watanabe, & Ohama, 2000; Ishida & Green, 2002; Saldarriaga, Taylor, Keeling, & Cavalier-Smith, 2001). A group of dinoflagellates called dinotoms (e.g. *Durinskia baltica*, *Glenodinium foliaceum*) (Chesnick, Morden, & Schmieg, 1996; Imanian & Keeling, 2007) harbor two distinct endosymbionts, one being the ancestral peridinin-pigmented plastid, the other being a newly obtained diatom-derived endosymbiont (Hehenberger, Imanian, Burki, & Keeling, 2014). Dinoflagellates of the *Karenia/Karlodinium* group possess plastids that derive from haptophytes (Tengs et al., 2000; Zapata, Fraga, Rodríguez, & Garrido, 2012). *Lepidodinium* spp., curiously, replaced the ancestral plastid with a chlorophyte-derived organelle (Matsumoto et al., 2011). Furthermore, there are examples of dinoflagellates that repeatedly extract plastids from their prey, maintain these plastids for a period of time (kleptoplasty), and therefore do not form a stable endosymbiotic relationship (Park, Kim, & Kim, 2014). As such, dinoflagellates represent an excellent platform to study the processes of plastid endosymbiosis and plastid replacement.

With respect to the arrangement of tetrapyrrole synthesis, a few differences in haptophytes, stramenopiles, and peridinin-pigmented dinoflagellates have been observed (Butterfield, Howe, & Nisbet, 2013; Cihlář et al., 2016; Kořený et al., 2011) compared to the primary algae (Fig. 4A). Although concentrated effort has not been made to localize tetrapyrrole enzymes in all lineages, predictions can be made with high credibility using algorithms trained on proteins with known localizations (Box 2). The entire pathway is plastid-localized. Apparently, all genes involved in the tetrapyrrole biosynthesis were transferred from the nucleus of the engulfed primary alga into the host nucleus and functionally replaced the genes for the mitochondrial/cytosolic pathway. There are, however, minor exceptions of this general arrangement. A new subfamily of putative plastid-targeted GSA-AT was identified in dinoflagellates, which clusters with proteobacteria and likely originates from non-endosymbiotic gene transfer (Cihlář et al., 2016). Haptophytes, stramenopiles and peridinin-pigmented dinoflagellates encode FeCH of different origins. Most dinoflagellates possess only a bacterial FeCH inherited from the rhodophyte

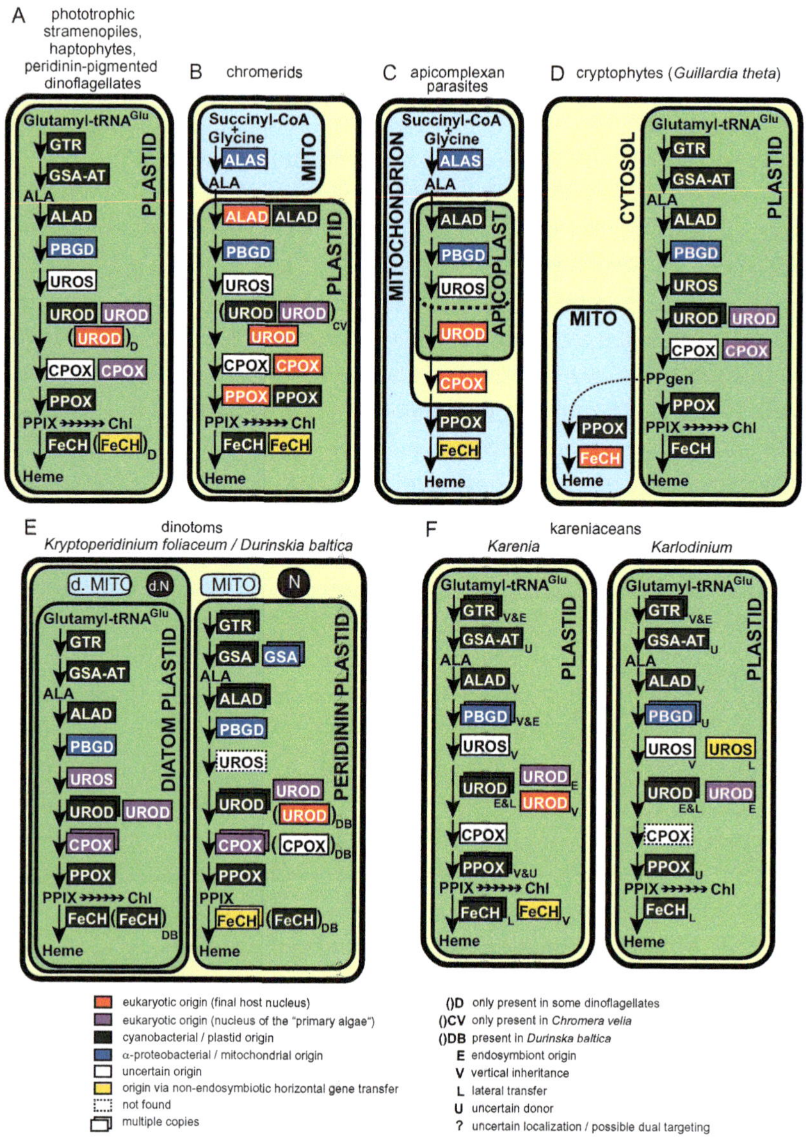

Fig. 4 Origins and subcellular locations of heme biosynthesis enzymes in complex algae with red-derived plastids. Enzyme abbreviations are listed in Fig. 1. Colored rectangles indicate the inferred origins of genes. The scheme of apicomplexan heme pathway is based on Kořený and Oborník (2011). The origins of chromerid, cryptophyte, dinotom, and kareniacean enzymes were determined by phylogenetic analyses (Cihlář et al., 2016; Kořený et al., 2011; Matsuo & Inagaki, 2018). The localizations are based on in silico predictions.

BOX 2 Targeting

Conventional routes for proteins translocated to the endomembrane system, mitochondria, and plastids include passing through protein complexes at their membranes, referred to as translocons. Targeting sequences recognized by the translocons are encoded at proteins' N-termini (i.e., they are presequences) and have a character of amphipathic or slightly hydrophobic helices (Kunze & Berger, 2015). These characteristics have been used by prediction algorithms to assess the probability of a given protein to be recognized for translocation into a membrane-bound compartment, when experimental data are lacking. There is a wide variety of localization predictors, and the most widely used include SignalP (Emanuelsson et al., 2007) for ER-targeted proteins, TargetP (Emanuelsson et al., 2007) for mitochondrial, plastid, and cytosolic proteins (invokes SignalP to predict signal peptides), TMHMM (Krogh, Larsson, von Heijne, & Sonnhammer, 2001) to predict membrane-anchored proteins and domains, MultiLoc2 to determine the highest probability of multiple localizations, including peroxisomes and Golgi, and HECTAR (Gschloessl, Guermeur, & Cock, 2008) and ASAFind (Gruber, Rocap, Kroth, Armbrust, & Mock, 2015) specifically trained to sort stramenopile proteins. Combined presequence analyses (signal and transit peptides) are needed to detect complex plastid-targeted proteins.

ancestor of the plastid, while haptophytes and stramenopiles encode an enzyme of cyanobacterial origin, and some dinoflagellate genera contain both copies (Fig. 5) (Cihlář et al., 2016; Kořený et al., 2011).

Cryptophytes based on predictions in *Guillardia theta* (Box 2), employ the typical red-derived arrangement of the pathway known from haptophytes, stramenopiles, and peridinin-pigmented dinoflagellates (Cihlář et al., 2016) (Fig. 4D). In comparison with the arrangement of the latter groups, some of the genes have been duplicated, including the originally cyanobacterial gene encoding UROD, and the nuclear (eukaryotic) CPOX copy of the endosymbiont. *G. theta* also possesses an additional FeCH of eukaryotic origin (passed on from the aplastidic cryptomonad ancestor), which is putatively mitochondrion-targeted like its closest homologs from heterotrophic eukaryotes. FeCH is functionally closely coupled with PPOX, so there is no obvious reason for *G. theta* to retain the mitochondrial FeCH if PPOX would be missing from this compartment. The protein sequence of *G. theta* PPOX possesses a putative signal peptide suggesting its localization in the plastid, but additional in silico predictions with Mitoprot II (Claros & Vincens, 1996) and TargetP (Emanuelsson, Brunak, von Heijne, & Nielsen, 2007) revealed

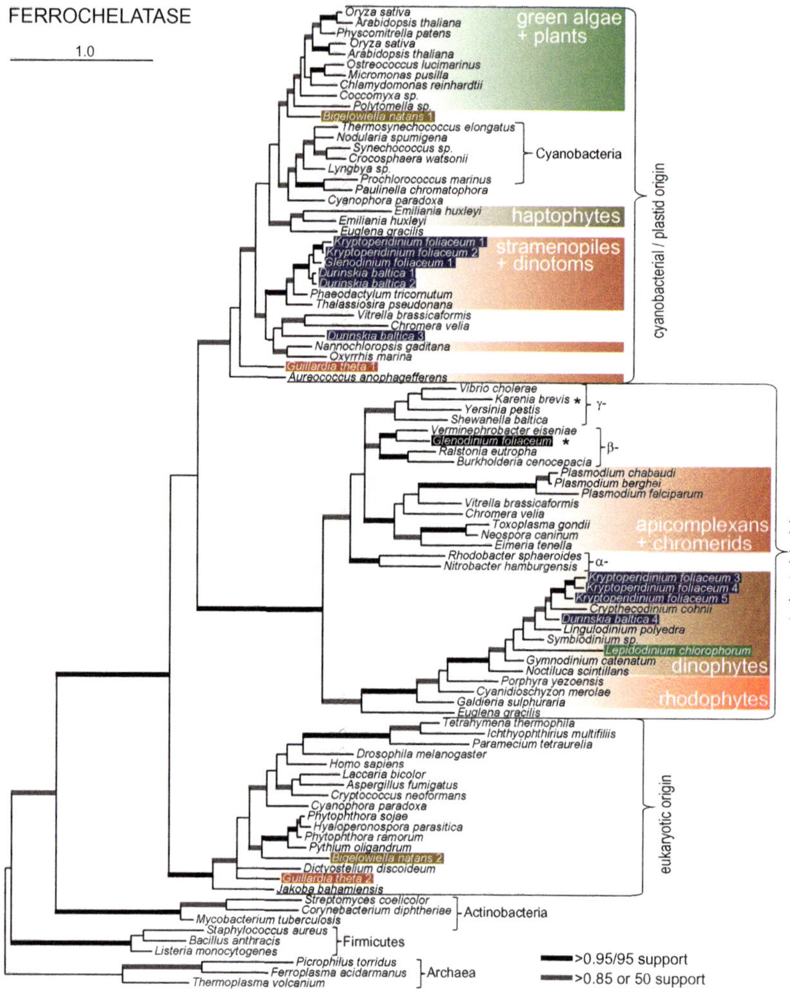

Fig. 5 Horizontal gene acquisitions of ferrochelatase in eukaryotes. The phylogenetic tree depicts independent origins of ferrochelatase sequences in algae, notably in apicomplexans and chromerids (directly from proteobacteria), dinoflagellates (from proteobacteria via rhodophytes), stramenopiles and haptophytes (from cyanobacteria via plastid endosymbiosis), and heterotrophic eukaryotes (sharing an ancestral eukaryotic gene with *Bigelowiella* and *Guillardia*). *Modified from Cihlář, J., Füssy, Z., Horák, A., & Oborník, M. (2016). Evolution of the tetrapyrrole biosynthetic pathway in secondary algae: Conservation, redundancy and replacement.* PLoS One, 11, *e0166338.*

that this presequence has also the characteristics of a mitochondrial transit peptide. PPOX may be therefore dual targeted to both the plastid and the mitochondrion (Cihlář, 2018). This allows to predict that *G. theta* is the sole rhodophyte-derived species that expresses a partial mitochondrial heme pathway. Such an arrangement may indicate a more recent plastid acquisition in cryptophytes (Burki et al., 2016; Cihlář et al., 2016), but indeed further experimental studies have to verify these predictions.

Other algae of the rhodophyte-derived lineage show various other arrangements of tetrapyrrole biosynthesis, reflecting more complicated endosymbiotic histories. Dinotoms, possessing diatom endosymbionts, engage two plastid-type tetrapyrrole pathways to supply independently tetrapyrroles to both the host and the endosymbiont (Cihlář et al., 2016; Hehenberger et al., 2014) (Fig. 4E). The pathway that is probably localized in the plastids of the diatom endosymbiont (Box 2), is encoded by the nucleomorph and consists of enzymes that cluster with homologs from free-living diatoms. The other pathway is host-encoded and represents the ancestral pathway of the peridinin-pigmented plastid, and likely supplies heme to the cytosol and mitochondria. In analogy, the enzymes of the plastid-type containing peridinin cluster together with other dinoflagellate sequences (Cihlář et al., 2016; Hehenberger et al., 2014). In kareniaceans, the tetrapyrrole pathway was probably retargeted to the newly acquired haptophyte-derived plastid, and is encoded by genes that share origin with homologs from peridinin-pigmented dinoflagellates but, at the same time, large influx of genes which were recruited from the endosymbiont or acquired horizontally, has been accented (Matsuo & Inagaki, 2018).

4.2.3 The unusual arrangement in chromerids and apicomplexans

Chromera velia and *Vitrella brassicaformis* are free-living or coral-associated algae with a rhodophyte-derived plastid and a close phylogenetic relationship to apicomplexan parasites (Janouškovec et al., 2015; Moore et al., 2008; Oborník et al., 2009, 2012, 2011; Woo et al., 2015). Unlike other phototrophs, *C. velia* and *V. brassicaformis* lack the enzymes of the C5 pathway. Instead, the mitochondrial C4 pathway supplies ALA for the synthesis of all tetrapyrroles (Füssy & Oborník, 2017a; Kořený et al., 2011; Oborník & Lukeš, 2013) as in aplastidic heterotrophs (Fig. 4B). Consistently, chromerid genomes lack the gene which encodes ferredoxin-dependent glutamate synthase, the plastid-localized enzyme synthesizing glutamate for the C5 pathway, which is ubiquitously present in green algae/plants, red algae, diatoms, and dinoflagellates (our unpublished data). Judging from the pathway arrangement in apicomplexans and closely related colpodellids (Fig. 4C; Janouškovec et al., 2015), ALA was synthesized exclusively from succinyl-CoA already in their phototrophic common ancestor.

Downstream of ALAS, all enzymes carry a well-predicted plastid-targeting presequence, suggesting their localization in the chromerid plastid stroma (Box 2). Although the spatial separation of the beginning and the end of the heme pathway was not found in any other organism, it seems that such an arrangement can satisfy the high demand of tetrapyrroles in the photosynthetic organelles of chromerids. This, on the other hand, necessitates efficient ALA transport to meet the demands of plastid tetrapyrrole synthesis. The feedback regulation of the pathway as known from other models (Czarnecki & Grimm, 2012; Masuda & Fujita, 2008; Vavilin & Vermaas, 2002) is probably not possible, so the regulation of tetrapyrrole synthesis in chromerids is an outstanding question to tackle.

Most of the apicomplexan parasites carry highly reduced non-photosynthetic plastids (termed apicoplasts), pointing out the photosynthetic abilities of their non-parasitic ancestor (Füssy & Oborník, 2017b; Janouškovec, Horák, Oborník, Lukeš, & Keeling, 2010; Köhler et al., 1997; Kořený et al., 2011; Maréchal & Cesbron-Delauw, 2001; McFadden, Reith, Munholland, & Lang-Unnasch, 1996; Moore et al., 2008; Wilson, 2002). The pathway is localized to three different cellular compartments and is a collection of enzymes of heterotrophic and photosynthetic origins, reflecting only a partial replacement of the host pathway genes by those of the endosymbiont (Kořený et al., 2013; Sato, Clough, Coates, & Wilson, 2004; van Dooren, Kennedy, & McFadden, 2012; Wu, 2006) (Fig. 4C). ALA is synthesized by ALAS as in chromerids (Sato et al., 2004; Varadharajan, Dhanasekaran, Bonday, Rangarajan, & Padmanaban, 2002; Wu, 2006). The following four steps (ALAD, PBGD, UROS and UROD) take place in the apicoplast (Nagaraj, Arumugam, et al., 2009; Nagaraj et al., 2008; Sato et al., 2004; Wu, 2006), except in *Toxoplasma gondii* where UROD is localized in the cytosol (Seeber & Soldati-Favre, 2010; Wu, 2006). An effective metabolic flow between the mitochondrion and the apicoplast is probably facilitated by an intimate association between these organelles (Botté et al., 2013; Hopkins et al., 1999; Waller & McFadden, 2005). CPOX is located in the cytosol of all apicomplexans investigated so far (Nagaraj, Prasad, Arumugam, Rangarajan, & Padmanaban, 2010; Wu, 2006) and the last two steps of heme synthesis (PPOX and FeCH) are again placed in the mitochondrion (Nagaraj, Arumugam, Prasad, Rangarajan, & Padmanaban, 2010; Nagaraj, Prasad, Rangarajan, & Padmanaban, 2009; van Dooren, Stimmler, & McFadden, 2006; Wu, 2006). The terminal part of the apicomplexan tetrapyrrole pathway hence differs from that of chromerids, which may have been triggered by the transition from the phototrophic to the parasitic lifestyle. The chromerid and apicomplexan orthologs involved in these steps share evolutionary origin but not localization being suggestive of the re-targeting of these proteins in Apicomplexa.

Despite the fact that the apicoplast is ultimately evolved from rhodophyte plastids, only a fraction of enzymes (ALAD, PBGD, UROS, and PPOX) originates in the endosymbiont. ALAS, UROD and CPOX were derived from the host pathway, and the FeCH-encoding sequence originates from proteobacteria (Cihlář et al., 2016; Kořený et al., 2011; Oborník & Green, 2005; Sato & Wilson, 2003). Notably, this FeCH is unrelated to that of rhodophytes and dinoflagellates suggesting it was independently acquired from bacteria (Kořený et al., 2011; Sato & Wilson, 2003). In comparison, the chromerid tetrapyrrole pathway is a composite of enzymes found in mitochondria, the apicoplast, and the red lineage plastids, especially the steps following UROS. A mosaic origin is particularly apparent on the example of UROD, CPOX, and PPOX (Fig. 4B). These enzymes are present in multiple copies. While one copy shares host origin with the apicomplexan orthologs, other copies have an origin in the endosymbiont nucleus (Kořený et al., 2011). The pathway differs very little among chromerids, *V. brassicaformis* UROD is encoded by a single copy gene and the endosymbiont-derived copies were lost (Woo et al., 2015). Chromerids may retain the additional copies (compared to apicomplexans) for regulatory reasons, one being constitutively expressed, while the other(s) responsive to stressful or environmental traits, similarly as *Arabidopsis* (Matsumoto et al., 2004), but currently there is no published quantitative analysis of their expression.

4.2.4 Complex algae with chlorophyte-derived plastids

The dinoflagellate *Lepidodinium chlorophorum* (Elbrachter & Schnepf, 1996; Takishita et al., 2008) contains a chlorophyte-derived plastid that replaced the ancestral peridinin-pigmented plastid. Genes for nearly all enzymatic steps of the plastid-located tetrapyrrole pathway were found in the available transcriptomic data. This led to an interesting discovery concerning the origin of particular enzymes (Cihlář et al., 2016; Matsuo & Inagaki, 2018) (Fig. 6). Surprisingly, some enzymes (GTR, PPOX, and FeCH) cluster with red and red-derived algae. Other enzymes (ALAD, GSA-AT, and PPOX) group in clades with green algae and plants at the base, but always sister to other dinoflagellate sequences (Cihlář et al., 2016). The close relationship of these *L. chlorophorum* enzymes to homologs from peridinin-pigmented dinoflagellates (that possess red-derived plastids) suggests that the pathway is orthologous in the green plastid of *Lepidodinium* and peridinin-pigmented

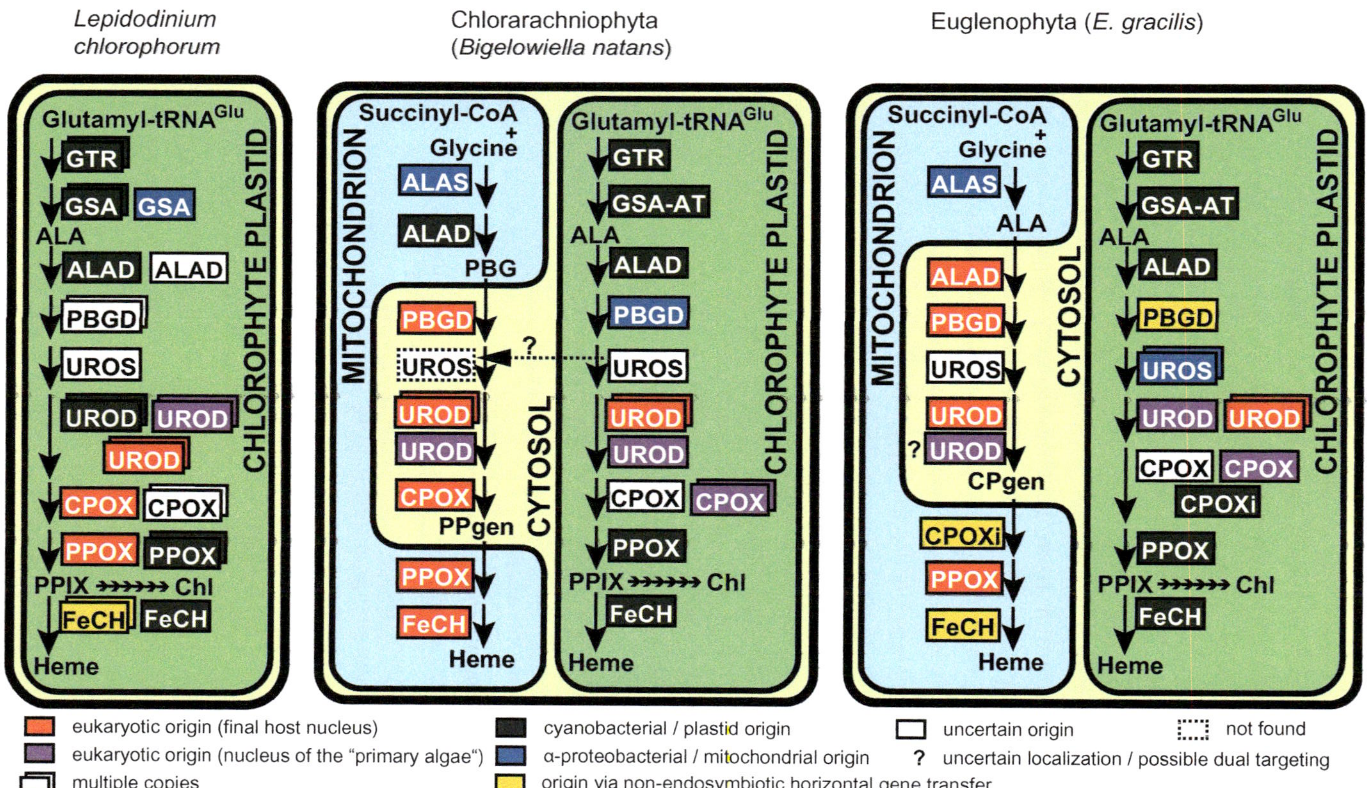

Fig. 6 Origins and subcellular locations of heme biosynthesis enzymes in complex algae with green-derived plastids. Enzyme abbreviations are listed in Fig. 1. Colored rectangles indicate the inferred origins of genes. The origins of *L. chlorophorum* and chlorarachniophyte enzymes are based on works of Cihlář et al. (2016) and Matsuo and Inagaki (2018), the localizations are based on our in silico predictions. Note that the scheme for *Lepidodinium* is a compilation of data from two strains. The scheme of *E. gracilis* heme pathways was modified according to Kořený and Oborník (2011) and Füssy and Záhonová et al. (unpublished).

plastids of dinoflagellates. In other words, *L. chlorophorum* appears to have employed some of the ancestral genes to function in the newly acquired plastid. The original tetrapyrrole pathway, introduced with the peridinin-containing plastid, hence remained functionally conserved in a novel organellar context (Cihlář et al., 2016). However, a study on a different strain of *Lepidodinium* found a large number of genes which were apparently recruited horizontally from non-endosymbiotic genetic sources (Matsuo & Inagaki, 2018).

The endosymbiotic events leading to the green-derived plastids of photosynthetic euglenids (euglenophytes) and chlorarachniophytes involved independent endosymbionts, unlike possibly the events forming the red-derived lineages (e.g., Burki, 2017; Jackson et al., 2018). Euglenophytes constitute an advanced monophyletic group and the presence of plastids or their remnants has never been confirmed in early-branching heterotrophic euglenozoans (trypanosomatids and diplonemids), except for *Rapaza viridis* that holds for an intermediary state of kleptoplasty among them (Hrdá, Fousek, Szabová, Hampl, & Vlček, 2012; Yamaguchi, Yubuki, & Leander, 2012). Biochemical experiments showed that *Euglena gracilis* synthesizes ALA by both the C4 and the C5 pathway, suggesting that *E. gracilis* engages two independent tetrapyrrole pathways (Weinstein & Beale, 1983). This was later confirmed on transcript sequence data (Füssy and Záhonová et al., unpublished; Kořený & Oborník, 2011; Lakey & Triemer, 2016) (Fig. 6). Similarly to *E. gracilis*, the chlorarachniophyte *Bigelowiella natans* also expresses two parallel tetrapyrrole pathways (Cihlář et al., 2016). In both *Euglena* and *Bigelowiella*, the arrangement of one of the pathways resembles the synthesis of tetrapyrroles in aplastidic heterotrophs and represents the ancestral host pathway, while the other pathway is plastid-localized and originates from the algal endosymbiont (Fig. 6).

Regardless of the independent origins of chlorarachniophyte (Rhizaria) and euglenophyte plastids (Excavata) (Rogers et al., 2007), they exhibit similar redirection of tetrapyrrole synthesis. In *Bigelowiella* the plastid-derived ALAD and UROS functionally replaced their heterotrophic counterparts (Cihlář et al., 2016), while paralogs of the host-derived UROD seem to be retargeted to the plastid in both algae. The mitochondrial localization of the cyanobacterial ALAD is notable in *B. natans*. Apart from these changes, the arrangement of the tetrapyrrole pathway in these algae corresponds to what was already found in other heterotrophs and phototrophs.

In spite of the chlorophyte origin of the *B. natans* plastid (Rogers et al., 2007), only the plastid copy of FeCH shows an expected chlorophyte origin and, strikingly, most genes of the plastid-localized pathway exhibit affinity to the red lineage. It has been reported that *B. natans* encodes a large number of rhodophyte-derived genes, including photosynthesis-related genes, suggesting multiple horizontal gene transfers from the red lineage (Archibald, Rogers, Toop, Ishida, & Keeling, 2003; Yang, Matsuzaki, Takahashi, Qu, & Nozaki, 2014). Nevertheless, the necessity to obtain a plastid-targeting presequence for several individual enzymes presumably complicates extensive functional replacements of genes that belong to a multi-step biochemical pathway. At the same time, the similarities of the origin of the tetrapyrrole biosynthesis pathway between *L. chlorophorum* and *B. natans* are conspicuous and open the possibility that a rhodophyte-derived plastid was maintained by early chlorarachniophytes. It is likely that this rhodophyte-derived plastid could be the donor of the tetrapyrrole pathway genes currently present in *B. natans*, as much as the tetrapyrrole enzymes of the peridinin-pigmented plastid operate in the current green plastid of *L. chlorophorum* (Cihlář et al., 2016).

4.3 Chlorophyll biosynthesis in algae and plants

Phylogenetic analyses of the Chl branch enzymes are scarce. An interesting evolutionary pattern was found for light-dependent protochlorophyllide oxidoreductase (POR) (Hunsperger, Randhawa, & Cattolico, 2015). Unexpectedly, POR is not directly related to the ancestral rhodophyte POR in the majority of rhodophyte-derived complex algae. Only in cryptophytes and two species of stramenopiles POR appears phylogenetically related to POR of rhodophytes, while most other stramenopiles, haptophytes, and several dinoflagellates encode only chlorophyte-related POR. Rhodophyte- and chlorophyte-derived PORs are structurally similar and supposedly functionally equivalent, so the pattern of their evolutionary distribution suggests that the ancestral POR gene of the rhodophyte plastid was replaced by a gene horizontally transferred from prasinophytes in most red-derived lineages (Hunsperger et al., 2015). Data addressing other Chl branch enzymes are even more problematic, a catchy example being the MgCH subunit H with two divergent homologs, neither of which follows the expected chlorophytes-to-green lineage and rhodophyte-to-red lineage evolutionary scenarios. In the phylogenies of other Chl branch enzymes (Matsuo & Inagaki, 2018), deep branches are also infrequently resolved, although crown groups tend to be monophyletic. This complicates inferences about the exact trajectory of genes participating in Chl synthesis, but may suggest several horizontal gene transfers have occurred at the time when major complex algal lineages were established.

4.4 What does tetrapyrrole synthesis reveal about the evolution of algae?

Understanding the way how biochemical pathway arrangements respond to endosymbiotic events allows us to look back in time at the general and lineage-specific factors that modulated these partnerships. Having some degree of conservancy, the collection of tetrapyrrole enzymes indeed testifies about the past events that come up during the evolution of individual lineages. The arrangement of the pathway directly reflects the trophic strategy of a given organism, allowing for differentiation between aplastidic heterotrophs, newly established endosymbioses, developed endosymbioses, and even non-photosynthetic algae, such as the apicomplexan parasites.

Heterotrophs rely on the mitochondrial/cytosolic tetrapyrrole synthesis, which is clearly distinct from the plastid-type synthesis typical for autotrophs. During early stages of endosymbiosis, which can be regarded as prolonged retention of algal prey within algivores (Bodył, 2017), plastids are not essential for the host survival and can be lost inadvertently (although specialist predators would starve without their prey). Indeed, temporary plastid sequestration (kleptoplasty) occurs in several lineages closely or distantly related to the recent phototrophic crown groups (Bodył, 2017; Yamaguchi et al., 2012; Oborník, Janouškovec, Chrudimský, & Lukeš, 2009). And based on phylogenetic evidence, some early branching lineages related to the algal crown groups are suggested to have had and lost a plastid (Janouškovec et al., 2010; Kořený et al., 2011). Hosts and their kleptoplastids retain their own metabolic pathways, though in some cases host-encoded factors might reinforce plastid function and longevity (Wisecaver & Hackett, 2010).

The bi-parallel arrangement of tetrapyrrole synthesis in euglenophytes and chlorarachniophytes is unique but probably represents a significant milestone in the evolutionary history of each endosymbiotic event. This arrangement reflects that both pathways may coexist for a period of time, and that the loss of the mitochondrial/cytosolic pathway as seen in other phototrophs may be gradual, and likely dependent on inter-compartmental exchange of tetrapyrroles or their intermediates. Each algal lineage passed through a similar stage of metabolic redundancy during the transformation from a heterotroph (Cihlář et al., 2016; Kořený & Oborník, 2011; Waller, Gornik, Kořený, & Pain, 2016). A bi-parallel organization is also expected in

Paulinella and its chromatophores (Fig. 2), and bi-parallel is also the arrangement in dinotoms and their highly independent diatom plastids (Fig. 4E).

In later stages of endosymbioses, plastids take over metabolic processes of the cytosol or the mitochondrion. Following such metabolic takeover, exemplified by the tetrapyrrole synthesis, plastids become indispensable for the survival of the host. This is often rationalized by metabolic and regulatory demands of the plastid compartment. Photosynthetic cells indeed require (and produce) more Chl and heme in the plastid in response to light (Papenbrock, Mock, Kruse, & Grimm, 1999). The plastid tetrapyrrole biosynthesis pathways, thus, developed a rather complex regulation that involves negative feedback loops on ALA synthesis from both the heme and Chl branch (Czarnecki & Grimm, 2012; Masuda & Fujita, 2008; Vavilin & Vermaas, 2002). Despite chromerids initiate tetrapyrrole synthesis in mitochondria, and must have developed a unique regulatory strategy, it was probably advantageous for phototrophs to employ a pathway entirely localized to the plastid, leading to the loss of importance for the mitochondrial/cytosolic pathway and an increased metabolic dependency on the plastid. The smaller demand for tetrapyrroles from other cellular compartments, especially the need for heme for the respiratory chain cytochromes in mitochondria, was then satisfied by tetrapyrrole export from the plastid by yet unknown mechanisms (Brzezowski et al., 2015; Zhang et al., 2015). Dependency on the plastid, i.e., on the production of not only tetrapyrroles but also other compounds, is most likely the reason why plastids remain, even in non-photosynthetic or parasitic lineages such as apicomplexans. Accordingly, many non-photosynthetic plastids house tetrapyrrole biosynthesis (reviewed in Hadariová, Vesteg, Hampl, & Krajčovič, 2018).

Presuming that enzyme homologs are functionally equivalent, they may complement or even replace one another during gene transfers to the host nucleus. Just as phylogenetic analyses report, some tetrapyrrole enzymes of "mitochondrial," "cytosolic," and "plastid" origin have indeed been relocated to a different compartment (see above). This is consistent with the mosaic evolution of not only the tetrapyrrole pathway in various presented organisms, but also of organellar proteomes in general (Gray, 2015; Qiu et al., 2013). Evolutionary mosaics develop as explained by the "shopping bag" or "plastid promiscuity" hypotheses (Larkum, Lockhart, & Howe, 2007), whereby host-encoded factors collected from various genetic sources can cooperate to maintain functions of an organelle of a defined, single origin. The mosaic complexity of organellar proteomes increases with the number and frequency of genetic interactions of organisms (Husník & McCutcheon, 2018). In case of serial plastid endosymbioses, this leads to an even higher mosaicism. The ancestral genes for tetrapyrrole pathway are quite conserved in the chlorophyte-derived plastid of *Lepidodinium* as well as the haptophyte-derived plastids of Kareniaceae (Cihlář et al., 2016; Matsuo & Inagaki, 2018), though a battery of new genes were acquired from various sources to function along with the original tetrapyrrole genes.

With a closer look on the pathway steps, we uncover the phylogenetic origin of each enzyme homolog, and the relationship among the host and the homolog donor. Besides ancestral, vertically inherited genes, homologs can be recruited from the endosymbiont, or horizontally from other sources, such as food and pathogens. While the contributions of these sources are hotly debated (Burki, 2017; Eliáš & Archibald, 2009; Moreira & Deschamps, 2014), it is clear that systematic recruitment of genes from a specific source indicates some biological interaction. A high contribution of ancestral genes in the tetrapyrrole pathways of *Lepidodinium* and kareniaceans points out an unpredicted resilience of the pathway to gene replacement, stemming from the necessity for its functionality throughout all stages of serial plastid endosymbioses. The lack of vertically inherited genes for the Chl branch, on the other hand, led to a suggestion that prior to ability to retain the current photosynthetic plastid, the ancestors of *Lepidodinium* and kareniaceans were heterotrophic (did not synthesize Chl), and that genes for this branch had to be reintroduced from the endosymbiont (Matsuo & Inagaki, 2018). Nevertheless, the presence of an ancestral POR in *Lepidodinium* questions this scenario.

A closer look at gene origin also points out that few genes may be recruited from a chlorophyte endosymbiont, if the ancestral plastid was a complex one. This is a direct result of differences of protein targeting into two-membrane and multi-membrane plastids, as proteins conveying to complex plastids carry a so-called bipartite targeting presequence, while proteins of primary algae carry a simpler presequence (Patron & Waller, 2007). Genes that encode proteins for primary plastids are therefore unlikely to be transferred to the host nucleus and correctly localize back to the multiple membrane-bound endosymbiont. This also explains why in *Lepidodinium* a majority of newly acquired tetrapyrrole genes came from sources other than chlorophytes, mostly complex algae (Cihlář et al., 2016; Matsuo & Inagaki, 2018). Analogously in *Bigelowiella*, which appears to encode a high number of rhodophyte- and red lineage-related genes (Archibald et al., 2003; Cihlář et al., 2016; Yang et al., 2014), we could expect some long-term interactions with rhodophyte(-derived) algae despite the clear chlorophyte descent of the chlorarachniophyte plastid (Rogers et al., 2007). Indeed, it would not be the first time we could not see the forest for the trees, and the organelles we see today may be the "lucky ones" that persisted the long run of attempts to form symbiotic relationships (Keeling, McCutcheon, & Doolittle, 2015).

The tetrapyrrole pathway represents a great model to illustrate the many factors that influence gene transfer and organelle endosymbiosis as such. Genes encoding the pathway enzymes are sufficiently resilient to replacement, but allow for a certain level of functional redundancy. As a result, phylogenetic analyses attest both ancient and more recent gene flows and bring evidence for evolutionary hypotheses that emerged to account for reductive evolution, genome mosaicism, plastid dependency, and other aspects related to endosymbioses. It is important to note that these principles are also reflected by other biological modules, such as the fatty acid and isopentenyl pyrophosphate synthesis, but also RNA biology (Dorrell & Howe, 2015; Lohr, Schwender, & Polle, 2012; Waller et al., 2016). However, the available data need to be carefully revised to find out if the evolution of these pathways may help to reliably answer the current issues that evolutionary protistologists face, including cryptic endosymbioses (such as that supposed in chlorarachniophytes), the position of the rhodophyte-derived algae on the tree of eukaryotes (mainly the position of Haptophyta and Cryptophyta), and the order of rhodophyte plastid transfers among the red-derived lineages.

5. Conclusion

The tetrapyrrole pathway plays a crucial role in the metabolism of the vast majority of organisms. With plastid endosymbiosis ongoing, organisms suddenly face metabolic redundancy that is in most algae resolved by plastid takeover of the tetrapyrrole supply for the whole cell. However, some algae possess parallel or hybrid pathways for tetrapyrrole synthesis. Dinotoms synthesize tetrapyrroles concurrently in both the tertiary diatom plastids and the remnant of peridinin plastids; *E. gracilis* and *B. natans* possess both the mitochondrial/cytosolic and the plastid tetrapyrrole pathways, thus probably represent an intermediate state in endosymbiosis; and chromerids use a hybrid pathway initiated in the mitochondrion via the C4 reaction and continuing in the plastid. From the phylogenetic point of view, the tetrapyrrole pathway of all eukaryotes is a true mosaic, an imaginary shopping bag of enzymes originating not only from eukaryotes, proteobacteria, and cyanobacteria, collected and combined throughout the history of mitochondrion and plastid endosymbioses, but also randomly obtained from various sources via horizontal gene transfer. Under some circumstances, the pathways appear to be evolutionarily well conserved even following serial endosymbioses, especially those seen in the dinoflagellate *Lepidodinium chlorophorum* and the chlorarachniophyte *Bigelowiella natans*, allowing for hypotheses about past evolutionary events.

Acknowledgments

The authors would like to thank the Czech Science Foundation (projects 15-17643S and 16-24027S granted to M.O.) and the European Regional Development Fund (ERDF) (No. CZ.02.1.01/0.0/0.0/16_019/0000759) for funding. We are also grateful to Luděk Kořený for helpful discussions.

References

Archibald, J. M. (2009). The puzzle of plastid evolution. *Current Biology*, *19*, R81–R88.

Archibald, J. M. (2015). Genomic perspectives on the birth and spread of plastids. *Proceedings of the National Academy of Sciences of the United States of America*, *112*, 10147–10153.

Archibald, J. M., Rogers, M. B., Toop, M., Ishida, K., & Keeling, P. J. (2003). Lateral gene transfer and the evolution of plastid-targeted proteins in the secondary plastid-containing alga *Bigelowiella natans*. *Proceedings of the National Academy of Sciences of the United States of America*, *100*, 7678–7683.

Beale, S. I. (1999). Enzymes of chlorophyll biosynthesis. *Photosynthesis Research*, *60*, 43–73.

Beale, S. I., & Weinstein, J. D. (1991). Biochemistry and regulation of photosynthetic pigment formation in plants and algae. In P. M. Jordan (Ed.), *Biosynthesis of tetrapyrroles* (pp. 155–235): Elsevier Science Publishers B.V.

Bodył, A. (2017). Did some red alga-derived plastids evolve via kleptoplastidy? A hypothesis. *Biological Reviews of the Cambridge Philosophical Society*, *93*, 201–222.

Bodył, A., Stiller, J. W., & Mackiewicz, P. (2009). Chromalveolate plastids: Direct descent or multiple endosymbioses? *Trends in Ecology & Evolution*, *24*, 119–121.

Botté, C. Y., Yamaryo-Botte, Y., Rupasinghe, T. W. T., Mullin, K. A., MacRae, J. I., Spurck, T. P., et al. (2013). Atypical lipid composition in the purified relict plastid (apicoplast) of malaria parasites. *Proceedings of the National Academy of Sciences of the United States of America*, *110*, 7506–7511.

Brzezowski, P., Richter, A. S., & Grimm, B. (2015). Regulation and function of tetrapyrrole biosynthesis in plants and algae. *Biochimica et Biophysica Acta*, *1847*(9), 968–985.

Burki, F., & Hirakawa, Y. (2017). The convoluted evolution of eukaryotes with complex plastids. In *Vol. 84. Secondary endosymbioses. Advances in botanical research* (pp. 1–30): Academic Press.

Burki, F., Kaplan, M., Tikhonenkov, D. V., Zlatogursky, V., Minh, B. Q., Radaykina, L. V., et al. (2016). Untangling the early diversification of eukaryotes: A phylogenomic study of the evolutionary origins of centrohelida, haptophyta and cryptista. *Proceedings of the Royal Society of London B: Biological Sciences*, *283*, 20152802.

Butterfield, E. R., Howe, C. J., & Nisbet, R. E. (2013). An analysis of dinoflagellate metabolism using EST data. *Protist*, *164*, 218–236.

Chen, M., & Blankenship, R. E. (2011). Expanding the solar spectrum used by photosynthesis. *Trends in Plant Science*, *16*, 427–431.

Chen, Z. X., Silva, H., & Klessig, D. F. (1993). Active oxygen species in the induction of plant systemic acquired-resistance by salicylic-acid. *Science, 262*, 1883–1886.

Chesnick, H. M., Morden, C. W., & Schmieg, A. M. (1996). Identity of the endosymbiont of *Peridinium foliaceum* (Pyrrophyta): Analysis of the rbcLC operon. *Journal of Phycology, 32*, 850–857.

Chiabrando, D., Mercurio, S., & Tolosano, E. (2014). Heme and erythropoieis: More than a structural role. *Haematologica, 99*, 973–983.

Chow, K. S., Singh, D. P., Roper, J. M., & Smith, A. G. (1997). A single precursor protein for ferrochelatase-I from *Arabidopsis* is imported in vitro into both chloroplasts and mitochondria. *Journal of Biological Chemistry, 272*, 27565–27571.

Chow, K. S., Singh, D. P., Walker, A. R., & Smith, A. G. (1998). Two different genes encode ferrochelatase in *Arabidopsis*: Mapping, expression and subcellular targeting of the precursor proteins. *Plant Journal, 15*, 531–541.

Cihlář, J. (2018). *Evolution of the heme biosynthetic pathway in eukaryotic phototrophs.* Ph.D. Thesis Series, České Budějovice, Czech Republic: University of South Bohemia, Faculty of Science, School of Doctoral Studies in Biological Sciences.

Cihlář, J., Füssy, Z., Horák, A., & Oborník, M. (2016). Evolution of the tetrapyrrole biosynthetic pathway in secondary algae: Conservation, redundancy and replacement. *PLoS One, 11*, e0166338.

Claros, M. G., & Vincens, P. (1996). Computational method to predict mitochondrially imported proteins and their targeting sequences. *European Journal of Biochemistry, 241*, 779–786.

Cornah, J. E., Roper, J. M., Pal Singh, D., & Smith, A. G. (2002). Measurement of ferrochelatase activity using a novel assay suggests that plastids are the major site of haem biosynthesis in both photosynthetic and nonphotosynthetic cells of pea (*Pisum sativum* L.). *Biochemical Journal, 362*, 423–432.

Czarnecki, O., & Grimm, B. (2012). Post-translational control of tetrapyrrole biosynthesis in plants, algae, and cyanobacteria. *Journal of Experimental Botany, 63*, 1675–1687.

Dailey, H. A., Gerdes, S., Dailey, T. A., Burch, J. S., & Phillips, J. D. (2015). Noncanonical coproporphyrin-dependent bacterial heme biosynthesis pathway that does not use protoporphyrin. *Proceedings of the National Academy of Sciences of the United States of America, 112*, 2210–2215.

Decker, K., Jungermann, K., & Thauer, R. K. (1970). Energy production in anaerobic organisms. *Angewandte Chemie International Edition in English, 9*, 138–158.

Delwiche, C. F. (1999). Tracing the thread of plastid diversity through the tapestry of Life. *American Naturalist, 154*, S164–S177.

Dorrell, R., & Howe, C. (2015). Integration of plastids with their hosts: Lessons learned from dinoflagellates. *Proceedings of the National Academy of Sciences of the United States of America, 112*, 10247–10254.

Duncan, R., Faggart, M. A., Roger, A. J., & Cornell, N. W. (1999). Phylogenetic analysis of the 5-aminolevulinate synthase gene. *Molecular Biology and Evolution, 16*, 383–396.

Elbrachter, M., & Schnepf, E. (1996). *Gymnodinium chlorophorum*, a new, green, bloom-forming dinoflagellate (Gymnodiniales, Dinophyceae) with a vestigial prasinophyte endosymbiont. *Phycologia, 35*, 381–393.

Eliáš, M., & Archibald, J. M. (2009). Sizing up the genomic footprint of endosymbiosis. *BioEssays, 31*, 1273–1279.

Emanuelsson, O., Brunak, S., von Heijne, G., & Nielsen, H. (2007). Locating proteins in the cell using TargetP, SignalP and related tools. *Nature Protocols, 2*, 953–971.

Ferreira, G. C., Andrew, T. L., Karr, S. W., & Dailey, H. A. (1988). Organization of the terminal two enzymes of the heme biosynthetic pathway. Orientation of protoporphyrinogen oxidase and evidence for a membrane complex. *Journal of Biological Chemistry, 263*, 3835–3839.

Ferreira, G. C., & Gong, J. (1995). 5-Aminolevulinate synthase and the first step of heme biosynthesis. *Journal of Bioenergetics and Biomembranes*, *27*, 151–159.
Frankenberg, N., Moser, J., & Jahn, D. (2003). Bacterial heme biosynthesis and its biotechnological application. *Applied Microbiology and Biotechnology*, *63*, 115–127.
Füssy, Z., & Oborník, M. (2017a). Chromerids and their plastids. *Advances in Botanical Research*, *84*, 187–218.
Füssy, Z., & Oborník, M. (2017b). Reductive evolution of apicomplexan parasites from phototrophic ancestors. In P. Pontarotti (Ed.), *Evolutionary biology: self/nonself evolution, species and complex traits evolution, methods and concepts* (pp. 217–236): Springer International Publishing AG.
Garg, S. G., & Gould, S. B. (2016). The role of charge in protein targeting evolution. *Trends in Cell Biology*, *26*, 894–905.
Gould, S. B., Maier, U. G., & Martin, W. F. (2015). Protein import and the origin of red complex plastids. *Current Biology*, *25*, R515–R521.
Gray, M. W. (2015). Mosaic nature of the mitochondrial proteome: Implications for the origin and evolution of mitochondria. *Proceedings of the National Academy of Sciences of the United States of America*, *112*, 10133–10138.
Gruber, A., Rocap, G., Kroth, P. G., Armbrust, E. V., & Mock, T. (2015). Plastid proteome prediction for diatoms and other algae with secondary plastids of the red lineage. *Plant Journal*, *81*, 519–528.
Gschloessl, B., Guermeur, Y., & Cock, J. M. (2008). HECTAR: A method to predict subcellular targeting in heterokonts. *BMC Bioinformatics*, *9*, 393.
Guernsey, D. L., Jiang, H., Campagna, D. R., Evans, S. C., Ferguson, M., Kellogg, M. D., et al. (2009). Mutations in mitochondrial carrier family gene SLC25A38 cause nonsyndromic autosomal recessive congenital sideroblastic anemia. *Nature Genetics*, *41*, 651–653.
Hadariová, L., Vesteg, M., Hampl, V., & Krajčovič, J. (2018). Reductive evolution of chloroplasts in non-photosynthetic plants, algae and protists. *Current Genetics*, *64*, 365–387.
Hamza, I., & Dailey, H. A. (2012). One ring to rule them all: Trafficking of heme and heme synthesis intermediates in the metazoans. *Biochimica et Biophysica Acta*, *1823*, 1617–1632.
Hedges, S. B., Blair, J. E., Venturi, M. L., & Shoe, J. L. (2004). A molecular timescale of eukaryote evolution and the rise of complex multicellular life. *BMC Evolutionary Biology*, *4*, 2.
Hehenberger, E., Imanian, B., Burki, F., & Keeling, P. J. (2014). Evidence for the retention of two evolutionary distinct plastids in dinoflagellates with diatom endosymbionts. *Genome Biology and Evolution*, *6*, 2321–2334.
Heinemann, I. U., Jahn, M., & Jahn, D. (2008). The biochemistry of heme biosynthesis. *Archives of Biochemistry and Biophysics*, *474*, 238–251.
Helliwell, K. E., Wheeler, G. L., Leptos, K. C., Goldstein, R. E., & Smith, A. G. (2011). Insights into the evolution of vitamin B12 auxotrophy from sequenced algal genomes. *Molecular Biology and Evolution*, *28*, 2921–2933.
Hey, D., Ortega-Rodes, P., Fan, T., Schnurrer, F., Brings, L., Hedtke, B., et al. (2016). Transgenic tobacco lines expressing sense or antisense ferrochelatase 1 RNA show modified ferrochelatase activity in roots and provide experimental evidence for dual localization of ferrochelatase 1. *Plant and Cell Physiology*, *57*, 2576–2585.
Hirakawa, Y. (Ed.), (2017). In *Advances in botanical research*: *Vol. 84*. *Secondary endosymbioses*. Academic Press.
Hopkins, J., Fowler, R., Krishna, S., Wilson, I., Mitchell, G., & Bannister, L. (1999). The plastid in *Plasmodium falciparum* asexual blood stages: A three-dimensional ultrastructural analysis. *Protist*, *150*, 283–295.

Hrdá, Š., Fousek, J., Szabová, J., Hampl, V., & Vlček, Č. (2012). The plastid genome of *Eutreptiella* provides a window into the process of secondary endosymbiosis of plastid in euglenids. *PLoS One*, *7*, e33746.

Hunsperger, H. M., Randhawa, T., & Cattolico, R. A. (2015). Extensive horizontal gene transfer, duplication, and loss of chlorophyll synthesis genes in the algae. *BMC Evolutionary Biology*, *15*, 16.

Husník, F., & McCutcheon, J. P. (2018). Functional horizontal gene transfer from bacteria to eukaryotes. *Nature Reviews Microbiology*, *16*, 67–79.

Imanian, B., & Keeling, P. J. (2007). The dinoflagellates *Durinskia baltica* and *Kryptoperidinium foliaceum* retain functionally overlapping mitochondria from two evolutionarily distinct lineages. *BMC Evolutionary Biology*, *7*, 172.

Inagaki, Y., Dacks, J. B., Doolittle, W. F., Watanabe, K. I., & Ohama, T. (2000). Evolutionary relationship between dinoflagellates bearing obligate diatom endosymbionts: Insight into tertiary endosymbiosis. *International Journal of Systematic and Evolutionary Microbiology*, *50*, 2075–2081.

Ishida, K., & Green, B. R. (2002). Second- and third-hand chloroplasts in dinoflagellates: Phylogeny of oxygen-evolving enhancer 1 (PsbO) protein reveals replacement of a nuclear-encoded plastid gene by that of a haptophyte tertiary endosymbiont. *Proceedings of the National Academy of Sciences of the United States of America*, *99*, 9294–9299.

Jackson, C., Knoll, A. H., Chan, C. X., & Verbruggen, H. (2018). Plastid phylogenomics with broad taxon sampling further elucidates the distinct evolutionary origins and timing of secondary green plastids. *Scientific Reports*, *8*, 1–10.

Janouškovec, J., Horák, A., Oborník, M., Lukeš, J., & Keeling, P. J. (2010). A common red algal origin of the apicomplexan, dinoflagellate, and heterokont plastids. *Proceedings of the National Academy of Sciences of the United States of America*, *107*, 10949–10954.

Janouškovec, J., Tikhonenkov, D. V., Burki, F., Howe, A. T., Kolísko, M., Mylnikov, A. P., et al. (2015). Factors mediating plastid dependency and the origins of parasitism in apicomplexans and their close relatives. *Proceedings of the National Academy of Sciences of the United States of America*, *112*, 10200–10207.

Jiroutová, K., Kořený, L., Bowler, C., & Oborník, M. (2010). A gene in the process of endosymbiotic transfer. *PLoS One*, *5*, e13234.

Jordan, P. M., & Shemin, D. (1972). D-aminolevulinic acid synthetase. In P. D. Boyer (Ed.), *Vol. 7. The enzymes* (pp. 339–356). New York/London: Academic Press.

Keeling, P. J. (2010). The endosymbiotic origin, diversification and fate of plastids. *Philosophical Transactions of the Royal Society of London Series B, Biological Sciences*, *365*, 729–748.

Keeling, P. J. (2013). The number, speed, and impact of plastid endosymbioses in eukaryotic evolution. *Annual Review of Plant Biology*, *64*, 583–607.

Keeling, P. J., McCutcheon, J. P., & Doolittle, W. F. (2015). Symbiosis becoming permanent: Survival of the luckiest. *Proceedings of the National Academy of Sciences of the United States of America*, *112*, 10101–10103.

Kim, S., & Park, M. G. (2016). *Paulinella longichromatophora* sp. nov., a new marine photosynthetic testate amoeba containing a chromatophore. *Protist*, *167*, 1–12.

Köhler, S., Delwiche, C. F., Denny, P. W., Tilney, L. G., Webster, P., Wilson, R. J., et al. (1997). A plastid of probable green algal origin in apicomplexan parasites. *Science*, *275*, 1485–1489.

Koonin, E. V. (2015). Origin of eukaryotes from within archaea, archaeal eukaryome and bursts of gene gain: Eukaryogenesis just made easier? *Philosophical Transactions of the Royal Society of London. Series B, Biological Sciences*, *370*, 20140333.

Kořený, L., Lukeš, J., & Oborník, M. (2010). Evolution of the haem synthetic pathway in kinetoplastid flagellates: An essential pathway that is not essential after all? *International Journal for Parasitology*, *40*, 149–156.

Kořený, L., & Oborník, M. (2011). Sequence evidence for the presence of two tetrapyrrole pathways in *Euglena gracilis*. *Genome Biology and Evolution*, *3*, 359–364.

Kořený, L., Oborník, M., & Lukeš, J. (2013). Make it, take it, or leave it: Heme metabolism of parasites. *PLoS Pathogens*, *9*, e1003088.

Kořený, L., Sobotka, R., Janouškovec, J., Keeling, P. J., & Oborník, M. (2011). Tetrapyrrole synthesis of photosynthetic chromerids is likely homologous to the unusual pathway of apicomplexan parasites. *Plant Cell*, *23*, 3454–3462.

Kořený, L., Sobotka, R., Kovářová, J., Gnipová, A., Flegontov, P., Horvath, A., et al. (2012). Aerobic kinetoplastid flagellate *Phytomonas* does not require heme for viability. *Proceedings of the National Academy of Sciences of the United States of America*, *109*, 3808–3813.

Krogh, A., Larsson, B., von Heijne, G., & Sonnhammer, E. L. (2001). Predicting transmembrane protein topology with a hidden Markov model: Application to complete genomes. *Journal of Molecular Biology*, *305*, 567–580.

Kunze, M., & Berger, J. (2015). The similarity between N-terminal targeting signals for protein import into different organelles and its evolutionary relevance. *Frontiers in Physiology*, *6*, 259.

Lakey, B., & Triemer, R. E. (2016). The tetrapyrrole synthesis pathway as a model of horizontal gene transfer in euglenoids. *Journal of Phycology*, *53*, 198–217.

Larkum, A. W., Lockhart, P. J., & Howe, C. J. (2007). Shopping for plastids. *Trends in Plant Science*, *12*, 189–195.

Layer, G., Reichelt, J., Jahn, D., & Heinz, D. W. (2010). Structure and function of enzymes in heme biosynthesis. *Protein Science*, *19*, 1137–1161.

Lermontova, I., Kruse, E., Mock, H. P., & Grimm, B. (1997). Cloning and characterization of a plastidal and a mitochondrial isoform of tobacco protoporphyrinogen IX oxidase. *Proceedings of the National Academy of Sciences of the United States of America*, *94*, 8895–8900.

Lhee, D., Yang, E. C., Kim, J. I., Nakayama, T., Zuccarello, G., Andersen, R. A., et al. (2017). Diversity of the photosynthetic *Paulinella* species, with the description of *Paulinella micropora* sp. nov. and the chromatophore genome sequence for strain KR01. *Protist*, *168*, 155–170.

Lohr, M., Schwender, J., & Polle, J. E. W. (2012). Isoprenoid biosynthesis in eukaryotic phototrophs: A spotlight on algae. *Plant Science*, *185-186*, 9–22.

Loughlin, P., Lin, Y., & Chen, M. (2013). Chlorophyll d and Acaryochloris marina: Current status. *Photosynthesis Research*, 116, 277–293.

Maréchal, E., & Cesbron-Delauw, M. F. (2001). The apicoplast: A new member of the plastid family. *Trends in Plant Science*, *6*, 200–205.

Martin, B., Nowack, E. C., & Melkonian, M. (2005). A plastid in the making: Evidence for a second primary endosymbiosis. *Protist*, *156*, 425–432.

Martin, W., Rujan, T., Richly, E., Hansen, A., Cornelsen, S., Lins, T., et al. (2002). Evolutionary analysis of Arabidopsis, cyanobacterial, and chloroplast genomes reveals plastid phylogeny and thousands of cyanobacterial genes in the nucleus. *Proceedings of the National Academy of Sciences of the United States of America*, *99*, 12246–12251.

Martin, W., Stoebe, B., Goremykin, V., Hapsmann, S., Hasegawa, M., & Kowallik, K. V. (1998). Gene transfer to the nucleus and the evolution of chloroplasts. *Nature*, *393*, 162–165.

Masuda, T. (2008). Recent overview of the Mg branch of the tetrapyrrole biosynthesis leading to chlorophylls. *Photosynthesis Research*, *96*, 121–143.

Masuda, T., & Fujita, Y. (2008). Regulation and evolution of chlorophyll metabolism. *Photochemical and Photobiological Sciences*, 7, 1131–1149.

Masuda, T., Suzuki, T., Shimada, H., Ohta, H., & Takamiya, K. (2003). Subcellular localization of two types of ferrochelatase in cucumber. *Planta*, *217*, 602–609.

Matsumoto, F., Obayashi, T., Sasaki-Sekimoto, Y., Ohta, H., Tekamiya, K.-I., & Masuda, T. (2004). Gene expression profiling of the tetrapyrrole metabolic pathway in Arabidopsis with a mini-array system. *Plant Physiology*, *135*, 2379–2391.

Matsumoto, T., Shinozaki, F., Chikuni, T., Yabuki, A., Takishita, K., Kawachi, M., et al. (2011). Green-colored plastids in the dinoflagellate genus *Lepidodinium* are of core chlorophyte origin. *Protist*, *162*, 268–276.

Matsuo, E., & Inagaki, Y. (2018). Patterns in evolutionary origins of heme, chlorophyll a and isopentenyl diphosphate biosynthetic pathways suggest non-photosynthetic periods prior to plastid replacements in dinoflagellates. *PeerJ*, *6*, e5345.

McFadden, G. I., Reith, M. E., Munholland, J., & Lang-Unnasch, N. (1996). Plastid in human parasites. *Nature*, *381*, 482.

Mochizuki, N., Tanaka, R., Grimm, B., Masuda, T., Moulin, M., Smith, A. G., et al. (2010). The cell biology of tetrapyrroles: A life and death struggle. *Trends in Plant Science*, *15*, 488–498.

Moore, R. B., Oborník, M., Janouškovec, J., Chrudimský, T., Vancová, M., Green, D. H., et al. (2008). A photosynthetic alveolate closely related to apicomplexan parasites. *Nature*, *451*, 959–963.

Moreira, D., & Deschamps, P. (2014). What was the real contribution of endosymbionts to the eukaryotic nucleus? Insights from photosynthetic eukaryotes. *Cold Spring Harbor Perspectives in Biology*, *6*, a016014.

Murphy, M. P. (2009). How mitochondria produce reactive oxygen species. *Biochemical Journal*, *417*, 1–13.

Nagai, S., Koide, M., Takahashi, S., Kikuta, A., Aono, M., Sasaki-Sekimoto, Y., et al. (2007). Induction of isoforms of tetrapyrrole biosynthetic enzymes, AtHEMA2 and AtFC1, under stress conditions and their physiological functions in *Arabidopsis*. *Plant Physiology*, *144*, 1039–1051.

Nagaraj, V. A., Arumugam, R., Chandra, N. R., Prasad, D., Rangarajan, P. N., & Padmanaban, G. (2009). Localisation of *Plasmodium* falciparum uroporphyrinogen III decarboxylase of the heme-biosynthetic pathway in the apicoplast and characterisation of its catalytic properties. *International Journal for Parasitology*, *39*, 559–568.

Nagaraj, V. A., Arumugam, R., Gopalakrishnan, B., Jyothsna, Y. S., Rangarajan, P. N., & Padmanaban, G. (2008). Unique properties of *Plasmodium falciparum* porphobilinogen deaminase. *Journal of Biological Chemistry*, *283*, 437–444.

Nagaraj, V. A., Arumugam, R., Prasad, D., Rangarajan, P. N., & Padmanaban, G. (2010). Protoporphyrinogen IX oxidase from *Plasmodium falciparum* is anaerobic and is localized to the mitochondrion. *Molecular and Biochemical Parasitology*, *174*, 44–52.

Nagaraj, V. A., Prasad, D., Arumugam, R., Rangarajan, P. N., & Padmanaban, G. (2010). Characterization of coproporphyrinogen III oxidase in *Plasmodium falciparum* cytosol. *Parasitology International*, *59*, 121–127.

Nagaraj, V. A., Prasad, D., Rangarajan, P. N., & Padmanaban, G. (2009). Mitochondrial localization of functional ferrochelatase from *Plasmodium falciparum*. *Molecular and Biochemical Parasitology*, *168*, 109–112.

Narita, S., Tanaka, R., Ito, T., Okada, K., Taketani, S., & Inokuchi, H. (1996). Molecular cloning and characterization of a cDNA that encodes protoporphyrinogen oxidase of *Arabidopsis thaliana*. *Gene*, *182*, 169–175.

Nowack, E. C., Melkonian, M., & Glöckner, G. (2008). Chromatophore genome sequence of *Paulinella* sheds light on acquisition of photosynthesis by eukaryotes. *Current Biology*, *18*, 410–418.

Nowack, E. C., Price, D. C., Bhattacharya, D., Singer, A., Melkonian, M., & Grossman, A. R. (2016). Gene transfers from diverse bacteria compensate for reductive genome evolution in the chromatophore of *Paulinella chromatophora*. *Proceedings of the National Academy of Sciences of the United States of America*, *113*, 12214–12219.

Oborník, M. (2018). The birth of red complex plastids: one, three, or four times? *Trends in Parasitology*, *34*, 923–925.

Oborník, M., & Green, B. R. (2005). Mosaic origin of the heme biosynthesis pathway in photosynthetic eukaryotes. *Molecular Biology and Evolution*, *22*, 2343–2353.

Oborník, M., Janouškovec, J., Chrudimský, T., & Lukeš, J. (2009). Evolution of the apicoplast and its hosts: from heterotrophy to autotrophy and back again. *International Journal for Parasitology*, *39*, 1–12.

Oborník, M., & Lukeš, J. (2013). Cell biology of chromerids: Autotrophic relatives to apicomplexan parasites. *International Review of Cell and Molecular Biology*, *306*, 333–369.

Oborník, M., Modrý, D., Lukeš, M., Černotíková-Stříbrná, E., Cihlář, J., Tesařová, M., et al. (2012). Morphology, ultrastructure and life cycle of *Vitrella brassicaformis* n. sp., n. gen., a novel chromerid from the Great Barrier Reef. *Protist, 163*, 306–323.

Oborník, M., Vancová, M., Lai, D. H., Janouškovec, J., Keeling, P. J., & Lukeš, J. (2011). Morphology and ultrastructure of multiple life cycle stages of the photosynthetic relative of Apicomplexa, *Chromera velia*. *Protist, 162*, 115–130.

Panek, H., & O'Brian, M. R. (2002). A whole genome view of prokaryotic haem biosynthesis. *Microbiology, 148*, 2273–2282.

Papenbrock, J., Mock, H. P., Kruse, E., & Grimm, B. (1999). Expression studies in tetrapyrrole biosynthesis: Inverse maxima of magnesium chelatase and ferrochelatase activity during cyclic photoperiods. *Planta, 208*, 264–273.

Parfrey, L. W., Lahr, D. J. G., Knoll, A. H., & Katz, L. A. (2011). Estimating the timing of early eukaryotic diversification with multigene molecular clocks. *Proceedings of the National Academy of Sciences of the United States of America, 108*, 13624–13629.

Park, M. G., Kim, M., & Kim, S. (2014). The acquisition of plastids/phototrophy in heterotrophic dinoflagellates. *Acta Protozoologica, 53*, 39–50.

Patron, N. J., & Waller, R. F. (2007). Transit peptide diversity and divergence: A global analysis of plastid targeting signals. *BioEssays, 29*, 1048–1058.

Pereira, I. A. C., Teixeira, M., & Xavier, A. V. (1998). Hemeproteins in anaerobes. *Structure and Bonding, 91*, 65–89.

Petersen, J., Ludewig, A. K., Michael, V., Bunk, B., Jarek, M., Baurain, D., et al. (2014). *Chromera velia*, endosymbioses and the rhodoplex hypothesis—Plastid evolution in cryptophytes, alveolates, stramenopiles, and haptophytes (CASH lineages). *Genome Biology and Evolution, 6*, 666–684.

Ponce-Toledo, R. I., Deschamps, P., López-García, P., Zivanovic, Y., Benzerara, K., & Moreira, D. (2017). An early-branching freshwater cyanobacterium at the origin of plastids. *Current Biology, 27*, 386–391.

Pratibha, P., Singh, S. K., Srinivasan, R., Bhat, S. R., & Sreenivasulu, Y. (2017). Gametophyte development needs mitochondrial coproporphyrinogen III oxidase function. *Plant Physiology, 174*, 258–275.

Qiu, H., Price, D. C., Weber, A. P. M., Facchinelli, F., Yoon, H. S., & Bhattacharya, D. (2013). Assessing the bacterial contribution to the plastid proteome. *Trends in Plant Science, 18*, 680–687.

Rodríguez-Ezpeleta, N., Brinkmann, H., Burey, S. C., Roure, B., Burger, G., Löffelhardt, W., et al. (2005). Monophyly of primary photosynthetic eukaryotes: Green plants, red algae, and glaucophytes. *Current Biology, 15*, 1325–1330.

Rogers, M. B., Gilson, P. R., Su, V., McFadden, G. I., & Keeling, P. J. (2007). The complete chloroplast genome of the chlorarachniophyte *Bigelowiella natans*: Evidence for independent origins of chlorarachniophyte and euglenid secondary endosymbionts. *Molecular Biology and Evolution, 24*, 54–62.

Roth, J. R., Lawrence, J. G., & Bobik, T. A. (1996). Cobalamin (coenzyme B12): Synthesis and biological significance. *Annual Review of Microbiology, 50*, 137–181.

Saldarriaga, J. F., Taylor, F. J., Keeling, P. J., & Cavalier-Smith, T. (2001). Dinoflagellate nuclear SSU rRNA phylogeny suggests multiple plastid losses and replacements. *Journal of Molecular Evolution, 53*, 204–213.

Sato, S., Clough, B., Coates, L., & Wilson, R. J. (2004). Enzymes for heme biosynthesis are found in both the mitochondrion and plastid of the malaria parasite *Plasmodium falciparum*. *Protist, 155*, 117–125.

Sato, S., & Wilson, R. J. (2003). Proteobacteria-like ferrochelatase in the malaria parasite. *Current Genetics, 42*, 292–300.

Scott, A. I., & Roessner, C. A. (2002). Biosynthesis of cobalamin (vitamin B12). *Biochemical Society Transactions, 30*, 613–620.

Seeber, F., & Soldati-Favre, D. (2010). Metabolic pathways in the apicoplast of apicomplexa. *International Review of Cell and Molecular Biology, 281*, 161–228.

Shih, P. M., & Matzke, N. J. (2013). Primary endosymbiosis events date to the later Proterozoic with cross-calibrated phylogenetic dating of duplicated ATPase proteins. *Proceedings of the National Academy of Sciences of the United States of America*, *110*, 12355–12360.

Sinclair, J., & Hamza, I. (2015). Lessons from bloodless worms: Heme homeostasis in *C. elegans*. *Biometals*, *28*, 481–489.

Singh, D. P., Cornah, J. E., Hadingham, S., & Smith, A. G. (2002). Expression analysis of the two ferrochelatase genes in *Arabidopsis* in different tissues and under stress conditions reveals their different roles in haem biosynthesis. *Plant Molecular Biology*, *50*, 773–788.

Smith, A. G., Marsh, O., & Elder, G. H. (1993). Investigation of the subcellular location of the tetrapyrrole-biosynthesis enzyme coproporphyrinogen oxidase in higher plants. *Biochemical Journal*, *292*, 503–508.

Sousa, F. L., Shavit-Grievink, L., Allen, J. F., & Martin, W. F. (2013). Chlorophyll biosynthesis gene evolution indicates photosystem gene duplication, not photosystem merger, at the origin of oxygenic photosynthesis. *Genome Biology and Evolution*, *5*, 200–216.

Storbeck, S., Rolfes, S., Raux-Deery, E., Warren, M. J., Jahn, D., & Layer, G. (2010). A novel pathway for the biosynthesis of heme in Archaea: Genome-based bioinformatic predictions and experimental evidence. *Archaea*, *2010*, 175050.

Sun, F. X., Cheng, Y. J., & Chen, C. Y. (2015). Regulation of heme biosynthesis and transport in metazoa. *Science China Life Sciences*, *58*, 757–764.

Suzuki, T., Masuda, T., Singh, D. P., Tan, F. C., Tsuchiya, T., Shimada, H., et al. (2002). Two types of ferrochelatase in photosynthetic and nonphotosynthetic tissues of cucumber: Their difference in phylogeny, gene expression, and localization. *Journal of Biological Chemistry*, *277*, 4731–4737.

Suzuki, K., & Miyagishima, S. Y. (2010). Eukaryotic and eubacterial contributions to the establishment of plastid proteome estimated by large-scale phylogenetic analyses. *Molecular Biology and Evolution*, *27*, 581–590.

Takishita, K., Kawachi, M., Noël, M. H., Matsumoto, T., Kakizoe, N., Watanabe, M. M., et al. (2008). Origins of plastids and glyceraldehyde-3-phosphate dehydrogenase genes in the green-colored dinoflagellate *Lepidodinium chlorophorum*. *Gene*, *410*, 26–36.

Tengs, T., Dahlberg, O. J., Schalchian-Tabrizi, K., Klaveness, D., Rudi, K., Delwiche, C. F., et al. (2000). Phylogenetic analyses indicate that the 19'hexanoyloxy-fucoxanthin-containing dinoflagellates have tertiary plastids of haptophyte origin. *Molecular Biology and Evolution*, *17*, 718–729.

Timmis, J. N., Ayliffe, M. A., Huang, C. Y., & Martin, W. (2004). Endosymbiotic gene transfer: Organelle genomes forge eukaryotic chromosomes. *Nature Reviews. Genetics*, *5*, 123–135.

Turrens, J. F. (2003). Mitochondrial formation of reactive oxygen species. *Journal of Physiology*, *552*, 335–344.

van Dooren, G. G., Kennedy, A. T., & McFadden, G. I. (2012). The use and abuse of heme in apicomplexan parasites. *Antioxidants and Redox Signaling*, *17*, 634–656.

van Dooren, G. G., Stimmler, L. M., & McFadden, G. I. (2006). Metabolic maps and functions of the *Plasmodium* mitochondrion. *FEMS Microbiology Reviews*, *30*, 596–630.

Varadharajan, S., Dhanasekaran, S., Bonday, Z. Q., Rangarajan, P. N., & Padmanaban, G. (2002). Involvement of delta-aminolaevulinate synthase encoded by the parasite gene in *de novo* haem synthesis by *Plasmodium falciparum*. *Biochemical Journal*, *367*, 321–327.

Vavilin, D. V., & Vermaas, W. F. J. (2002). Regulation of the tetrapyrrole biosynthetic pathway leading to heme and chlorophyll in plants and cyanobacteria. *Physiologia Plantarum*, *115*, 9–24.

Vothknecht, U. C., & Soll, J. (2007). The endosymbiotic origin of organelles: An ancient process still very much in fashion. *Biological Chemistry*, *388*, 877.

Waller, R. F., Gornik, S. G., Kořený, L., & Pain, A. (2016). Metabolic pathway redundancy within the apicomplexan-dinoflagellate radiation argues against an ancient chromalveolate plastid. *Communicative & Integrative Biology*, *9*, e1116653.

Waller, R. F., & McFadden, G. I. (2005). The apicoplast: A review of the derived plastid of apicomplexan parasites. *Current Issues in Molecular Biology*, *7*, 57–79.

Wang, P., Li, C., Wang, Y., Huang, R., Sun, C., Xu, Z., et al. (2014). Identification of a geranylgeranyl reductase gene for chlorophyll synthesis in rice. *Springerplus*, *3*, 201.

Watanabe, N., Che, F. S., Iwano, M., Takayama, S., Yoshida, S., & Isogai, A. (2001). Dual targeting of spinach protoporphyrinogen oxidase II to mitochondria and chloroplasts by alternative use of two in-frame initiation codons. *Journal of Biological Chemistry*, *276*, 20474–20481.

Weinstein, J. D., & Beale, S. I. (1983). Separate physiological roles and subcellular compartments for two tetrapyrrole biosynthetic pathways in *Euglena gracilis*. *Journal of Biological Chemistry*, *258*, 6799–6807.

Williams, P., Hardeman, K., Fowler, J., & Rivin, C. (2006). Divergence of duplicated genes in maize: Evolution of contrasting targeting information for enzymes in the porphyrin pathway. *Plant Journal*, *45*, 727–739.

Wilson, R. J. (2002). Progress with parasite plastids. *Journal of Molecular Biology*, *319*, 257–274.

Wisecaver, J. H., & Hackett, J. D. (2010). Transcriptome analysis reveals nuclear-encoded proteins for the maintenance of temporary plastids in the dinoflagellate *Dinophysis acuminata*. *BMC Genomics*, *11*, 366.

Wolf, Y. I., & Koonin, E. V. (2013). Genome reduction as the dominant mode of evolution. *BioEssays*, *35*, 829–837.

Woo, Y. H., Ansari, H., Otto, T. D., Klinger, C. M., Kolísko, M., Michálek, J., et al. (2015). Chromerid genomes reveal the evolutionary path from photosynthetic algae to obligate intracellular parasites. *eLife*, *4*, e06974.

Wu, B. (2006). *Heme biosynthetic pathway in apicomplexan parasites*. Dissertation available from ProQuest. AAI3246266. https://repository.upenn.edu/dissertations/AAI3246256.

Yamaguchi, A., Yubuki, N., & Leander, B. S. (2012). Morphostasis in a novel eukaryote illuminates the evolutionary transition from phagotrophy to phototrophy: Description of *Rapaza viridis* n. gen. et sp. (Euglenozoa, Euglenida). *BMC Evolutionary Biology*, *12*, 29.

Yang, Y., Matsuzaki, M., Takahashi, F., Qu, L., & Nozaki, H. (2014). Phylogenomic analysis of "red" genes from two divergent species of the "green" secondary phototrophs, the chlorarachniophytes, suggests multiple horizontal gene transfers from the red lineage before the divergence of extant chlorarachniophytes. *PLoS One*, *9*, e101158.

Yin, L., & Bauer, C. E. (2013). Controlling the delicate balance of tetrapyrrole biosynthesis. *Philosophical Transactions of the Royal Society of London. Series B, Biological Sciences*, *368*, 20120262.

Yoon, H. S., Hackett, J. D., Ciniglia, C., Pinto, G., & Bhattacharya, D. (2004). A molecular timeline for the origin of photosynthetic eukaryotes. *Molecular Biology and Evolution*, *21*, 809–818.

Yoon, H. S., Nakayama, T., Reyes-Prieto, A., Andersen, R. A., Boo, S. M., Ishida, K., et al. (2009). A single origin of the photosynthetic organelle in different *Paulinella* lineages. *BMC Evolutionary Biology*, *9*, 98.

Zapata, M., Fraga, S., Rodríguez, F., & Garrido, J. L. (2012). Pigment-based chloroplast types in dinoflagellates. *Marine Ecology Progress Series*, *465*, 33–52.

Zhang, Z. W., Zhang, G. C., Zhu, F., Zhang, D. W., & Yuan, S. (2015). The roles of tetrapyrroles in plastid retrograde signaling and tolerance to environmental stresses. *Planta*, *242*, 1263–1276.